The Heliosphere through the Solar Activity Cycle

André Balogh, Louis J. Lanzerotti
and Steven T. Suess

The Heliosphere through the Solar Activity Cycle

Springer

Published in association with
Praxis Publishing
Chichester, UK

PRAXIS

Professor André Balogh
The Blackett Laboratory
Imperial College
London
UK

Professor Louis J. Lanzerotti
Center for Solar–Terrestrial Research
New Jersey Institute of Technology
Newark
New Jersey
USA

Dr Steven T. Suess
National Space Science & Technology Center
NASA Marshall Space Flight Center
Huntsville
Alabama
USA

SPRINGER–PRAXIS BOOKS IN ASTRONOMY AND SPACE SCIENCES
SUBJECT *ADVISORY EDITOR*: John Mason B.Sc., M.Sc., Ph.D.

Springer is part of Springer-Science + Business Media (springer.com)

ISBN 978-3-642-09368-5 e-ISBN 978-3-540-74302-6

Cover design: Jim Wilkie
Project management: Originator Publishing Services Ltd, Gt Yarmouth, Norfolk, UK

Printed on acid-free paper

Contents

Preface

Since its inauguration in 1979 and launch in 1990, the joint European Space Agency (ESA)/National Aeronautics and Space Administration (NASA) solar polar Ulysses mission has produced transformational new insights into the dynamics of the heliosphere. The motivation for this book is the desire to provide a unique record of the heliospheric environment through a complete 11-year solar activity cycle, from the Sun to the orbit of Jupiter. This is now possible, thanks to opportunities provided by observations of the Sun using ground-based techniques as well as important vantage points in space, including the unique out-of-ecliptic orbit of Ulysses. The close connection between the solar cycle and the state of the heliosphere is well recognized; however, the just completed solar cycle 23 resulted in much important progress in gathering and combining solar observations and *in situ* observations in space. Although the Editors and contributing authors of this volume are associated principally with the Ulysses mission, the book is intended to provide a status report on contemporary understanding of the heliosphere that has been achieved using the many sources of data and observations available since the early to mid-1990s.

The story of the heliosphere is longer than that of space sciences: 2007 commemorates 50 years of space research, but almost 100 years of heliospheric research. Heliospheric research was born of cosmic ray research that started in 1912; the connection of the variations in cosmic ray intensity, as well as associations of sudden decreases in cosmic ray intensity following solar flares, was recognized before the first measurements made in space. In the first decade and more following the launch of Sputnik during the International Geophysical Year (5 October 1957), space around the Earth was the new frontier, ever expanding as space probes moved farther and farther from their Earth origin. The wealth of data acquired by numerous spacecraft in Earth orbit, but which probed the medium beyond the Earth's own volume of space (the magnetosphere), gave a more and more detailed view of the interplanetary medium and its connection with the Sun.

It was a cosmic ray physicist, Leverett Davis, who in 1955 named the volume of space around the Sun the heliosphere. Davis expressed in this way the connection of the variations in the intensity of cosmic rays (requiring a very large volume, comparable with the solar system) to solar activity as measured by the nearly periodic variations in the number of sunspots. The approximately 11-year periodicity in both cosmic ray intensity and sunspot numbers strongly hinted at a connection. It was some time, however, before the connection was correctly identified: by linking the constantly outpouring plasma from the Sun that formed the bluish, ionic tails of comets (as noted by Biermann) with the supersonically expanding solar atmosphere proposed by Eugene Parker in 1958. Finally, in the early 1960s the first well-instrumented space probes measured a solar wind and its embedded magnetic field and laid the basis for quantitatively developing the concept of the heliosphere.

It was another cosmic ray physicist, John Simpson, who already in 1959 proposed a space mission that would chart the interplanetary medium outside the plane in which the planets circle the Sun. He and others had recognized that the intensity of cosmic rays can only be affected if a three-dimensional volume breathes at the same rhythm as the Sun. The space mission that Simpson advocated in the late 1950s became a reality when, after a long wait, the Ulysses space probe was launched in 1990. Before then, in the 1970s and 1980s, there were several key interplanetary space missions—Helios 1 and 2, Pioneers 10 and 11, Voyagers 1 and 2—that paved the way to the inner and outer heliosphere, but still only close to the ecliptic plane. The exploration of the third dimension began with Ulysses.

Over the decades, from the early 20th century onward, the Sun has been observed with increasingly sophisticated instruments from both Earth- and space-based telescopes. The 11-year activity cycle became a subject of intense research as its potential effects on the Earth and its technologies were recognized. The complexity of the Sun, from its interior to its atmosphere, is now bewildering, as is the recognition that many or even most key physical processes operate at spatial and temporal scales that may need further generations of observatories (and observers) to resolve. But in terms of the phenomenology of the solar cycle, the richness and variety of observations have brought their fruit: the development of the "activity" cycle can be observed and catalogued from one solar minimum to the next and successive cycles can be compared to probe the underlying causes of solar variability. The best solar observations through this most recent activity cycle have come from the SOHO spacecraft, with important contributions from Yohkoh and many complementary observations from ever more capable ground-based facilities.

The last complete solar cycle, cycle number 23, has been the best observed 11-year period in the Sun's history. In December 2004, just as that cycle was approaching its close, the Voyager 1 spacecraft crossed the first outer boundary of the heliosphere, the termination shock, at about 100 times the distance of the Sun to Earth. Throughout the cycle, spacecraft closer to Earth, such as ACE in orbit around Lagrange Point 1 of the Sun–Earth system, as well as WIND and even the aging IMP 8 spacecraft, provided steady streams of observations about the properties of the interplanetary medium near the ecliptic plane. The Ulysses spacecraft, which, as we write, is about to celebrate its 17th launch anniversary, has provided data covering more than an entire solar activity cycle. Solar cycle 23 has been unique, and will remain so, in that it is the only cycle to date that is covered in three dimensions, by Ulysses.

The three-dimensional heliosphere in the context of the Ulysses mission has been covered by four prior books. The first took stock of understanding of the heliosphere before Ulysses (Marsden, 1986), while the next three (Marsden, 1995; Balogh, Marsden, and Smith, 2001; and Marsden, 2001) covered the first results and the state of the heliosphere in three dimensions at solar minimum and solar maximum, respectively. The current volume has a different perspective, providing a more integrated approach to the important questions concerning solar activity and the state of the heliosphere. We and the chapter authors look forward to this volume becoming an important heliosphere reference, especially as new missions are developed in the future.

REFERENCES

Balogh, A., Marsden, R. G., and Smith, E. J. (eds.) (2001), *The Heliosphere near Solar Minimum: The Ulysses Perspective*, Springer-Praxis, Chichester, U.K.

Marsden, R. G. (ed.) (1986), *The Sun and the Heliosphere in Three Dimensions*, Astrophysics & Space Science Library Vol. 125, D. Reidel, Dordrecht, The Netherlands.

Marsden, R. G. (ed.) (1995), *The High Latitude Heliosphere*, Kluwer Academic, Dordrecht, The Netherlands.

Marsden, R. G. (ed.) (2001), *The 3-D Heliosphere at Solar Maximum*, Kluwer Academic, Dordrecht, The Netherlands.

Acknowledgments

The Editors wish to thank first the contributing authors of this volume for enthusiastically bringing their expertise and experience to this undertaking. It is natural for us to acknowledge the worldwide solar and heliospheric scientific community for exploiting the opportunities provided by the space missions and ground-based observations to achieve remarkable advances in understanding the heliosphere over the past decades. We also wish to express our indebtedness specifically to the Ulysses community, the project scientists, Richard Marsden of ESA and Edward Smith of NASA/JPL. Since launch, the Ulysses project managers (Derek Eaton, Peter Wenzel, and Richard Marsden at ESA, and Willis Meeks and Ed Massey at NASA/JPL), together with the Mission Operation Team led formerly by Peter Beech and now by Nigel Angold, have worked wonders to keep the spacecraft operational and the data stream flowing. Similar thanks are due to the other missions such as SOHO, Yohkoh, ACE, WIND and their scientists and operations staffs for the excellent observations which have been extensively used to generate the scientific results presented here.

A.B. wishes to thank his colleagues at Imperial College London and at the International Space Science Institute for the help and facilities provided to pursue research activities beyond the formal retirement age. L.J.L. thanks his Ulysses/HI-SCALE team members for their collegial and friendly collaborations and support since the beginning of the program nearly 30 years ago. S.T.S. thanks the Ulysses/SWOOPS instrument team and the Ulysses Project for their support.

André Balogh, Louis J. Lanzerotti, Steven T. Suess
25 August 2007

Figures

TABLE

Abbreviations and acronyms

ACE	Advanced Composition Explorer
ACR	Anomalous Cosmic Ray
AMPTE	Active Magnetospheric Particle Tracer Explorer
AT	Anisotropy Telescope
CELIAS	Charge, ELement, and Isotope Analysis System (SOHO instrument)
CH	Coronal Hole
CHEM	CHarge Energy Mass spectrometer
CIR	Corotating Interaction Region
CME	Coronal Mass Ejection
COSPIN	Cosmic Ray and Solar Particle INvestigation
CRNC	Cosmic Ray Nuclear Composition
CRR	Corotating Rarefaction Region
DFG	Deutsche Forschungsgemeinschaft
EC	Ecliptic Crossing
EPAC	Energetic Particle Composition Experiment
EPAM	Electron, Proton, and Alpha Monitor
FIP	First-Ionization Potential
FLS	Fast-Latitude Scan
FS	Forward Shock
GAS	Interstellar Neutral Gas Experiment (Ulysses instrument)
GCR	Galactic Cosmic Rays
GHO	Great Heliospheric Observatory
GMIR	Global Merged Interaction Region
GOES	Geostationary Operational Environmental Satellites
HCS	Heliospheric Current Sheet
HELIOS	A pair of interplanetary spacecraft in the inner heliosphere (1970s)

HET	High-Energy Telescope
HI-SCALE	Heliosphere Instrument for Spectra, Composition, and Anisotropy at Low Energies
HMF	Heliospheric Magnetic Field
IBEX	Interstellar Boundary EXplorer
ICE	International Cometary Explorer
ICI	Ion Composition Instrument
ICME	Interplanetary CME
IGY	International Geophysical Year
IMF	Interplanetary Magnetic Field
IMP	Interplanetary Monitoring Platform
IPS	InterPlanetary Scintillation
ISEE	International Sun–Earth Explorer
ISPM	International Solar Polar Mission
ISSI	International Space Science Institute
JE	Jovian Encounter
KET	Kunow Electron Telescope (Ulysses instrument)
LASCO	Large Angle and Spectrometric COronagraph Experiment (SOHO instrument)
LIC	Local Interstellar Cloud
LIS	Local Interstellar Spectrum
LISM	Local InterStellar Medium
LSPF	Least Squares Planar Fit
MHD	MagnetoHydroDynamic
MIR	Merged Interaction Region
MPV	Most Probable Value
MTOF	Mass Time Of Flight spectrometer (SOHO instrument)
MVA	Minimum Variance Analysis
NASA	National Aeronautics and Space Administration
NOAA	National Oceanic and Atmospheric Administration
NRF	South African Research Foundation
NSO	National Solar Observatory
NSSDC	National Space Science Data Center (NASA)
OOE	Out Of the Ecliptic
PCH	Polar Coronal Hole
PFSS	Potential Field Source Surface
PVO	Pioneer Venus Orbiter
RD	Rotational Discontinuity
RHESS	Ramaty High-Energy Solar Spectroscopic Imager
RS	Reverse Shock
RTG	Radioisotope Thermoelectric Generator
RTN	Heliospheric coordinate system with components R, T, and N
S/B	Sector Boundary
SAMPEX	Solar, Anomalous, and Magnetospheric Particle EXplorer

SB	Sector Boundary
SCR	Solar Corpuscular Radiation
SDO	Solar Dynamics Observatory
SEP	Solar Energetic Particles
SI	Stream Interface
SMM	Solar Maximum Mission
SOHO	SOlar Heliospheric Observatory
SSNL	Source Surface Neutral Line
STEREO	Solar TErrestrial Relations Observatory
SUMER	Solar Ultraviolet Measurements of Emitted Radiation (SOHO instrument)
SWICS	Solar Wind Ion Composition Spectrometer
SWOOPS	Solar Wind Observations Over the Poles of the Sun (Ulysses instrument)
TD	Tangential Discontinuity
TGO	The Great Observatory
TRACE	Transition Region And Coronal Explorer
TS	Termination Shock
ULECA	Ultra Low-Energy Charge Analyzer (ISEE-3/ICE instrument)
ULYSSES	The International Solar-Polar Space Mission
UV	UltraViolet
VELA	Large-radius Earth-orbiting satellites for monitoring nuclear tests (1960s)
WIND	Spacecraft in NASA's Global Geospace Program to monitor the solar wind

1

The heliosphere: Its origin and exploration

A. Balogh and L. J. Lanzerotti

1.1 INTRODUCTION

The heliosphere exists because of the presence of the solar wind, the expanding hot upper atmosphere (the corona) of the Sun, which excludes the local interstellar medium (LISM) from the vicinity of the Sun and planets. The size and boundaries of the heliosphere are determined through the interaction between the solar wind and the LISM. The internal properties, structure, and dynamics of the heliospheric medium are defined by spatial and temporal variability of the regions of origin of the solar wind in the solar corona. The variability of the solar wind leads to evolving, dynamic phenomena throughout the heliosphere on all spatial and temporal scales. The most important timescale is imposed by the approximately 11-year solar activity cycle and the approximately 22-year solar magnetic cycle (the Hale cycle). Regions of the origin and of the properties of the solar wind undergo considerable change on this 11-year timescale, the dominant parameter in the description of the global heliosphere. Due to the rotation of the Sun and the interaction of this rotation with the generation of the internal and external magnetic fields of the Sun, the 11-year periodicity is most significant in the heliosphere in the solar meridian, and thus as a function of heliolatitude.

The chapters that follow contain a synopsis of the new understandings that have been achieved about the behavior and physics of the heliosphere through more than a solar cycle. In particular, solar cycle 23 (1997 to 2007) as measured by instruments carried by the Ulysses spacecraft is the focal point of the chapters. The Ulysses mission, due to its near-polar orbit around the Sun, has provided the first global, three-dimensional view of the heliosphere following its launch in October 1990. Together with several other robotic space missions, including Voyagers 1 and 2, SOHO, WIND, ACE, and now STEREO, knowledge and understanding of the heliosphere has dramatically increased in the last nearly two decades. These missions together, and taken especially with the three-dimensional views provided by Ulysses, have contributed to an integrated systems view of the Sun, the heliosphere, and the LISM. A brief introduction to the physical processes that are involved in the heliosphere is provided in the following sections. The subsequent five chapters detail different aspects of the behavior of the Sun and the heliosphere over solar cycle 23. Finally, Chapter 7 presents a concluding overview of the contributions of the Ulysses mission to heliospheric science.

1.2 THE PRE–SPACE AGE HELIOSPHERE

1.2.1 The expanding hot solar atmosphere

Early evidence for the existence of a volume of space controlled by the Sun came first from suggestions of a "medium" that communicated to the near-Earth environment information about the solar activity cycle and the solar rotation period through periodicities and fluctuations in geomagnetic phenomena. Variations in the appearance of aurora and large excursions of the geomagnetic field were observed to approximately coincide with the appearance of sunspots and the sunspot cycle (e.g., historical reviews in Chapman and Bartels, 1940). Interestingly, the growth of electrical technologies for communications, beginning with the telegraph in the mid-19th century, stimulated much work toward the understanding and possible "prediction" of the solar activity that appeared to be causally related to disturbances in the technologies.

Second, indications for the existence of a continuously outflowing solar wind came from the study of the orientation of comet tails (Biermann, 1951). A third argument that implied a large volume of space controlled by the Sun was based on the quasi-11-year "modulation" cycle of galactic cosmic rays, as well as sudden decreases in cosmic ray intensity (Forbush decreases) that often followed large solar flares. Once the high temperature of the solar corona was recognized by measurements in the 1940s, theoretical models of the "connecting" material in the heliosphere were developed (Chamberlain, 1960; Parker, 1963). Two conflicting concepts were proposed, one for a subsonic "breeze" and one for a supersonic "wind" that encompassed Earth, and distances beyond. The opposing theories were not resolved in favor of a supersonic wind, until *in situ* observations were made by robotic spacecraft in the early 1960s.

The volume of space filled with the expanding solar wind from the Sun is what is now called the heliosphere. The solar wind originates in the solar corona, the upper atmosphere of the Sun that can be seen visually at the time of solar eclipses. The solar wind continuously expands into space with speeds that can vary between about 250 km/s to more than 1,000 km/s. The corona itself is a rarefied, hot gas, with temperatures well in excess of a million degrees K. At these temperatures, most of the electrons around atomic nuclei are stripped away. Why the corona is so hot—when the surface of the Sun, the visible photosphere, is only about 6,000 degrees K—has remained a mystery since the high coronal temperatures were originally identified. It is now known that there is enough energy emerging from the Sun in the form of convective motions (mass transport from the solar interior to the surface) to supply the energy needed to heat the corona, but the way that this convective mechanical energy is transmitted to the gas in the corona remains a topic of intense observational investigation and theoretical debate. An important factor in the heating must be the magnetic flux that is transported with the material from the solar interior to the solar surface. In the lower corona (the layers closest to the solar surface) magnetic fields emerging from the solar photosphere form complex magnetic loops that can be observed by space-based solar telescopes or even from the ground at the times of total solar eclipses. It is likely that some form of waves and magnetic dissipation are the main contributors to the heating of the corona.

Looking in more detail at the evidence implying the existence of a medium connecting the Sun to Earth, quasi-periodic perturbations were observed in the Earth's magnetic field that approximately matched the synodic period of the solar rotation (about 27 days). These approximately 27-day periodicities were associated with the passage, as the Sun rotates, of specific regions on the Sun that appeared to cause a higher level of auroral activity and other manifestations of geomagnetic disturbances. These regions were called "M-regions" because they were thought to be somehow more magnetically active that other parts of the Sun (Chapman and Bartels, 1940). Their occurrence also changed with the 11-year solar cycle; the M-regions appeared to be more active (i.e., caused more terrestrial disturbances) not at the time of maximum in solar activity, but rather away from it. It is now known that such periodic disturbances are actually produced by corotating interaction regions (CIRs) in the heliosphere. These CIRs are large-scale structures in the solar wind caused by the collision of faster and slower solar wind streams that originate from different solar regions. Since the solar regions with which these structures are associated remain reasonably stable over several solar rotations, the interaction regions in the solar wind that cause geomagnetic disturbances also recur at approximately the same time in each solar rotation. This is then observed as an approximately 27-day periodicity in the terrestrial effects.

As mentioned above, intense geomagnetic storms were often found to occur following large solar flares in sunspot regions. These geomagnetic storms are manifest in the records as large, fast depressions in the geomagnetic field intensity, followed by a recovery that can last from a few hours to a few days. This behavior can be understood by the compression of the geomagnetic field by some large-scale wave front that travels from the Sun after large solar flares. These are now known to be coronal mass ejections (CMEs), the explosive expulsion of coronal mass into the heliosphere. In summary, both the matching periodicities at the rate of the solar rotation and the nature of the geomagnetic disturbances following solar flares pointed to some agent that brought solar phenomena and disturbances to the vicinity of the Earth.

An investigation that indicated a continuous emission of particles from the Sun was developed in the 1950s. Ludwig Biermann (1951) determined that the bluish tails of comets (now called the plasma tail) that always point radially away from the Sun (see Figure 1.1) can only be produced by particles constantly streaming also radially away from the Sun. Even though the orientation of comet tails had been known for

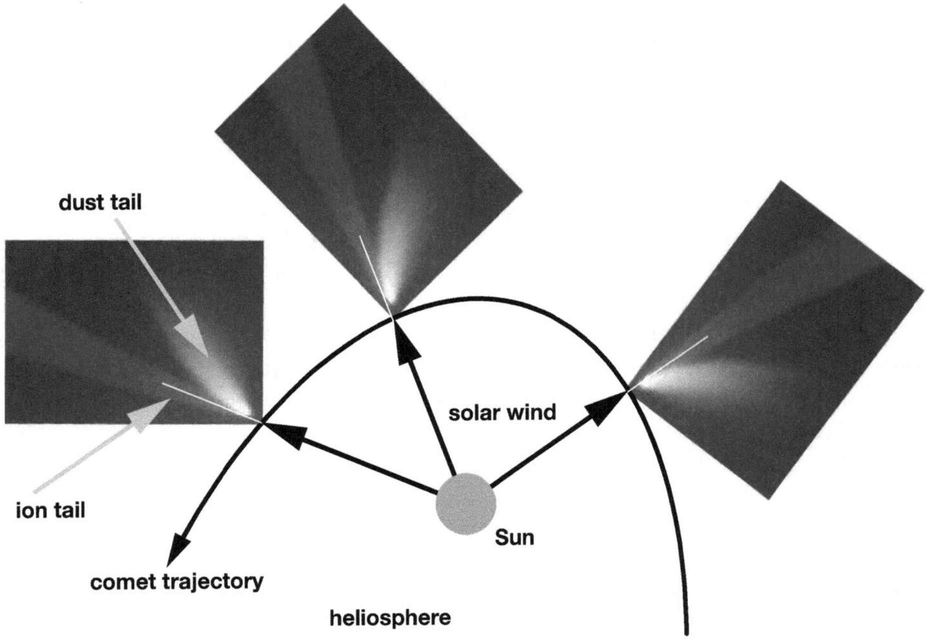

Figure 1.1. Biermann proposed that ion tails arise from atomic particles in the coma that are ionized by solar ultraviolet radiation which are then "entrained" by a "continuously flowing" corpuscular radiation from the Sun (now known to be the solar wind).

centuries, it was always thought that the pressure of radiation (the visible light) from the Sun was the responsible cause. Biermann's key contribution was to demonstrate that radiation pressure was insufficient, and that particles traveling from the Sun at hundreds of km/s were necessary to create these plasma tails. The other tail, the dust tail that curves away from the comet (still in a generally anti-Sunward direction), is in fact generated by solar radiation pressure; that is, photons from the Sun striking the micron-sized dust particles that emanate from the comet. Both tails are very visible in the photograph of comet Hale–Bopp in Figure 1.1.

In 1958 a highly controversial idea was put forward by Eugene Parker, a young researcher at the University of Chicago (Parker, 1958). He calculated the consequences that would result from the million-degree solar corona above the solar photosphere. While his theoretical solution included many simplifications, it nevertheless provided a sophisticated mathematical model for a solar wind that would escape from the upper solar atmosphere at supersonic speeds. In a magnetized plasma (unlike in an ordinary gas like air) three kinds of waves can propagate: the so called Alfvén wave (a wave that propagates along a magnetic field line as a wave would along a stretched string) and two kinds of sound (longitudinal, compressional) waves—slow-mode and fast-mode waves. Parker demonstrated that the solar wind, as it escapes the Sun's corona, travels at speeds in excess of the speed of the fast-mode sound wave. This makes the solar wind a supersonic flow of plasma in interplanetary space.

Parker's idea was controversial at the time since the scientific establishment favored a different solution for the solar atmosphere. This was the so-called solar "breeze" that existed due to "evaporation" at the outer edges of the corona. Nevertheless, Parker's theory soon found vindication. Measurements from NASA's Mariner 2 probe to Venus in 1962 returned measurements that showed that the solar wind was continuously present with speeds of a few hundred km/s, close to the values predicted by Parker. The density of the solar wind close to the Earth's orbit was found to be less than in Parker's original theory, about 7 cm^{-3} versus the predicted 30 to 50 cm^{-3}. This discrepancy principally arose from the simplifying assumption made by Parker that the temperature of the corona is constant. Since this time, knowledge gained about the solar wind shows it to be a more complex phenomenon than was originally treated by Parker. Nevertheless, many of the original conclusions remain valid, and Parker's ideas have continued to shape the way the solar wind and the heliosphere are viewed today.

Since the solar wind flows from the Sun in all directions with varying speeds and densities, its time-varying pressure will alter the location of the boundary of the heliosphere with the LISM. Information about the properties of the LISM is difficult to obtain, as no space mission has up to now reached it to provide direct information. Indirect inferences and remote sensing are used to obtain such parameters as the density, temperature, composition, and magnetic field in the LISM. Very sophisticated remote-sensing techniques have shown that even in the neighborhood of the Sun the medium is not uniform but rather lumpy on the scale of a few parsecs (1 parsec $= 3.26$ light years $\approx 2.06 \times 10^5$ AU where 1 AU $\approx 1.5 \times 10^8$ km) and even less. This means that calculating the size of the heliosphere from a balance of pressures between the solar wind and the interstellar medium is not easy. In Section 1.3.1, the size of the heliospheric cavity is estimated based on the best current estimates of the parameters of the LISM.

After more than four decades of observations, the properties of the solar wind have been well documented. The wind has an average density of 7 particles/cm^3 at the orbit of Earth (but is highly variable from about 1 to 100 particles/cm^3), with a speed that varies from less than 300 km/s to more than 1,000 km/s. The wind is composed principally of ionized hydrogen (protons), with a few percent of doubly ionized helium (alpha particles). The wind also contains detectable amounts of fully and partially ionized heavier elements that comprise the Sun, such as carbon, oxygen, silicon, magnesium and iron. The total density of these heavier ions is very small but these elements provide vital information on the temperature conditions in the regions of the corona in which the solar wind originates. The density of the solar wind decreases as the inverse square of the distance from the Sun (since the wind is a spherically expanding gas), with a speed that varies very little all the way out to the outer boundary of the heliosphere.

1.2.2 Energetic particles in the heliosphere

During the depths of the Second World War, on the same date in February 1942 (the 28th) when British radars that were being used to track enemy airplanes were suddenly blacked out by a large burst of electromagnetic (radio) noise from the Sun, a large increase was measured in ground-based detectors of cosmic rays (Forbush, 1946). It was only after the war, when publication restrictions were eliminated, that the association between the two solar-produced events was evident. From that time until the advent of the space age in 1957, during the International Geophysical Year (IGY), only five similar occurrences of particles with energies sufficient to penetrate Earth's atmosphere were measured by instruments on the ground. However, these occurrences were sufficient to demonstrate that the Sun was capable of producing energetic charged particles whose time behavior and whose energy characteristics required explanation. The explanations for the time behavior, generally a rapid rise to a peak flux and an extended decay in intensity, involved both acceleration processes at the Sun, propagation processes in the heliosphere between the Sun and Earth, and a finite extent to the size of the solar system (see Section 1.3).

The use of balloons, rockets, and satellites for the study of solar particle phenomena was greatly accelerated by the IGY, whose first major discovery was of the Van Allen radiation belts using the Explorer 1 satellite. These experimental techniques encouraged the development of instruments for measuring solar-emitted particles to ever-lower energies. And such measurements demonstrated the ever-increasing complexities of the time dependencies and of the directions of arrival of solar-produced particles as the energies that were able to be measured continued to be pushed lower and lower by new instrumentation developments. As significant was the new information obtained on the types and intensities of solar activity that could produce energetic particle enhancements in the heliosphere near Earth.

The decision in 1961 to send humans to the Moon provided a large impetus for solar energetic particle research. It was widely recognized that such solar emissions could be significant health threats to astronauts *en route* to the Moon and on its surface. The solar particle event of August 1972 occurred between the last two Apollo missions to the Moon. If the event had occurred at the time of one of the missions, it has been stated that the crew could have faced serious illness or death. Measurements on spacecraft and on the ground of solar energetic particles continues to date, both for new scientific returns as well as for the practical implications that these events (and their possible predictability) can have on sensitive spacecraft electronics and for human exploration of space.

Forbush (1954) also discovered that the intensity of galactic cosmic rays at Earth depends upon the general level of solar activity. This is because the size of the heliosphere and its internal physical conditions determine the access of galactic rays to the entire solar system. The variation of galactic cosmic ray intensities and sunspot numbers (an indication of solar activity and thus of the solar disturbances that affect the size of the heliosphere) is shown in Figure 1.2. These cosmic rays are generated throughout the galaxy by various processes (e.g., supernova explosions, shock waves), and therefore are present everywhere in the galaxy. As they reach the volume of space around the Sun their propagation is impeded by the outward flowing solar wind and the diverse magnetic structures that are carried in the solar wind. Since the solar wind and its structures change with the level of activity on the Sun (such as in response to the 11-year activity cycle), then cosmic rays will be more or less impeded

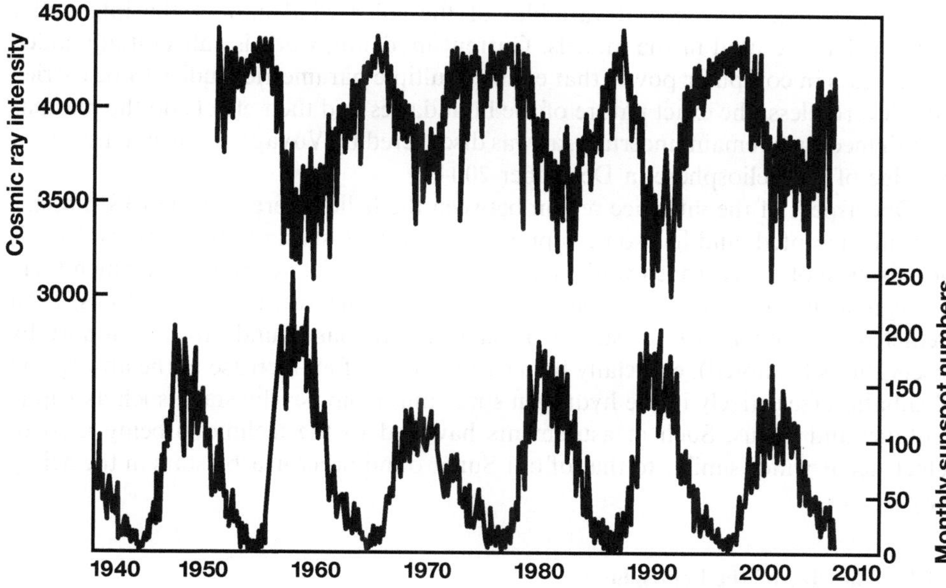

Figure 1.2. The 11-year variation in the intensity of cosmic radiation at Earth, out of phase with the sunspot cycle, implies a large volume of space around the Sun to which the access of galactic cosmic rays is somehow remotely controlled by the Sun.

in reaching the Earth: when the Sun is in its more active state, with the largest number of sunspots, fewer cosmic rays can reach the Earth than when solar activity is low. Another way of looking at this is that at high solar activity a larger amount of energy needs to be expended by cosmic rays to reach Earth, but as there are fewer cosmic rays of such higher energies, the number detected at Earth at such times is lower. Conversely, during low solar activity levels, lower energy cosmic rays can reach Earth, and as they are more numerous, their intensity increases.

1.3 THE HELIOSPHERE AND ITS BOUNDARIES

Given the established properties of the solar wind, the size of the heliosphere and the nature of its boundaries can be estimated, taking into account properties of the LISM. The parameters (density, temperature, composition, magnetic field) of the LISM cannot be directly measured, but some can be deduced indirectly, with some accuracy.

The models of the heliosphere that take into account the known characteristics of the solar wind as it propagates to large distances from the Sun and the characteristics of the Local Interstellar Cloud (deduced from remote and indirect observations) tend to agree on basic parameters, such as the approximate distance of the outer boundaries, the termination shock, and the heliopause. Modeling in detail is quite difficult, however, as there are several variables of the solar wind and of the interstellar medium that are used in the models. Current modeling work is substantially aided by increases in computer power that enable multiple parametric studies to be carried out. Nevertheless, the exact nature of the boundaries and their effects on the plasmas of both mediums remain uncertain, as was discovered by Voyager 1 when it arrived at the edge of the heliosphere in December 2004.

One aspect of the interface region between the heliosphere and the LISM seems generally accepted, and has some support from remote-sensing observations. This is the existence of a "hydrogen wall", a region in front of the heliopause in which there is a significant increase in the density of interstellar neutral hydrogen atoms. First predicted by models, the existence of such a wall has found strong support by observations (as noted), especially by measurements of an increase in the absorption of radiation (selectively in the hydrogen spectrum) from nearby stars, such as Alpha Centauri and Sirius. Such measurements have led to the technique being used to detect stellar winds similar to that of our Sun around other nearby stars in the Milky Way.

1.3.1 The size of the heliosphere

The best indication of the size of the heliosphere at the present time was obtained when NASA's Voyager 1 spacecraft (launched in 1977) crossed the termination shock, one of its key outer boundaries, in December 2004 (Fisk, 2005). The distance of Voyager from the Sun was then 94 AU. After four decades of theoretical speculation and modeling, the measurement of the distance to this outer heliospheric boundary established a firm foundation for future modeling work.

The extent of a heliosphere around the Sun was a subject of theoretical discussion, using known data at the time, over many years prior to the launch of the Voyager spacecraft in 1977. Davis (1955) concluded that "solar corpuscular emission[s]" would reach a balance at about 200 AU with a local galactic magnetic field strength of 10^{-5} gauss. Meyer, Parker, and Simpson (1956) in a classic study and analysis of the large solar particle event of February 1958 concluded that the decay time of the event implied that the boundary of the heliosphere was beyond the orbit of Earth, but certainly less than the orbit of Jupiter ($\sim$5 AU).

The size of the galactic cosmic ray modulation region (and therefore the heliosphere) was placed at less than 5 AU by Simpson and Wang (1967) whereas Lanzerotti and Schulz (1969) suggested that an apparent solar cycle variation in the appearance of Jovian decametric radio emissions indicated that the heliosphere boundary was likely near 5 AU and varied with the solar cycle, being beyond Jupiter's orbit at solar maximum and closer to the orbit at solar minimum. Experimental studies of the radial gradient of cosmic rays (e.g., O'Gallagher, 1967) and theoretical examination of interplanetary neutral hydrogen intensities (Hundhausen, 1968) tended to place the heliosphere boundary within about 5 AU.

Many of the early estimates of the heliospheric boundary assumed that the density of the solar wind, as it becomes rarefied with increasing distance from the Sun, will simply drop to a sufficiently low level that the solar wind dynamic pressure will just balance the encountered low pressure of the interstellar medium. Axford, Dessler, and Gottlieb (1963) calculated that this interaction would occur at a distance of about 20 AU.

The parameters of the LISM are poorly known. It is now recognized that interstellar space is not at all uniform and has large spatial variations in density and temperature, as well as in the relative speed between different interstellar regions. Primarily through measurements of starlight from different stars located in directions all around the Sun, it has been deduced that the Sun's neighborhood in space (about 100 parsec in dimension), beyond the heliosphere, is rather emptier than interstellar space in general. In fact, inside this "bubble" there are smaller irregular regions that are considerably cooler and denser than the average of the bubble. Even so, the Sun's immediate neighborhood is usually described as a "warm, partially ionized diffuse interstellar cloud"; this is the Local Interstellar Cloud (LIC) whose properties are those that define, together with those of the solar wind, the size of the heliosphere and the nature of its boundaries.

Current estimates suggest that the heliosphere has been immersed in this cloud for perhaps 10,000 to 100,000 years. The key parameters of the LIC are: the density of neutral hydrogen atoms ($0.24\,cm^{-3}$); the density of electrons ($0.09\,cm^{-3}$); the ratio of ionized hydrogen (or number of electronless protons) to hydrogen atoms (about 23%); the density of helium atoms ($0.014\,cm^{-3}$); the ratio of ionized to neutral helium (about 45%); the temperature (about 6,400 K, similar to the temperature of the photosphere, but with many orders of magnitude difference in their respective densities). For comparison, it is estimated that the temperature of the large local interstellar bubble is about a million degrees, consisting mostly of very low density ionized hydrogen, about 0.005 particles cm^{-3}. The heliosphere is moving through the LIC at $\sim$25 km/s (Izmodenov, 2004).

The physical principles of the interaction between the solar wind and the LIC are needed to estimate the size of the heliosphere using these parameters of the LIC. As the highly supersonic solar wind encounters the near-stationary LIC medium, the wind slows to subsonic speeds. The shock wave that is formed from this slowing of the solar wind is called the termination shock (the boundary crossed by Voyager 1 in late 2004), and is the locale where the solar wind becomes subsonic.

Beyond the termination shock, the plasma medium consists of the slowed and heated solar wind that extends out to the heliopause, the ultimate boundary between the solar wind and the LIC. In space plasmas, the plasmas of different origin do not easily mix because of the commonly entrained magnetic fields in each of the plasmas (except in very special circumstances), so the LIC and solar wind plasmas probably have a distinct boundary that separates them.

However, there are important complications. The LIC (unlike the solar wind) is not a fully ionized plasma; rather it also contains neutral hydrogen, helium, and other atoms which are not affected by the presence of a magnetic field. Neutral atoms from the LIC can penetrate into the heliosphere, some traveling even inward to Earth's orbit. These neutral atoms have been measured directly *in situ*, providing important information on properties of the LIC. A portion of the neutral atoms become ionized when the atoms encounter the solar wind (primarily by charge exchange with the solar wind protons). Once ionized, these particles are "picked up" by the solar wind and can, if they are numerous, influence the local properties of the solar wind. Such pick-up ions were long believed by theory to be energized by their interaction with the termination shock once they were carried there from the interior heliosphere. At the higher energies, these energized pick-up ions then become part of the cosmic ray population. As they are generally quite recognizable by their composiition and energy spectra as a different population from the more generally observed galactic cosmic rays, they are called "anomalous" cosmic rays (ACRs). Voyager 1 data taken at the termination shock showed that the more than three-decade-old theoretical consensus on the acceleration region for ACRs did not apply at the location of the crossing. Measurements by the Voyager 1 instruments showed that the ACR populations before and after the shock crossing were the same. Thus, there remains a major puzzle in understanding as to just where the energization of these anomalous cosmic rays occurs.

The distance to the termination shock cannot be a constant, as the properties of the solar wind vary significantly in time, in particular with the 11-year cycle of solar activity. As a result, the termination shock (and all other shock waves in space, including the Earth's and other planetary bow shocks) is in constant motion, moving in and out at speeds that are probably of the order of about 100 km/s. When Voyager 1 first encountered the shock, it was moving toward the Sun with about that velocity. The distance of travel (inward or outward) of the termination shock probably varies, depending on timescales: it undoubtledly makes large excursions (perhaps as much as 10 AU or more) in response to solar cycle variations; smaller distances of movement likely occur in response to the always changing conditions in both the solar wind and, presumably, in the LIC as well.

The possibility of an outer shock wave, outside the heliopause and surrounding the whole heliosphere, is still an open question in the context of the models. The existence of such an outer shock depends partly on the relative velocities of the Sun and the LIC (measured to be about 25 km/s), but partly also on the other physical parameters of the two colliding media, such as their densities and temperatures.

From studies of the relative motion of the Sun and the heliosphere in the LISM and the rest of the Milky Way, it has been estimated that occasionally (but many times during the lifetime of the solar system) conditions in the LIC can change significantly when large, cool, and dense interstellar molecular clouds are encountered. The densities of these clouds can be 10 particles cm^{-3} or more, on the order of 30 to 40 times greater than the density in the present LIC. Under such conditions the heliosphere would be significantly compressed, to perhaps less than a half or a third of its current size. How such conditions might affect the Earth and its space environment has not yet been fully simulated with heliosphere models.

In the present era, with the termination shock near 100 AU, the Earth is deep within the heliosphere, only 1% of the way to the shock. However, so are Jupiter (at about 5.5 AU) and Saturn (at about 10 AU). Even the other gas giants, Uranus (19 AU) and Neptune (30 AU) are well within the inner half of the heliosphere, while Pluto (with a highly elliptical orbit) is, when farthest away from the Sun, only halfway to the current location of the termination shock. The heliosphere is, on the solar

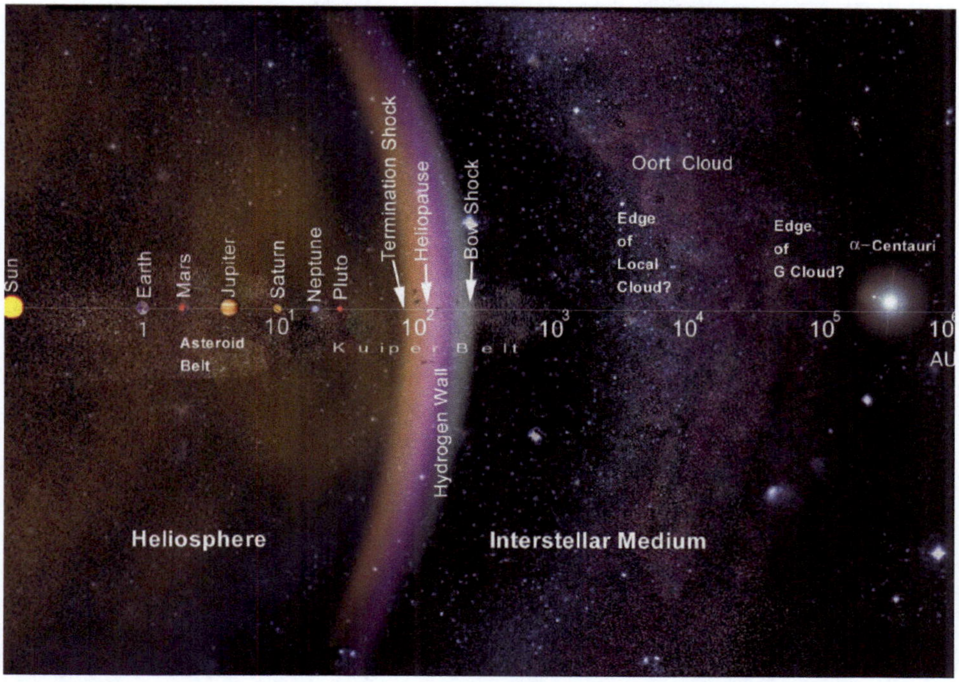

Figure 1.3. The solar system and its nearby Galactic neighborhood are illustrated here on a logarithmic scale extending from <1 to 10^6 AU. Our Sun and its planets are shielded by a bubble of solar wind—the heliosphere—that is $\sim 10^2$ AU in size. Beyond this bubble is the local interstellar medium. Threaded through the boundaries of the heliosphere is the Kuiper Belt—the source of short-period comets. The Oort Cloud is a spherical shell of comets extending to ~10,000 AU. Coincidentally, the edge of the Local Interstellar Cloud that surrounds our solar system may also lie at a similar distance. Alpha Centauri, the best known member of our nearest star system, lies well beyond at ~300,000 AU.

system scale, truly enormous. But on astrophysical scales, whether in the local interstellar bubble or in the Milky Way itself, it is dwarfed by the vastness of the universe (Figure 1.3).

1.3.2 The termination shock and beyond: Voyager 1 results

In heliospheric research, once Voyager 1 passed ~50 AU, modelers and data analysts began to converge on the conclusion that the distance to the termination shock would be in the vicinity of 80–100 AU, depending on the time during the activity cycle of the Sun. Prior to encountering the termination shock, instruments on Voyager 1 detected what were apparently sporadic upstream energetic particles coming from the shock. Finally, in December 2004, at a distance of 94 AU, particle and magnetic field instruments on board Voyager 1 unanimously indicated the crossing of a shock front, clearly the termination shock where the solar wind becomes subsonic. (Since the solar wind instrument on Voyager 1 had not been operative for several years, direct measurement of the subsonic wind was not made. Deductions of a subsonic wind flow were made by analysis of the directionality of the low-energy charged particle fluxes that were measured on the spacecraft. The solar wind instrument on the Voyager 2 spacecraft is still operative, so direct measurements of the subsonic flow can be expected when Voyager 2 crosses the shock sometime in the next few years.)

At first sight, the set of observations appeared to match many theoretical predictions, but as Voyager 1 mover farther out into the heliosheath, the expected increase in the ACR population was not observed. The existence of ACRs has long been considered a strong indication of the strength of the termination shock, but with no change detected in the ACR flux long after crossing the termination shock, the production process for ACRs is now uncertain. Suggestions include either that it occurs beyond the shock or only at some locations on the shock.

A current, somewhat *ad hoc* suggestion to explain the "anomalous" ACR observations is that the geometry of the termination shock is not suitable for the energization process at the expected location. That is, ACRs actually originate around the flanks of the heliosphere, where the geometry of the termination shock more readily enables the energization to be carried out. Obviously, substantially more theory and measurements are required.

1.4 HELIOSPHERIC STRUCTURE AND DYNAMICS OVER THE SOLAR CYCLE

1.4.1 The solar wind through the solar activity cycle

The principal changes in the solar wind, as a consequence of changing solar activity levels, occur in three dimensions, mainly out of the ecliptic plane which contains the Earth's (and other planets') orbit(s) and that is inclined at 7.25° to the solar equatorial plane. Ground- and space-based solar observations have shown that there are many important changes in the Sun and its corona that accompany the sunspot cycle. One of the main discoveries concerning the solar wind in the 1970s was that there are two kinds of solar wind, fast streams (at speeds in excess of about 650 km/s) and slow solar wind (with speeds less than about 550 km/s). In addition to speed, most of the parameters that describe the solar wind, such as its density, temperature, and composition, are significantly different in the two kinds of wind. Also in the 1970s, solar observers identified dark, relatively cool regions in the corona as being the origin sites of the fast solar wind streams. These regions are known as coronal holes. One of the characteristics of coronal holes is that the solar magnetic fields that they contain are open to interplanetary space. The magnetic field lines that are dragged into the heliosphere by the solar wind originate mostly from these coronal holes.

The origin of the slow solar wind is less well understood. Slow wind is far more variable in its parameters than is fast solar wind. It is likely to be generated near or at the edges of hot regions (streamers) in the solar corona.

The magnetic field of the Sun that threads the solar corona is carried into space by the solar wind from the coronal holes. The Sun, unlike the Earth, does not have simple north and south magnetic poles. Around the minimum activity phase, there are large coronal holes covering the heliographic poles of the Sun; the magnetic fields that emanate from these holes are of opposite polarities in the north and the south. This is the closest the Sun ever gets to exhibiting a magnetic dipole structure such as do the magnetized planets. As solar activity increases, these polar coronal holes shrink and fragment, so that there is apparently much less open magnetic flux. The polarities in the corona become much more mixed.

As the magnetic field from the coronal holes is carried out in the solar wind, the magnetic polarities are separated by a so-called neutral line, separating the inward- and outward-pointing magnetic fields. In interplanetary space, the surface that separates the polarities is called the Heliospheric Current Sheet (HCS); near solar minimum this vast surface is close to the equatorial plane of the Sun, while near solar maximum it becomes very complex and highly inclined with respect to the solar equator. Pictures of the solar corona, taken at the time of total eclipses and now more routinely from space-based observatories such as SOHO, show long, bright streamers that are confined near the solar equator near solar minimum, but point in various directions at high heliographic latitudes when the sunspot number is high. It is now known that these coronal streamers are the roots of the HCS close to the Sun. With increasing distance from the Sun (measured in distances of a few solar radii), the streamers become fainter as the density of the material becomes thinner. The "folds" that occur in the HCS have likened it to a ballerina's skirt; this is the largest structure in the heliosphere and is likely to play a key role in the modulation of galactic cosmic rays.

The two kinds of solar wind are well delineated around solar minimum: all the fast solar wind comes from the two large polar coronal holes and the slow wind originates from above the equatorial regions where the magnetic field remains in the form of loop systems. The HCS is always embedded in slow solar wind, just as the streamers are observed to originate above the hot-loop system in the equatorial corona. A further effect that shapes the heliospheric medium is what is called the over-expansion of the fast wind: even though coronal holes above the poles have a large-scale angular extent of only about 30° or so away from the poles of the Sun, the fast wind actually fills the inner heliosphere to much lower heliolatitudes. It is as though at least at the edges of coronal holes the solar wind flows bend in a direction away from the solar poles. Around solar minimum, the polar coronal holes remain approximately the same for many 27-day periods of solar rotation.

At solar maximum, coronal holes are small and can be found everywhere on the Sun, not just near the poles. They are also generally short-lived, often appearing and disappearing within a single solar rotation. As a result, the solar wind is rarely fast, but generally mixed, with some not-so-fast wind and slow wind streams mingling at all heliolatitudes. The contrast between the solar wind at solar minimum and solar maximum is well illustrated by the observations of Ulysses.

1.4.2 Close to solar-minimum activity: corotating interaction regions

The formation of polar coronal holes, following solar-maximum activity, is not a simple process. It involves the migration, over many solar rotations, of magnetic regions of opposite polarities towards the heliographic poles. During the time when these large coronal holes form, portions of them can reach low latitudes, towards the equator. This means that, at equatorial solar latitudes, both fast and slow solar wind streams can be generated in successive solar longitude ranges. As the Sun rotates, fast and slow solar wind streams are therefore emitted alternately into a given radial direction. But then a parcel of fast wind will catch up with the slow stream ahead of it. In the interaction between the fast and slow winds, the solar wind plasma is compressed and heated. As the compressed plasma travels out away from the Sun, it forms traveling shock waves at its leading and trailing edges. If, as happens close to solar minimum, the flow pattern of fast and slow streams remains approximately the same over successive solar rotation periods, the outward traveling interaction regions appear to rotate with the Sun, hence the name CIR. Such CIRs can persist for a year or more, usually in the period between solar maximum and minimum activity and constitute the major structuring process at those times in the inner heliosphere. In addition, the leading- and trailing-edge shock waves constitute an important source of energetic particles in the heliosphere, not competing with galactic cosmic rays, but still providing an important example of how energetic particles can be produced in the universe.

1.4.3 Around solar-maximum activity: coronal mass ejections

Solar-maximum activity has been historically defined by the number of sunspots that can be observed and measured. While this remains a useful descriptive criterion, it is now known that sunspots themselves are only a symptom of considerable changes in the Sun's upper layer, the convection zone (just below the visible photosphere). Solar activity in fact increases in response to the complexity of magnetic fields that emerge from the convection zone to the surface of the Sun; the increased complexity restructures the magnetic fields in the solar corona. In the course of this restructuring process, very large and occasionally explosive amounts of magnetic energy are transformed into sudden heating of the corona in localized areas. During this sudden heating, very large amounts of coronal material are expelled into space, embedded in the solar wind. These events are the CMEs; their number increases from close to zero at solar minimum to several per day close to solar maximum.

In many cases, CMEs appear to be closed magnetic structures, unlike the ordinary solar wind that has open magnetic fields embedded in it. CMEs, when directed towards the Earth, often carry enough hot plasma and strong magnetic fields to cause major disturbances in the terrestrial magnetic field, producing magnetic storms on Earth. The strongest of these storms, usually before or after solar activity maximum, depress the Earth's magnetic field and cause an increase in the intensity of trapped charged particle radiation in Earth's magnetosphere. These particles can damage spacecraft and indirectly even affect large sections of terrestrial power supply grids. As the CMEs propagate towards the outer heliosphere, they often amalgamate and the largest among them form what has been called a Global Merged Interaction Region (GMIR). Such a region acts as a barrier between the inner- and outer-heliospheric regions and significantly impedes the access of galactic cosmic rays to the vicinity of Earth.

One of the effects of large-scale, probably merged CMEs is the effect they can have on the outer boundaries of the heliosphere. Some intriguing radio noise observations by the Voyager spacecraft in 1983 and 1992–1993 have been interpreted as the signs of radio emissions from the heliopause when large-scale CMEs reach it and disturb it. This provides yet another tentative measure of the distance of the heliopause: estimating the travel time of the CMEs that perhaps produce the radio emissions gives a distance of about 150 AU to the heliopause, a figure that is largely consistent with theoretical expectations and the currently measured termination shock location at about 100 AU.

1.4.4 Energetic solar particles

The intensity levels of energetic particles originating at or near the Sun in the heliosphere depend importantly upon the stage in the 11-year solar activity cycle in which measurements are made. This is simply because the level of solar activity determines the number of events that can produce energetic particles, either directly from activity in the corona itself or by interactions of solar-produced CMEs and CIRs with the ambient heliosphere population.

Since the initial measurements at ground level of solar-produced energetic particles, as well as pre-spacecraft era measurements of the blackouts of polar region radio transmissions by lower energy particles, it has been known that appearances of solar energetic particles roughly tracked the solar activity cycle. Energetic particle events are less prevalent during solar-minimum than during solar-maximum conditions. Nevertheless, solar particle events do occur during solar minimum if there is solar activity at these times. The ground level solar event of January 2005 and the intense solar activity of December 2005 are recent examples of this, although such examples exist from past solar cycles as well.

Instruments on the Ulysses mission have been critical in defining the three-dimensional spatial extent of solar energetic particles in the heliosphere as a function of solar activity and the solar cycle. During solar minimum, Ulysses has shown that energetic particles are most often associated with CIRs, and not with specific solar activity. These CIR-associated particle enhancements are confined approximately to the region of the heliospheric current sheet, not extending much beyond about 30° to 40° in heliolatitude, in general. When a solar event does occur, solar particles from the event itself and from CME-produced particle acceleration can be found up to the highest latitudes that Ulysses traveled.

During solar-maximum conditions, solar energetic particles can be found at essentially all heliolatitudes. In both solar-maximum and solar-minimum conditions, solar energetic particles have been detected by both Voyagers out to the outer edges of the heliosphere.

1.4.5 Large-scale structures and the modulation of cosmic rays

The cause of the observed modulation of galactic cosmic rays (i.e., the increase in intensities during solar minimum and decrease in intensities during solar maximum) is the increase in complexity of heliospheric structures from solar-minimum to solar-maximum activity. However, the precise processes that control the access of cosmic rays into the inner heliosphere are not fully understood. Almost certainly, several factors jointly play a role in this process. One such factor is the formation of GMIRs; it has been noted that cosmic ray intensity decreases significantly, almost in a step-like way, following the passage of a GMIR. But the modulation process as measured at the Earth is smoother and almost certainly involves higher levels of turbulence (or simply disorder) in the heliospheric magnetic field around maximum activity. The access of galactic cosmic rays to the heliosphere is thus largely controlled by a combination of GMIRs, turbulence, and the heliospheric boundary. The relative importance of each likely depends upon the phase of the solar cycle.

1.5 THE EXPLORATION OF THE HELIOSPHERE

The space era has provided critical data that has permitted quantitative modeling of the global heliosphere. The Sun, its interior, and its atmosphere are much better understood, as are properties of the solar wind and solar energetic particles and their dependence on conditions in the solar corona. Nevertheless, significant gaps remain. Importantly, only a few measurement points are available in critical parameter ranges and very sparse measurement sites exist in the vast extent of the heliosphere. It can certainly be expected that major surprises will be encountered as further space missions slowly fill the gaps in the future.

Several of the key missions that have explored the heliosphere and that provided the main milestones in achieving new understanding are briefly summarized in Sections 1.5.1–1.5.4. This discussion is not exhaustive, meant principally as a guide to the past as well as a look toward the future. References to the missions mentioned are widely available through search services on the World Wide Web.

1.5.1 Inner heliosphere

Early in the space age, significant uncertainties as to the quality and validity of data returned by instruments flown on the Lunik 2 and 3 spacecraft (1959) and on the Explorer 10 spacecraft (1961) left open the question of the existence of a solar wind. It is now recognized that the first successful interplanetary mission, the Mariner 2 mission to Venus (1962), conclusively discovered the solar wind and returned the first details of its properties.

The joint German/U.S. dual spacecraft mission, Helios 1 and 2, with perigees at ~0.3 AU and apogees at Earth orbit, are the only missions to date to travel so near to the Sun and to return such important solar wind data on the deep inner heliosphere. The Pioneer Venus Orbiter mission (1978), with its elliptical trajectory around the planet, provided an additional platform at the orbit of Venus (~0.7 AU) for studies of the solar wind inside 1 AU.

1.5.2 Earth-orbiting missions

There have been numerous spacecraft that have circled Earth and have traveled to the Moon in the first half-century of the space age. Many Earth-circling missions had orbits outside the normal dayside magnetosphere or highly elliptical orbits that took the spacecraft beyond the magnetosphere. Most of these missions had instruments that provide data on the solar wind and *in situ* measurements of solar particle events. Such missions included spacecraft in the Interplanetary Monitoring Platform (IMP) series (1960s), the International Sun–Earth Explorer (ISEE) set of two spacecraft (1970s). In many ways, the ISEE 1, 2 continued the exploration mission of the IMP spacecraft series.

Also important for heliosphere studies near Earth were the near-circular orbit Vela series of spacecraft (1960s and 1970s) that were implemented for the monitoring and detection of clandestine nuclear testing in space. In addition to its monitoring responsibilities and studies of the solar wind, instruments on the Vela 5 spacecraft first detected gamma ray bursts from astronomical sources.

1.5.3 L1 spacecraft

The L1 Lagrangian point, at ~0.01 AU from Earth along the Earth–Sun line, is an ideal location from which the solar wind and energetic particles can be measured—a location that rotates around the Sun at the same rate as does Earth. A number of important interplanetary missions have been flown to L1 for studies of the solar wind and of the Sun. The first of these, the ISEE 3 mission, was designed for the L1 location to provide upstream solar wind data to interpret the solar wind energy input to the magnetosphere as was measured by the Earth-orbiting dual ISEE 1, 2 pair. (ISEE 3 was targeted later in its life to a study of the deep magnetotail of Earth and an encounter with the comet Giacobini–Zinner; it was thus renamed the International Cometary Explorer—ICE.)

Other important L1 missions over the last two decades have been the Solar and Heliospheric Observatory (SOHO; launched in 1995) and the Advanced Composition Explorer (ACE) mission (launched in 1997). SOHO carries instruments that monitors the Sun, the solar wind, and energetic particles and has been exceptional in bringing new insights into solar activity and its effects on the heliosphere.

In addition to its primary mission of measuring the atomic and isotopic composition of primary cosmic rays and solar particles, ACE also carries a solar wind instrument, a magnetometer, and an energetic particle instrument. These latter three instruments supply real-time "space weather" data to the U.S. NOAA Space Environment Center in Boulder, Colorado.

1.5.4 Outer heliosphere

The first spacecraft to travel beyond the orbit of Mars were the Pioneer 10 and 11 missions to Jupiter (launched in 1972 and 1973, respectively). Passing Mars, data from the instruments on Pioneer 10 showed definitively that the boundary of the heliosphere was clearly and considerably beyond this planet's orbit, firmly contradicting some of the speculations and theories of a decade and more earlier.

Following encounters with all four giant planets, it was finally instruments on the Voyager 1 spacecraft (both Voyagers were launched within a month of each other in 1977) that "discovered" the boundary to the solar system in December 2004. If the power sources on the Voyagers permit, both Voyagers may be able to traverse the heliosheath region and reach interstellar space to return data from there.

Shortly following the dawn of spacecraft flight, Simpson *et al.* (1959) proposed that a mission be designed and flown out of the ecliptic plane of the planets. Such a mission could provide important data concerning cosmic rays in the solar system. Two decades later (1978) such a mission consisting of two spacecraft, Out-of-the-Ecliptic (OOE; later renamed Ulysses), was born from a joint agreement between NASA and the European Space Agency. Two years later the U.S. spacecraft was canceled and OOE continued to its launch in 1990 with the single spacecraft that continues to return information on the three-dimensional heliosphere. The orbit and objectives of Ulysses are unique in the annals of space exploration to date, and no other such heliosphere mission to such high heliolatitudes is planned.

1.5.5 Future heliosphere missions

Several heliosphere missions are in various stages of planning and/or execution. The dual STEREO mission, launched in late 2006, will provide stereoscopic views of solar activity that give rise to interplanetary events, as well as two-point *in situ* measurements of the solar wind and energetic particle events.

The Interstellar Boundary Explorer spacecraft (IBEX, launch in 2008), from its location in a highly eccentric and high-altitude Earth orbit, will return energetic neutral atom images of the distant boundary of the heliosphere termination shock and the interstellar medium beyond.

The ESA-planned Solar Orbiter mission will rise over about a 7-year period from the ecliptic plane to a latitude of $\sim 35^\circ$ with respect to the solar equator and will approach as close as 35 solar radii to the Sun. The mission will take images of solar activity with unprecedented precision and measure the solar wind and energetic particles. Solar Orbiter is still in a definition phase, with a possible launch in the middle of the second decade of this century.

The Interplanetary Sentinels mission, consisting of four spacecraft at and within 1 AU and designed in the context of NASA's Living with a Star program, will enable the understanding and modeling of the connection between solar phenomena (including on the far side of the Sun) and heliosphere disturbances, particularly those impacting Earth's space environment. This mission is still in its very initial definition phase with a possible launch in the coming decade. Currently, discussions are underway that are exploring possible collaboration between the Sentinels and the Solar Orbiter missions.

1.5.6 Summary

1. Heliosphere research with spacecraft began at the outset of the space age, motivated largely by the early ground-based cosmic ray studies, and has continued vigorously to date.

2. Ever more sophisticated spacecraft and spacecraft instrumentation have returned ever more detailed views of the Sun-influenced space environment in which Earth and all solar system planets are embedded.

3. A number of missions in the near-execution phase as well as in more distant planning for the future will continue the past heritage of clever and insightful investigations of the heliosphere.

4. The possibilities for continued advances in heliosphere research look bright going forward.

1.6 REFERENCES

Axford, W. I., A. J. Dessler, and B. Gottlieb (1963), Termination of Solar Wind and Solar Magnetic Field, *Astrophys. J.*, **137**, 1268.

Biermann, L. (1951), Kometenschweife und solare Korpuskularstrahlung, *Z. Astrophys.*, **29**, 274–286.

Chamberlain, J. W. (1960), Interplanetary gas II. Expansion of a model solar corona, *Astrophys. J.*, **131**, 47–56.

Chapman, S., and J. Bartels (1940), *Geomagnetism*, Oxford University Press.

Davis, L. (1955), Interplanetary Magnetic Fields and Cosmic Rays, *Phys. Rev.*, **109**, 1440–1444.

Fisk, L. A. (2005), Journey into the Unknown Beyond, *Science*, **309**(5743), 2015.

Forbush, S. E. (1946), Three Unusual Cosmic-Ray Increases Possibly due to Charged Particles from the Sun, *Phys. Rev.*, **70**, 771.

Forbush, S. E. (1954), World-Wide Cosmic-Ray Variations, 1837–1952, *J. Geophys. Res.*, **59**, 525.

Hundhausen, A. J. (1968), Interplanetary Neutral Hydrogen and the Radius of the Heliosphere, *Planet. Space Sci.*, **16**, 783–793.

Izmodenov, V. V. (2004), The Heliospheric Interface: Models and Observations, in *The Sun and the Heliosphere as an Integrated System* (G. Poletto & S. T. Suess, eds.), pp. 23–64, Kluwer Academic Publishers.

Lanzerotti, L. J., and M. Schulz (1969), Interaction between the Boundary of the Heliosphere and the Magnetosphere of Jupiter, *Nature*, **222**, 1054.

Meyer, P., E. N. Parker, and J. A. Simpson (1956), Solar Cosmic Rays of February 1956 and Their Propagation through Interplanetary Space, *Phys. Rev.*, **104**, 768–783.

O'Gallagher, J. J. (1967), Cosmic-Ray Radial Density Gradient and Its Rigidity Dependence Observed at Solar Minimum on Mariner IV, *Astrophys. J.*, **150**, 675.

Parker, E. N. (1958), Dynamics of the interplanetary gas and magnetic field, *Astrophys. J.*, **128**, 664.

Parker, E. N. (1963), *Interplanetary Dynamical Processes*, Interscience Publishers, New York.

Simpson, J. A., and J. R. Wang (1967), Dimension of the Cosmic-Ray Modulation Region, *Astrophys. J. Lett.*, **149**, L73.

Simpson, J. A., B. Rossi, A. R. Hibbs, R. Jastrow, F. L. Whipple, T. Gold, E. Parker, N. Christofilos, and J. A. Van Allen (1959), Round table discussion, *J. Geophys. Res.*, **64**, 1691.

2

Solar cycle 23

David H. Hathaway and Steven T. Suess

2.1 INTRODUCTION

Ulysses' launch in October of 1990 was at the maximum of solar activity cycle 22. The first passages through the polar regions of the heliosphere came in 1994 and 1995, very near the minimum of activity between cycles 22 and 23. The second orbit then took Ulysses through the polar regions in 2000 and 2001, at the maximum of solar activity for cycle 23, and its third orbit will again sample the polar regions at near-minimum conditions (Figure 7.1). Ulysses has thus observed heliospheric conditions through a complete solar cycle, solar cycle 23. How typical was cycle 23? In this chapter we will examine the characteristics of this cycle, its noteworthy events, and compare it with other cycles.

Solar cycle 23 has been, by far, the best observed solar activity cycle. The Ulysses spacecraft was joined in space by WIND in 1994, the Solar and Heliospheric Observatory (SOHO) in 1995, and by the Advanced Composition Explorer (ACE) in 1997. These three spacecraft continue to hover around the L1 point between the Earth and Sun and observe the Sun and solar wind from that vantage point. The Transition Region And Coronal Explorer (TRACE) was launched in 1998 and provides images at high spatial and temporal resolution of photospheric and coronal structures from low Earth orbit. The Ramaty High Energy Solar Spectroscopic Imager (RHESSI) was launched in 2002 and provides spectroscopic imaging of hard X-rays and gamma rays from solar events. Although the Yohkoh satellite failed in 2001 it did provide X-ray observations of the Sun's corona during the rise of cycle 23. This array of space-based observatories is complemented by a similar array of improved ground-based instruments. The combined observations from these observatories and their instruments provide a detailed view of solar activity through cycle 23.

2.2 SOLAR ACTIVITY CYCLES

Sunspots are the traditional indicators of solar activity, with good reason. First, the sunspot record extends back nearly 400 years to the time of Galileo. This long record includes the Maunder Minimum of solar activity and the recent high-amplitude cycles, with a broad range of behavior in between (Figure 2.1). Second, the sunspot number is well correlated with other indicators of solar activity and thus provides a good measure of solar activity. The smoothed sunspot number is correlated with sunspot areas, the 10.7 cm radio flux, and the total solar irradiance at the 99% level (Figures 2.2a, b, and c). There is a 93% correlation between smoothed sunspot numbers and the number of M- and X-class flares (Figure 2.2d) and a 91% correlation with the basal level of geomagnetic activity (Figure 2.2e). The geomagnetic field is also buffeted by high-speed solar wind streams during the declining phase of the solar cycle that contribute to high geomagnetic activity at low levels of solar activity. Cosmic ray fluxes have an anti-correlation with sunspot number (Figure 2.2f). This anti-correlation is strongest for the time lag of about 8 months between sunspots (solar activity) and low-energy cosmic rays. The lag presumably reflects the time it takes for solar disturbances to propagate through the heliosphere, where it takes $\sim$1 year for solar wind to reach the termination shock in the upstream direction relative to the local interstellar medium.

Studies of the sunspot number record reveal several significant characteristics of the average sunspot cycle. The average cycle has a period from minimum to minimum of nearly 11 years (131 ± 14 months) with a normal distribution about that mean (Hathaway and Wilson, 2004). The level of activity is asymmetric with respect to the time of maximum (Figure 2.3a). It usually takes 4 or 5 years to rise from minimum to maximum and 6 or 7 years to fall from maximum back to minimum again. This asymmetry is accentuated for large-amplitude cycles (as measured by their maximum

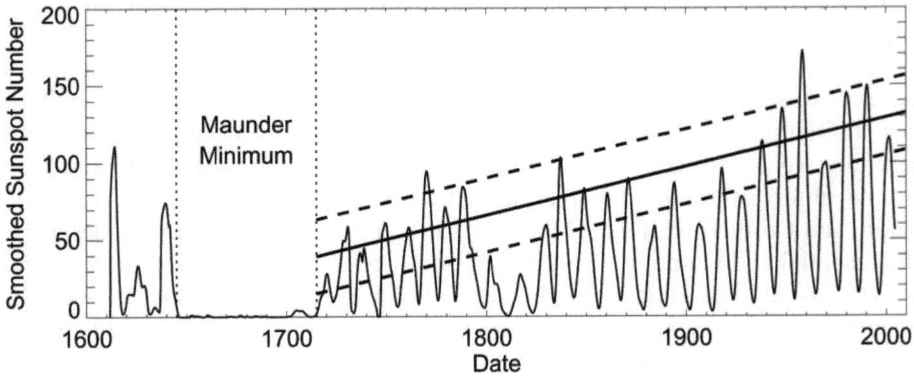

Figure 2.1. The 400-year record of sunspot numbers shows the Maunder Minimum and the trend toward bigger cycles in recent decades. Cycle 23 is the last cycle in this series, a slightly larger-than-average cycle. The solid line is a linear fit to the maxima since the Maunder Minimum, with the $\pm 1\sigma$ uncertainty shown by the two dashed lines.

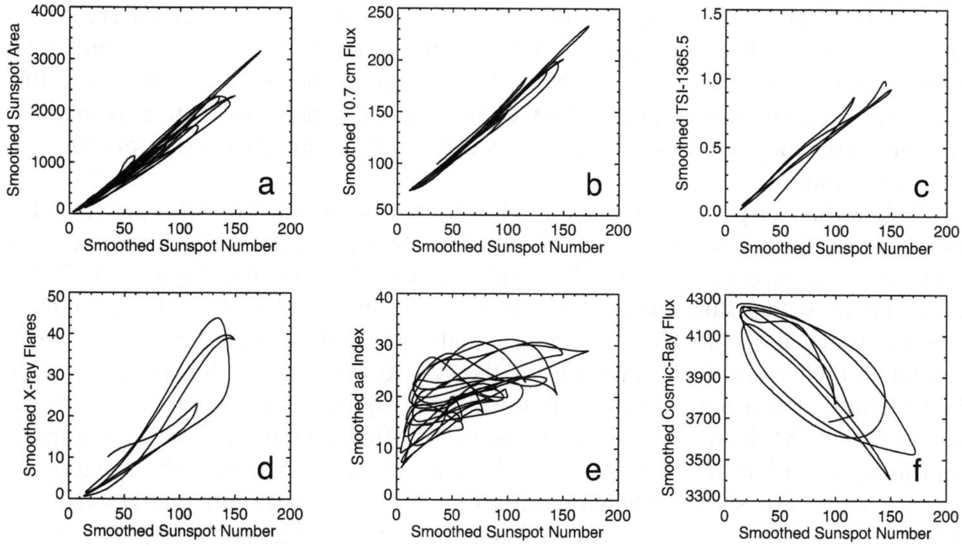

Figure 2.2. The number of sunspots on the Sun is tightly correlated with sunspot areas (a), 10.7 cm radio flux (b), and total solar irradiance (c). Strong correlations also exist between sunspot numbers and the number of M- and X-class flares (d) and the basal level of geomagnetic activity (e). There is a strong anti-correlation between sunspot numbers and low-energy, ground-level cosmic ray neutrons (f). (NOAA/National Geophysical Data Center)

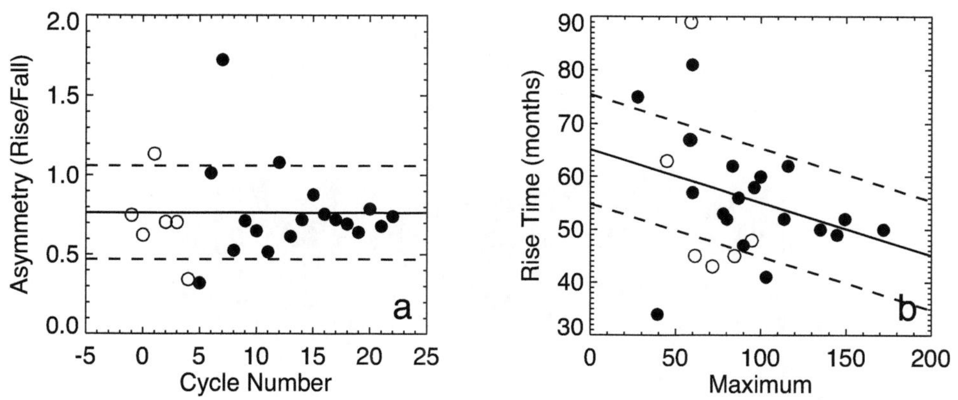

Figure 2.3. Sunspot cycles are asymmetric with respect to the time of cycle maximum. The ratios of the cycle rise time to the cycle fall time for each cycle since 1797 are shown by the filled circles in panel (a). Earlier (less reliable) data are shown by open circles. On average the rise time is only about 75% as long as the fall time. The decrease in rise time with cycle amplitude (the "Waldmeier Effect") is shown in panel (b). Larger cycles take less time to rise to maximum.

sunspot number). This was first noted by Waldmeier (1939) and is often referred to as the "Waldmeier Effect" (Figure 2.3b). Large-amplitude cycles are usually preceded by short period cycles and high minima. These characteristics are well captured by the idea that large-amplitude cycles start early and rise rapidly to their maxima. By starting early they cut short the previous cycle and the added overlap produces a high minimum.

While the sunspot number is a good indicator of the level of solar activity, the positions of the sunspots are also significant. Sunspot positions with respect to latitude are represented in "butterfly diagrams" like that shown in Figure 2.4. These diagrams show that sunspots appear in two bands, one on either side of the equator. The bands develop at about 30° from the equator at the start of each cycle. They then spread to higher and lower latitudes as the activity rises. As each cycle progresses these bands of sunspots slowly drift toward the equator but appear to avoid the equator itself. At the time of minimum the cycles often overlap, with new cycle spots appearing at mid-latitudes while old cycle spots are still found near the equator.

Sunspots also show systematic variations with longitude but this behavior is much more subtle than the latitudinal behavior seen in the butterfly diagrams. Indi-

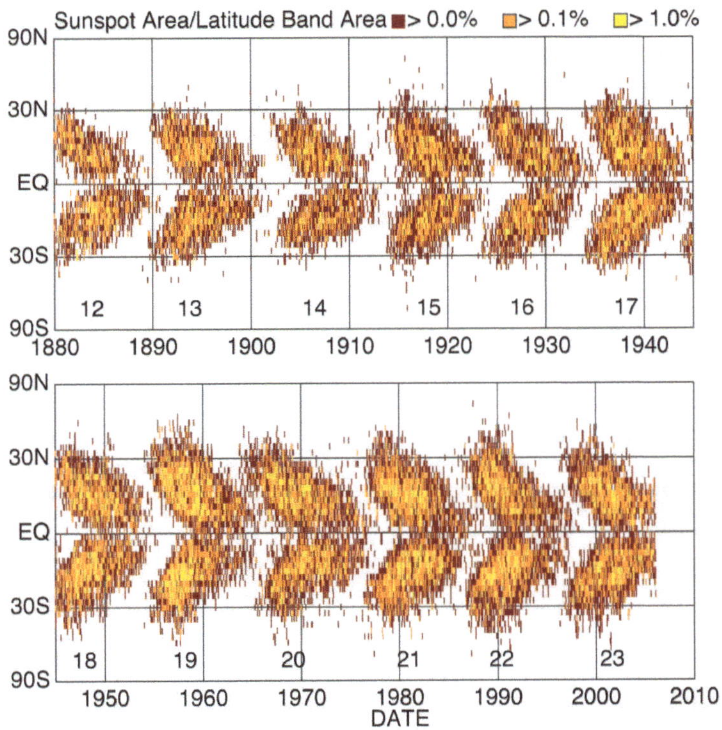

Figure 2.4. The sunspot butterfly diagram shows the latitudinal position of sunspots on the Sun as a function of time. The active latitudes drift toward the equator as each cycle progresses. Cycles often overlap significantly near times of activity minima. (After Hathaway *et al.*, 2004)

vidual sunspots typically last for days and the active regions they are embedded in may last for weeks or even months. Yet, there appear to be active longitudes or "nests" where active regions tend to emerge recurrently for years or even decades. This phenomenon is complicated by the Sun's differential rotation because active regions and active longitudes commonly drift with respect to the longitudes given by the Carrington rotation rate. Nonetheless, simply averaging the sunspot areas in longitude bins fixed in the Carrington system indicates the presence of significantly overactive longitude positions (Hathaway and Wilson, 2004; Henney and Harvey, 2002; Castenmiller, Zwaan, and van der Zalm, 1986).

Preferred-longitude effects are also found to map into the solar wind at a synodic period of 27.03 ± 0.02 days over a multi-decade data set. Other periodicities are often more prominent for shorter intervals, such as a single solar cycle or less. On average, solar magnetic field lines in the ecliptic plane point outward on one side of the Sun and inward on the other, reversing direction approximately every 11 years while maintaining the same phase. This is consistent with a model in which the solar magnetic dipole returns to the same longitude after each reversal (Neugebauer *et al.*, 2000). The equatorial dipole component is also visible in Ulysses data at solar maximum, during the time of polar field reversal (Jones, Balogh, and Smith, 2003), although it is often masked in photospheric data by strong active-region magnetic fields.

Measurements of the Sun's magnetic field reveal far more about solar activity than do sunspot observations alone. However, systematic daily observations of the magnetic field strength over the full solar disk only cover three solar cycles, three cycles with similar amplitudes. The daily magnetograms from ground-based observatories (NSO/Kitt Peak, Wilcox Solar Observatory, and Mount Wilson Observatory) and from SOHO can be used to construct synoptic maps of the Sun's magnetic field. Since only the Earth-facing side of the Sun is observed, these maps do not represent a "snapshot" of the Sun's magnetic field distribution but instead represent a convolution with different times associated with different longitudes. Nonetheless, these maps do provide our best estimation of the global distribution of magnetic field elements over the surface of the Sun. When we average the magnetic field in longitude over each solar rotation the resulting distribution (Figure 2.5) again reveals a butterfly pattern for the active latitudes but with many additional details.

Active regions have a characteristic tilt (Joy's Law) in that the leading polarity spots tend to be closer to the equator than the opposite polarity following spots (Hale *et al.*, 1919; Howard, 1996). This tilt leaves a predominance of leading polarity field closer to the equator and following polarity field toward the poles. A poleward meridional flow with a speed of about 10 m/s (Komm *et al.*, 1993) carries these field elements to the poles where they cancel the existing opposite polarity field and ultimately reverse the polar field polarity at about the time of cycle maximum (Figure 2.6).

The global dipolar nature of the Sun's magnetic field and its reversal at the maximum of each solar cycle brings up the point that the solar magnetic cycle takes up two solar activity cycles, or about 22 years. This magnetic cycle has been given the name Hale cycle, which represents the true physical cycle. Figure 2.6 shows the polar

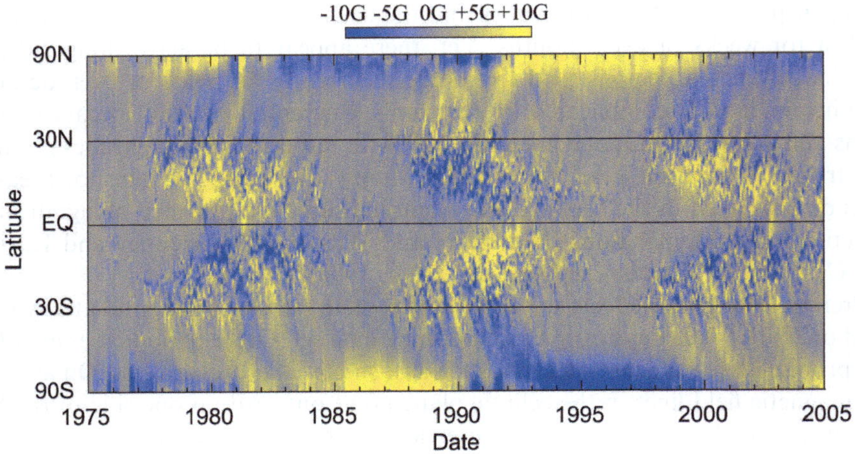

Figure 2.5. The magnetic butterfly diagram shows the latitudinal distribution of weak magnetic fields on the Sun obtained by averaging the observed field strength at each latitude over each solar rotation. The butterfly wing pattern in the active latitudes shows evidence of both Hale's Law and Joy's Law, the equatorward edges are dominated by leading polarity field while the poleward edges are dominated by following polarity field. These fields flip from cycle to cycle and from hemisphere to hemisphere. A poleward meridional flow of predominantly following polarity fields accumulates a field at the poles that reverses the polar field polarities at about the time of cycle maximum.

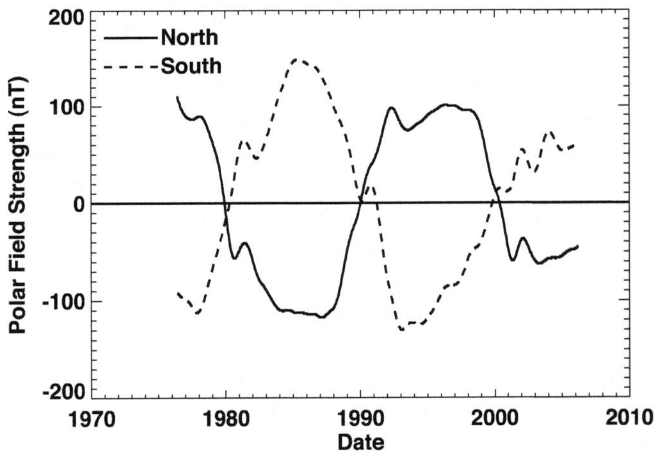

Figure 2.6. Polar field strengths from the Wilcox Solar Observatory. Daily measurements of the field strength in the polemost aperture (latitudes above about 55°) are smoothed to remove seasonal effects. The fields reversed in 1980, 1990, and 2000, about the time of sunspot cycle maximum for each cycle. Both hemispheres are not perfectly synchronized. The northern hemisphere reversed first during cycles 21 and 22 but was preceded by the southern hemisphere in cycle 23.

fields for slightly more than one Hale cycle. The reversals need not be symmetric and it is seen here that the north and south reversals occurred simultaneously only in 1980, at the maximum of sunspot cycle 21. Generally, the reversals are not simultaneous. The field reversed first in the north in cycle 22 and first in the south in cycle 23—several months earlier than in the north. However, the field strength in the north changed sign much more rapidly than in the south in cycle 23, with the reversed (negative) field gaining strength more rapidly than in the south. So far, the polar field strength between cycles 23 and 24 has only been about half as large as the minimum between cycles 22 and 23.

2.3 CYCLE 23

Cycle 23 began with a controversy concerning the "official" time of minimum. The traditional method for determining the levels and times of cycle minima and maxima uses the 12-month running mean of the monthly averages of the daily sunspot numbers. The 12-month running mean is actually a weighted average over 13 months with weights of 1/24 for the first and last month and weights of 1/12 for the 11 months in between. According to this measure, cycle minimum occurred in May of 1996. This date corresponded with the date of first appearance of new cycle spots, an occurrence which usually comes before minimum. The sunspot activity over the following few months was dominated by old-cycle spots rather than the rising number of new-cycle spots and a secondary minimum occurred later that year. This has led many to accept September of 1996 as the date of minimum between cycles 22 and 23 (Harvey and White, 1999).

Cycle 23 was predicted to be a much-bigger-than-average cycle (Joselyn *et al.*, 1997). This prediction was primarily based on the level of geomagnetic activity near the time of cycle minimum, something that had been a reliable predictor in the past. The initial rise of cycle 23 activity seemed at first to support this prediction. By September of 1998 the monthly average of the daily sunspot number reached 93, well on its way to the predicted maximum of 160 ± 30, but activity faltered during the next few months and did not rise above this level again until May of 1999. Cycle 23 ultimately was larger than average but still significantly smaller than predicted and smaller than four out of the previous five cycles (Figure 2.1). Its maximum was characterized by two peaks, with the smoothed sunspot number reaching 121 in April of 2000, dropping to 104 in February of 2001, and reaching its second maximum of 115 in November of 2001 (Figure 2.7). Double peaks such as this are not uncommon when using the 12-month running mean.

While cycle 23 did not reach its expected amplitude, it was in many respects a typical cycle. Its amplitude was above average and within the range of variability expected from the secular increase in cycle amplitudes shown in Figure 2.1. Its shape was typical, with a rapid rise to maximum and slower decline toward minimum, giving the characteristic asymmetry for a cycle of its size. Even the double peak is fairly typical. There were, however, several atypical aspects to cycle 23.

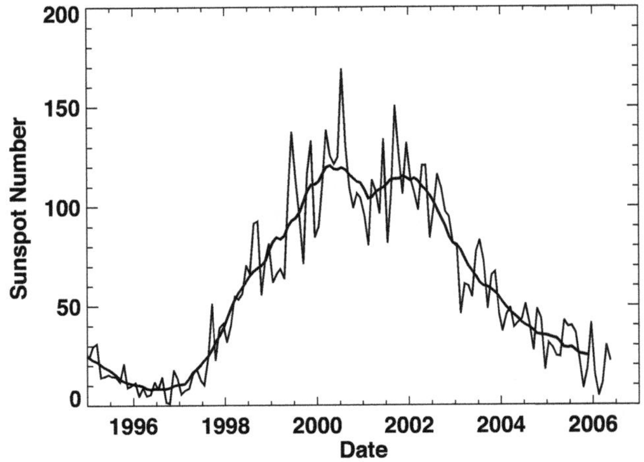

Figure 2.7. Cycle 23 as defined by the sunspot number. The monthly averages of the daily sunspot numbers are shown by the thin line. The smoothed (12-month running mean) values are shown by the thick line. Cycle 23 was double-peaked with the first, and higher, peak reaching 121 in April of 2000.

Flares have been observed by X-ray monitors on the NOAA GOES satellites continuously since 1975 (*http://www.ngdc.noaa.gov/*). The number of X-ray flares is generally proportional to the number of sunspots (Figure 2.8) but with a higher proportionality constant for the declining phase of the sunspot cycle. Cycle 23 was exceptional in the number of large flares that came late in the cycle. As the smoothed sunspot number dropped below 50 the number of flares remained relatively constant,

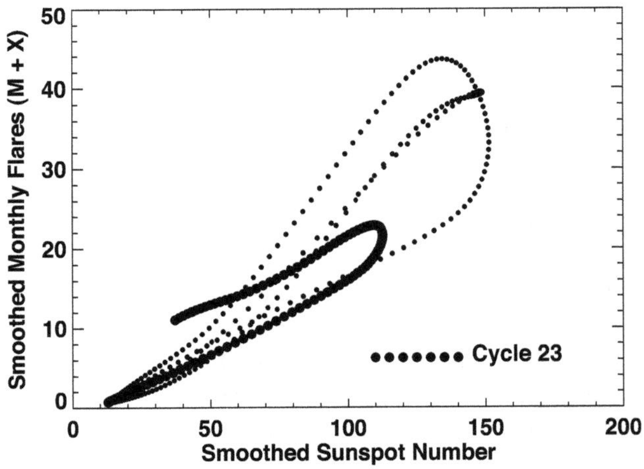

Figure 2.8. X-ray flares per month as a function of smoothed sunspot number. For a given sunspot number there tend to be more flares after cycle maximum. Cycle 23 is exceptional in that well after maximum it had twice as many flares per month as the previous two cycles had at the same sunspot number level.

giving nearly twice the number of flares than were seen in the previous two cycles at the same sunspot number level. Late cycle flare activity for cycle 23 certainly "pushes the envelope" for our experience with X-ray flare data.

Two intervals of late cycle solar activity were so remarkable as to be given names: the Bastille Day storms of July 2001 and the Halloween Storms of October–November 2003 (Gopalswamy *et al.*, 2003). The former came between the first and second sunspot maxima, while the latter was well into the declining phase of the cycle. The Halloween Storms spawned aurorae that were seen over most of North America and extensive satellite problems were reported. The 4 November flare was one of the most powerful X-ray flares ever detected. The events were far away from Ulysses in heliographic longitude, but were nevertheless detected in magnetic field, plasma, and energetic particle data. The resulting merged interaction region produced unusually large pressures at Voyagers 1 and 2 and apparently significantly moved the termination shock at ~100 AU (Webber, 2005).

The excess level of flaring activity late in cycle 23 is not obviously reflected in ground-level cosmic ray neutron monitor data. In general, the cosmic ray flux is anti-correlated with sunspot number, the higher the sunspot number the lower the cosmic ray flux. The correlation is strongest with an 8-month lag for the cosmic ray flux signal. As with flares and geomagnetic activity through the cycles, there is a stronger effect during the declining phase of the sunspot cycles. Cycle 23 showed the same effect (Figure 2.9) but the lower cosmic ray fluxes after cycle maximum, while low, were very much in line with those seen during previous cycles.

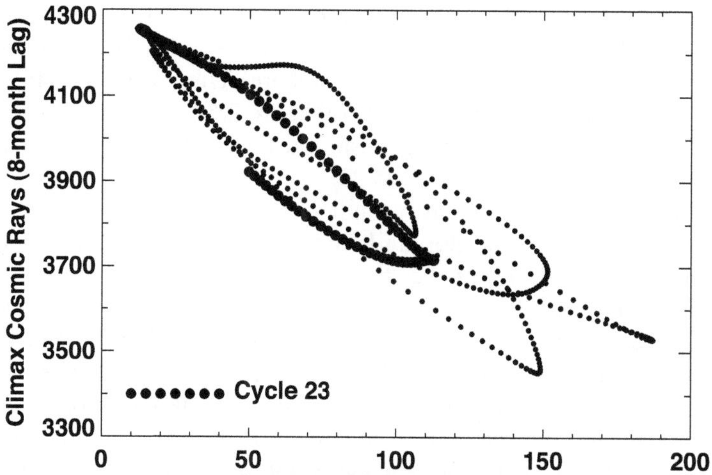

Figure 2.9. Cosmic ray flux measured at Climax, Colorado as a function of smoothed sunspot number. The quantities are anti-correlated but for a given sunspot number there tends to be even lower fluxes after cycle maximum. Cycle 23, while at the lower extreme for cosmic ray flux after maximum, has behaved very much like previous cycles.

In two aspects important for the structure of the heliosphere, cycle 23 was extreme. The polar field strength (Figure 2.6) as measured at the Wilcox Solar Observatory was significantly lower than during the previous two cycles (the full extent of our experience with polar field strength). It also appears that the strength of the meridional flow in the Sun's convection zone was by far the weakest it has been in the last 130 years.

The Sun's meridional circulation is now thought to play two key roles in producing the 11-year sunspot cycle (cf. Dikpati and Charbonneau, 1999). The poleward flow measured in the surface layers by direct Doppler measurements (Hathaway, 1996) and by helioseismology (Giles *et al.*, 1997) carries the weak magnetic elements poleward in the surface layers. Magnetic reconnections across the equator eliminate some of the leading polarity magnetic flux, leaving a surplus of following polarity flux to be carried poleward. This poleward flux of following polarity magnetic elements ultimately reverses the polar fields by the time of cycle maximum and then builds up and maintains the new polar fields over the declining phase of each cycle (this is evident in Figure 2.5). The second role played by the meridional circulation is to slowly transport the sheared and strengthened magnetic field at the base of the convection back toward the equator. This slow return flow at the base of the convection zone produces the equatorward drift of the active latitudes seen in the sunspot butterfly diagram (Figure 2.4) and in the magnetic butterfly diagram (Figure 2.5). The speed of the return flow sets the length of the cycle as well as the equatorward drift rate of the active latitudes.

Measurements of the equatorward drift rate of the active (sunspot) latitudes since 1874 give flow speeds commensurate with those expected for the meridional flow speed at the base of the convection zone (Hathaway *et al.*, 2003, 2004). The drift rate at the time of cycle maximum for previous cycles is about 2.1°/yr with a range from 1.5° to 2.7°/yr. Cycle 23 had the lowest drift rates on record—1.4°/yr in the northern hemisphere and 0.9°/yr in the southern hemisphere. These exceptionally slow drift rates and the weak polar fields in 2005–2007 (Figure 2.6) are linked through the flux transport process. Both of these features should have influenced the structure and evolution of the heliosphere during cycle 23.

Coronal hole initial development was not symmetric at the beginning of cycle 23, with the northern polar hole developing before that in the south. The field reversed first in the south (Figure 2.6), but the reversal in the north was more rapid and the field gained greater strength there than in the south. Figure 2.10 shows the early north coronal hole development in terms of coronal hole boundaries determined from He 10830 Å observations. The coronal hole map from late 2000 on the left shows no polar coronal holes, as is often the case at solar maximum and for several months on either side of maximum. The map from late 2001 on the right shows a prominent coronal hole in the north, with the new cycle polarity, and no coronal hole in the south. The He 10830 Å image in the center from 18 November 2001 exhibits this new cycle coronal hole, which continued to grow as Ulysses moves south after its north polar passage in 2001. At the highest northern latitudes reached, Ulysses was continuously embedded in high-speed flow from this coronal hole, even though the secondary sunspot peak of cycle 23 had not yet occurred (Figure 2.7).

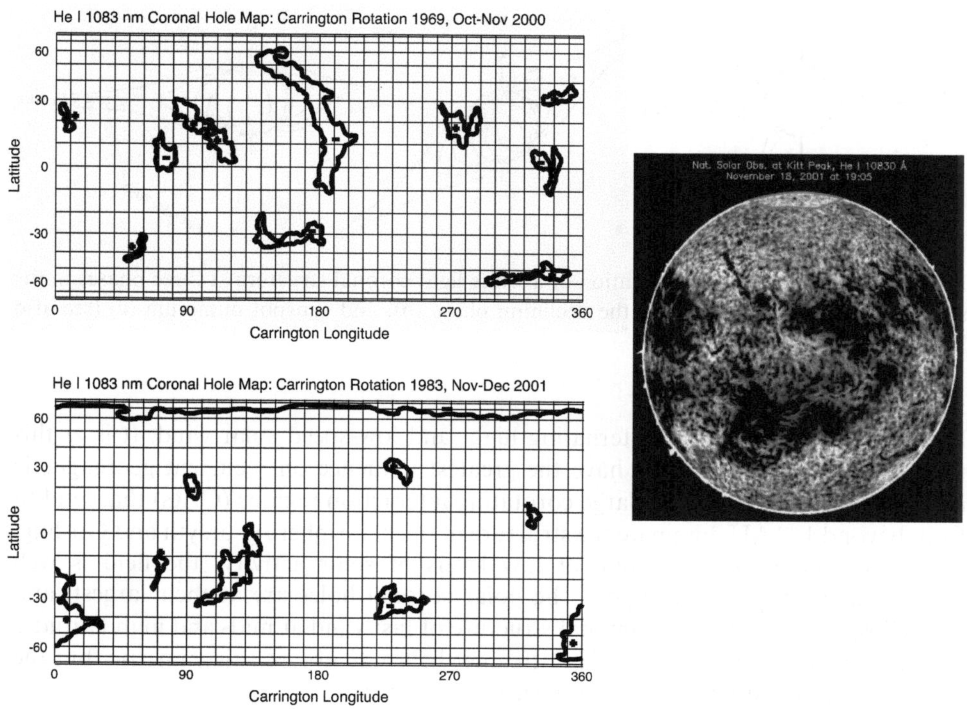

Figure 2.10. He 10830 Å synoptic maps from late 2000 and late 2001, and an image from 18 November 2001 showing a prominent northern polar coronal hole (outlined by a gray line) and no apparent southern coronal hole at the time just after the maximum of cycle 23. (National Solar Observatory/Kitt Peak)

2.4 THE EXTENSION OF CYCLE 23 INTO THE INTERPLANETARY MEDIUM

The coronal magnetic field controls the configuration of the heliosphere and the structure of the solar corona evolves through the solar cycle in direct relation to the way the photospheric magnetic field is changing. This is already seen in Figure 2.10, in relation to Figure 2.6, in terms of polar coronal hole development. Figure 2.11 is a simple schematic representation of the general appearance of white light coronal evolution over the solar cycle. Figure 2.11a indicates that the corona is highly structured and spiky around solar maximum. Polar coronal holes are small, or absent altogether. There may be many small and short-lived coronal holes distributed elsewhere over the Sun. Solar wind emanates from these small coronal holes, active regions, and quiet corona (Neugebauer *et al.*, 2002). In the declining phase of the solar cycle (Figure 2.11b) the magnetic field and corona become more organized but generally not symmetrically about the solar rotation axis. Therefore, at these times the Sun's magnetic field can often be approximated as a tilted dipole with concurrent off-axis polar coronal holes. Because of solar rotation, this exposes the mid- and

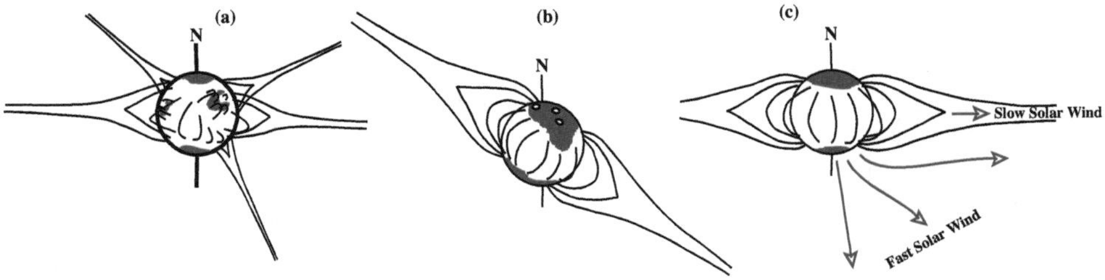

Figure 2.11. Schematic representation of a white light coronal structure at three phases in the solar cycle: solar maximum (a), the declining phase (b), and sunspot minimum (c) (see also Figure 7.1).

low-latitude heliosphere to alternating high- and low-speed solar wind. It is at this time that high-speed streams have the greatest influence on recurrent geomagnetic activity (Section 2.2) and the large corotating interaction regions (CIRs) that tend to form beyond 1–2 AU dominate the structure of the heliosphere (Forsyth and Gosling, 2001). Around solar minimum, at low levels of solar activity, the field is well approximated by an axial dipole, the polar coronal holes are at their largest, and the heliosphere exhibits the bimodal structure of fast wind at mid- and high latitudes and slow wind at low latitudes. The streamer belt may not be axisymmetric, but the coronal holes themselves are often nearly axisymmetric.

Recurrent geomagnetic activity during the declining phase of the cycle is a continuing topic of research. The connection to high-speed solar wind comes directly through currents on the magnetopause being proportional to the square root of the solar wind dynamic pressure and is characterized in standard geoeffectiveness parameters. The dynamic pressure is affected by whether a CIR has evolved far enough to produce shock waves, although this usually does not happen until beyond 1 AU. There is, in addition, a dependence on southward turnings in the heliospheric magnetic field (HMF) which, in turn, are influenced by proximity to the "heliospheric current sheet" (HCS) and whether a "$+ \rightarrow -$" or "$- \rightarrow +$" magnetic polarity change sector boundary is being encountered. Finally, there is probably a correlation between active longitudes and sector boundary locations, with active longitudes being the location of many of the large flares that tend to occur in the declining phase of the solar cycle (Burton, McPherron, and Russell, 1975; Vasyliunas, 2006; Forsyth and Gosling, 2001).

Coronal structure maps in a very direct way into the heliosphere and models can estimate or approximate this structure (see Chapter 4). The best such models are fully three-dimensional and time-dependent (Riley *et al.*, 2003). But, an excellent approximation that has been used for many years is the "potential field source surface" (PFSS) model that has been improved over the years to include current sheets outside the source surface (Zhao and Hoeksema, 1995). This is the type of model used to interpret solar wind sources at solar maximum (Neugebauer *et al.*, 2002). The models use the line-of-sight photospheric magnetic field to build synoptic maps of the photospheric field through the solar cycle. These maps are then used as a photospheric

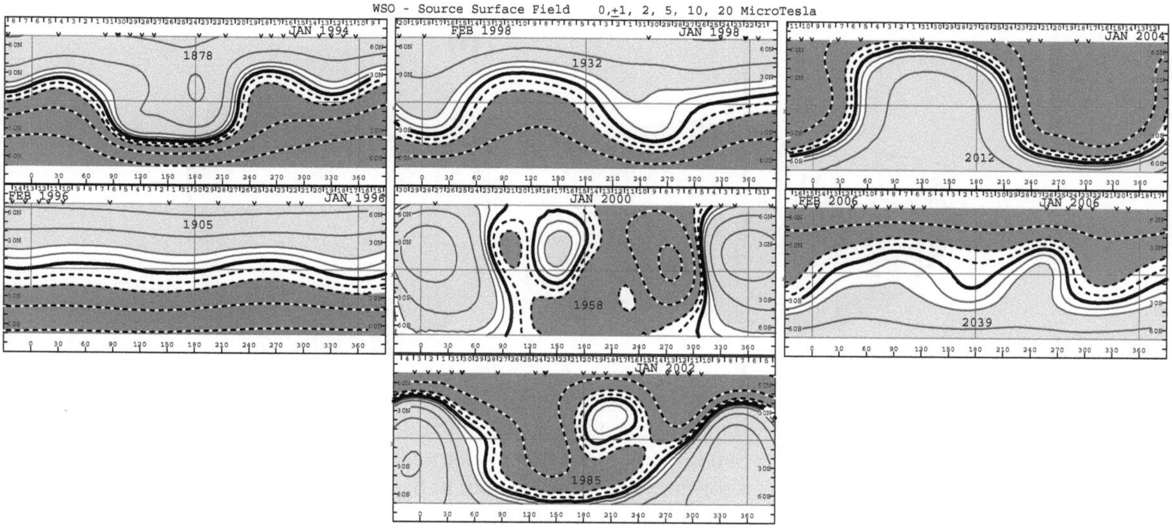

Figure 2.12. Solar magnetic field source surface synoptic charts at 2-year intervals throughout cycle 23. The coronal magnetic field is calculated from photospheric field observations with a potential field model where the field is forced to be radial at the source surface at 2.5 solar radii. There is a polar field correction to closely match the observations of the IMF at Earth. Light gray and solid contours indicate positive or outward fields. (Wilcox Solar Observatory data and maps, Zhao and Hoeksema, 1995)

boundary condition for a potential magnetic field model of the corona that assumes the magnetic field becomes radial at some height, the "source surface", that is usually chosen as $2.5R_\odot$. They are a substitute for the lack of a method to directly observe magnetic fields in the corona.

Figure 2.12 is a selection of source surface synoptic charts at 2-year intervals throughout cycle 23. Each plot shows the sign and magnitude of the field on the $2.5R_\odot$ radius source surface in coordinates of Carrington longitude and heliographic latitude. The field is not plotted above $\sim 70°$ due to the large uncertainties in observing near the poles. Seven Carrington rotations are shown, from January 1994 to January 2006. The main feature in these plots is the neutral line that divides outward from inward magnetic field. This neutral line maps directly into the HCS that is the surface dividing outward from inward polarity in the interplanetary medium. The HCS is one of the most easily identified features in the solar wind and interplanetary magnetic field. The light and dark shadings represent positive (+) and negative (−) polarities on the source surface. As suggested by Figure 2.6, the polarity is reversed on the source surface between 1994 and 2006. The reversal takes place around the year 2000, at the same time as the photospheric polarity reverses.

At solar minimum (January 1996) the neutral line lies very near the heliographic equator, which is also shown schematically in Figure 2.11c. As solar activity increases, the neutral line becomes more contorted until, in January 2000, it appears to pass directly over the pole. This is precisely what was observed at Ulysses (Jones *et*

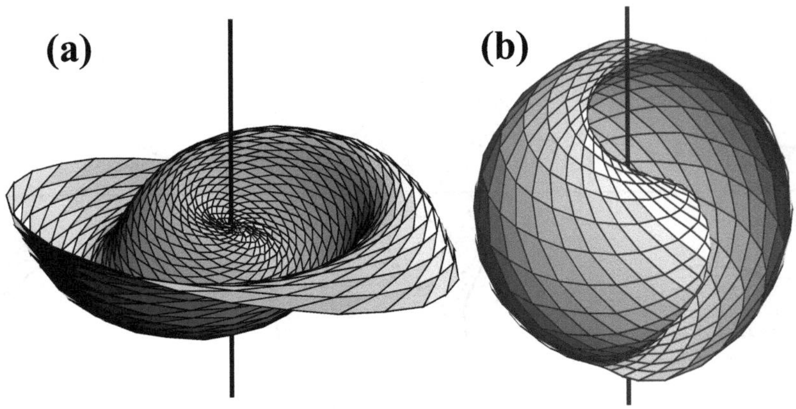

Figure 2.13. The heliospheric current sheet for a simple tilted dipole and 400 km/s wind. The calculation is carried from the Sun to ~6 AU, not to the same scale in both panels. This distance corresponds to the time it takes for the Sun to revolve once at the Carrington rate. (a) Tilted at 22.5° to the equator. (b) Tilted at 70° to the equator.

al., 2003). Then, after solar maximum, the neutral line slowly moves back towards the equator (January 2004) at the same time as the corona appears as if it were organized by a tilted dipole magnetic field (Figure 2.11b).

The magnetic field on the source surface in Figure 2.12 is transported out into the heliosphere by the solar wind. The neutral line maps into the HCS which, in turn, eventually is transported into the heliosheath (Suess, 2004). Figure 2.13 shows the idealized HCS out to about 6 AU for a dipolar field tilted to the heliographic equator by 22.5° in (a) and 70° in (b). These figures show how the HCS is distorted into a warped surface that was compared to a ballerina skirt by Alfvén. However, the shape becomes more reminiscent of a sea shell for large tilts.

The poleward reach of the HCS is defined as the "tilt" in the general case, even though the solar field is not precisely dipolar. The tilt varies from near zero at minimum to 90° at maximum. Figure 2.14 shows a plot of the tilt through approximately the past two solar cycles. Values appear to saturate at ~75° due to the restrictions on knowledge of the polar fields. This is confined to near the heliographic equator at minimum but extends to the polar regions of the heliosphere at maximum. This has an influence on the transport of cosmic rays in the heliosphere and, furthermore, cosmic ray transport has a different characteristic behavior with time through the solar cycle depending on whether the northern hemisphere is of positive polarity ("A+ cycle") or negative polarity ("A− cycle") (see Chapter 6). In cycle 23 the tilt maximized in the southern hemisphere around maximum and stayed large for 2–3 years. Conversely, the tilt in the north began to decrease within a year of solar maximum, along with the formation of the larger fields in the north and the northern polar coronal hole.

The shape of the HCS in the interplanetary medium is not simply an outward mapping of the neutral line shown in Figure 2.12. The magnetic field energy density is small compared with the kinetic energy of the solar wind. Therefore, the HCS is

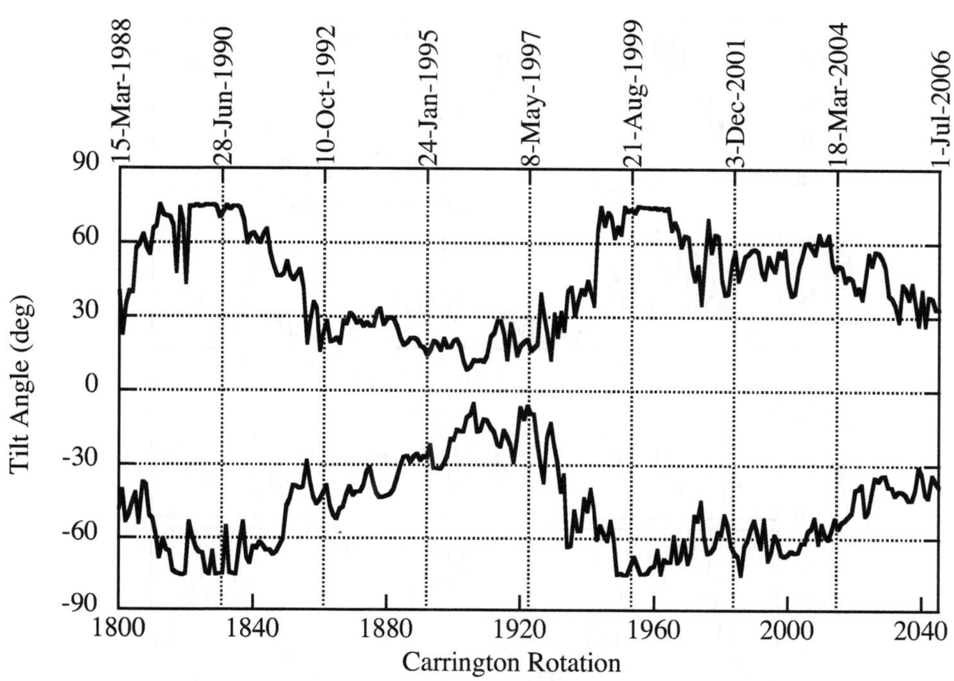

Figure 2.14. The maximum extent of the HCS for the indicated dates/Carrington rotations, over approximately two solar cycles. The extent is called the "tilt", although this angle may differ in opposite hemispheres. It is taken directly from the location of the neutral line on source surface plots from Wilcox Solar Observatory (see Figure 2.12).

carried by the solar wind. The dominant effect is that the field is drawn into a backward Archimedian spiral (see Chapter 4) since the solar wind does not corotate with the Sun beyond a few solar radii (Parker, 1963). In addition CIRs further distort the large-scale structure of the HCS. The HCS further develops extensive fine structure due to small-scale fluctuations in solar wind speed. Finally, the structure of the magnetic field shown in Figure 2.12 is a dramatic simplification of coronal physical processes. There are probably many embedded small-scale current sheets, especially above active regions, and there is no reflection of the motion of field line footpoints in the photosphere that maps out into the solar wind and has an important influence on the motion of energetic particles throughout the heliosphere.

PFSS models are tools to estimate the locations of coronal holes, which tend to occur in large, unipolar magnetic field regions. Using this knowledge, they have been used to look for relationships between properties of the models and solar wind speed. Such a relation has been found between the ratio of magnetic field strength on a specific magnetic field line at the source surface to the field strength at the photospheric footpoint of that field line. Taking out spherical r^2 divergence, this ratio is known as the spreading factor. What has been found is that there is a reliable empirical relationship between the spreading factor and mass and energy flux in

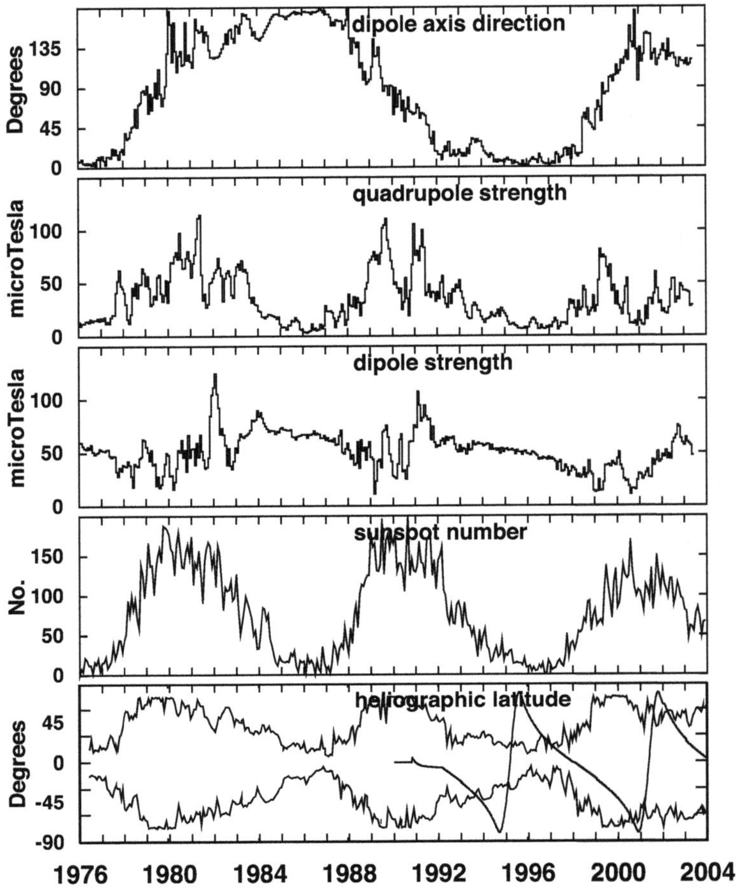

Figure 2.15. Maximum extent of the HCS (as in Figure 2.12, Wilcox Solar Observatory data), together with the spacecraft heliographic latitude, sunspot number, dipole and quadruple strength, and dipole axis direction for the 28-year period starting on 1 January 1976.

the solar wind (Wang, 1995). This result greatly expanded the utility of the PFSS model to predict the imprint of the coronal magnetic field on the HMF and the solar wind. The strength and size of CIRs can be predicted.

Figure 2.15 draws some of these concepts together. The top three panels are taken from the data used to make the source surface plots shown in Figure 2.12. These PFSS models are computed in terms of spherical harmonics, with the lowest order being the dipole components, the next the quadrupole components, and so on. Figure 2.15 shows the dipole axis direction and the quadrupole and dipole strengths for the full Wilcox Solar Observatory data set. The fourth panel is the sunspot number and the bottom panel shows the same "tilt angle" plotted in Figure 2.4. This figure first illustrates what was described earlier, that the solar magnetic field always has a significant dipolar component and that the reversal at solar maximum is partly achieved through rotation of the dipole through the equator. In addition, the quadrupolar and higher order components become large around solar maximum, reflecting the large number of active regions. Superimposed on the tilt angle at the bottom is the latitude of Ulysses. Obviously, Ulysses will be exposed to CIRs when it is below the tilt angle. Revealing the details of this exposure is one of the objectives of the mission and the result will help to understand the large-scale dynamics of the heliosphere.

2.5 SUMMARY

1 Solar activity cycle 23 is characterized here in the context of typical solar cycles. There are some important differences that are reflected in phenomena observed in the solar wind. There were several instances of large activity in the declining phase of the cycle. This is not unusual in itself but was unusual in that the activity was even larger than usual and had measurable consequence throughout the heliosphere and even out to the heliosheath. Polar coronal holes formed asymmetrically and the tilt of the HCS was also asymmetric. This will have resulted in asymmetric streams in the north and south.

2 There was a large-scale asymmetry in the heliosphere at the last solar minimum that was discovered by Ulysses, which was discovered in cosmic ray measurements but then shown to exist in almost all properties of the solar wind (McKibben, 2001). Solar activity completely obscures any possibility to measure the same asymmetry at solar maximum. Nevertheless, the results shown in Figures 2.6, 2.10, and 2.13 imply that the Sun was also north–south asymmetric at solar maximum. It remains to be seen whether the asymmetry persists into the 2006–2007 solar minimum.

3 Ulysses passed over the north pole of the Sun in 2001 and began moving south as the north polar hole was forming. Then, as the tilt decreased and CIRs became confined closer to the equator, Ulysses was following this evolution southward. Through the declining phases, Ulysses was continuously in the vicinity of the HCS and moving in and out of the resulting CIRS. The solar imprint on the large-scale heliosphere is particularly strong at this time of the solar cycle.

2.6 ACKNOWLEDGMENTS

The writing of this chapter was supported by the Ulysses Mission, the Ulysses/SWOOPS instrument team, and grants from NASA headquarters.

2.7 REFERENCES

Burton, R. K., R. L. McPherron, and C. T. Russell (1975), An empirical relationship between interplanetary conditions and *Dst*, *J. Geophys. Res.*, **80**, 4204.

Castenmiller, M. J. M., C. Zwaan, and E. B. J. van der Zalm (1986), Manifestations of sequences in magnetic activity, *Solar Phys.*, **105**, 237.

Dikpati, M., and P. Charbonneau (1999), A Babcock–Leighton flux transport dynamo with solar-like differential rotation, *Astrophys. J.*, **518**, 508–520.

Forsyth, R. J., and J. T. Gosling (2001), Corotating and transient structures in the heliosphere, in *The Heliosphere near Solar Minimum* (A. Balogh, R. G. Marsden, and E. J. Smith, eds.), p. 107, Springer/Praxis.

Giles, P. M., T. L. Duvall, P. H. Scherrer, and R. S. Bogart (1997), A subsurface flow of material from the Sun's Equator to its poles, *Nature*, **390**, 52–54.

Gopalswamy, N., L. Barbieri, E. W. Cliver, G. Lu, S. P. Plunkett, and R. Skoug (2003), Introduction to violent Sun–Earth connection events of October–November 2003, *J. Geophys. Res.*, **110**, A09S00.

Hale, G. E., F. Ellerman, S. B. Nicholson, and A. H. Joy (1919), Magnetic Polarity of Sunspots, *Astrophys. J.*, **49**, 153–178.

Harvey, K. L., and O. R. White (1999), What is solar cycle minimum?, *J. Geophys. Res.*, **104**(A9), 19759–19764.

Hathaway, D. H. (1996), Doppler measurements of the Sun's meridional flow, *Astrophys. J.*, **460**, 1027–1033.

Hathaway, D. H., and R. M. Wilson (2004), What the Sunspot Record Tells Us About Space Climate, *Solar Phys.*, **224**, 5–19.

Hathaway, D. H., D. Nandy, R. M. Wilson, and E. J. Reichmann (2003), Evidence that a deep meridional flow sets the sunspot cycle period, *Astrophys. J.*, **589**, 665–670.

Hathaway, D. H., D. Nandy, R. M. Wilson, and E. J. Reichmann (2004), Erratum: Evidence that a deep meridional flow sets the sunspot cycle period, *Astrophys. J.*, **602**, 543.

Henney, C. J., and J. W. Harvey (2002), Phase coherence analysis of solar magnetic activity, *Solar Phys.*, **207**, 199.

Howard, R. F. (1996), Axial tilt of active regions, *Solar Phys.*, **169**, 293–301.

Jones, G. H., A. Balogh, and E. J. Smith (2003), Solar magnetic field reversal as seen at Ulysses, *Geophys. Res. Lett.*, **30**(19), 8028.

Joselyn, J. A., J. B. Anderson, H. Coffey, K. Harvey, D. Hathaway, G. Heckman, E. Hildner, and W. Mende (1997), Panel achieves consensus prediction of solar cycle, *EOS, Trans. AGU*, **78**(20), 205, 211–212.

Komm, R. W., R. F. Howard, and J. W. Harvey (1993), Meridional flow of small photospheric magnetic features, *Solar Phys.*, **147**, 207–223.

McKibben, R. B. (2001), Cosmic rays at all latitudes in the inner heliosphere, in *The Heliosphere near Solar Minimum* (A. Balogh, R. G. Marsden, and E. J. Smith, eds.), p. 327, Springer/Praxis.

Neugebauer, M., E. J. Smith, A. Ruzmaikin, J. Feynman, and A. H. Vaughn (2000), The solar magnetic field and the solar wind: Existence of preferred longitudes, *J. Geophys. Res.*, **105**(A2), 2315.

Neugebauer, M., P. C. Liewer, E. J. Smith, R. M. Skoug, and T. H. Zurbuchen (2002), Sources of the solar wind at solar activity maximum, *J. Geophys. Res.*, **107**(A12), 1488.

Parker, E. N. (1963), *Interplanetary Dynamical Processes*, Interscience Publishers, New York.

Riley, P., Z. Mikić, and J. A. Linker (2003), Dynamical evolution of the inner heliosphere approaching solar activity maximum: Interpreting Ulysses observations using a global MHD model, *Annales Geophysicae*, **21**, 1347.

Suess, S. T. (2004), The magnetic field in the outer heliosphere, in *Physics of the Outer Heliosphere* (V. Florinski, N. V. Pogorelov, and G. P. Zank, eds.), p. 10, American Institute of Physics Conf. Proc., vol. 719.

Vasyliunas, V. M. (2006), Reinterpreting the Burton–McPherron–Russell equation for predicting *Dst*, *J. Geophys. Res.*, **111**, A07S04.

Waldmeier, M. (1939), Über die Struktur der Sonnenflecken, *Astron. Mitt., Zurich*, **14**(138), 439–485.

Wang, Y. M. (1995), Empirical relationship between the magnetic field and the mass and energy flux in the source regions of the solar wind, *Astrophys. J. Lett.*, **449**, L157.

Webber, W. R. (2005), An empirical estimate of the heliospheric termination shock location with time with application to the intensity increases of MeV protons seen at Voyager 1 in 2002–2005, *J. Geophys. Res.*, **110**, A10103.

Zhao, X., and J. T. Hoeksema (1995), Prediction of the interplanetary magnetic field strength, *J. Geophys. Res.*, **100**(A1), 19.

3

The solar wind throughout the solar cycle

Rudolf von Steiger

3.1 INTRODUCTION: THE PRE-ULYSSES PICTURE

The existence of solar corpuscular radiation (SCR) was conjectured by Biermann (1951) based on the fact that the ion tails of comets always point radially away from the Sun. Earlier it had been thought that this was due to solar radiation pressure, but when the relevant cross-sections were measured it became clear that these were far too small. This is visible in Figure 3.1, where stars can be seen shining through the ion tail of comet Hale–Bopp, one of the more spectacular sights in the sky of the 20th century. Parker (1958) provided the first theoretical description of the SCR in terms of a supersonic magnetized fluid. He coined the term "solar wind" in order to set it apart from other ideas of a (subsonic) solar breeze that were around at the time. The solar wind was ultimately observed in the early 1960s by the Soviets and independently with the American Mariner 2 mission to Venus (Gringauz *et al.*, 1961; Neugebauer and Snyder, 1962). An excellent account of these early developments is given by Parker (2001).

The first generation of solar wind instruments were Faraday cups with stepped-potential retarding grids or curved-plate electrostatic analyzers, both providing energy-per-charge (E/q) spectra. This revealed the basic constituents of protons with an admixture of a few percent (by number) of alpha particles (Neugebauer and Snyder, 1966). It is remarkable how much could be gleaned from the first few months of observations: The existence of alternating high-speed streams with (slow) inter-stream solar wind, the rough proportionality of the proton temperature with the bulk speed, the approximate equality of the proton and alpha thermal speeds (i.e., the mass-proportionality of their kinetic temperatures), and more (cf. Neugebauer and von Steiger, 2001). With E/q sensors of increasing sophistication, Bame *et al.* (1970)

Figure 3.1. Comet Hale–Bopp as seen in April 1997. The straight, bluish ion tail, as opposed to the curved, white dust tail, points radially away from the Sun. Since it is transparent to starlight this must be due to solar corpuscular radiation (© W. Pacholka).

discovered heavy ions[1] such as oxygen, silicon, and iron in charge states that gave direct proof of the million-degree temperatures in the solar corona. Around the same time the noble gases Ne and Ar were first measured using the foil collection technique on the Apollo lunar missions (Geiss *et al.*, 2004, and references therein). Bame *et al.* (1977) also found that the solar wind from the newly discovered coronal holes (Krieger, Timothy, and Roelof, 1973) was structure-free and thus a distinctly different state of the phenomenon. With the plasma instruments on the two Helios missions (1974–1982) the distribution functions of protons and alpha particles could be characterized in full detail and as a function of heliocentric distance from 0.3 to 1 AU. Anisotropies perpendicular to the magnetic field direction were found to be common at small distances; with increasing distance they could be seen to isotropize gradually while at the same time another anisotropy along the magnetic field directed away from the Sun would build up (Marsch, 1991, and references therein).

A second generation of solar wind instruments combined an electrostatic E/q analyzer with either magnetic deflection to form a Wien filter thus providing a mass-per-charge (m/q) measurement such as the ICI sensor on the ISEE-3 mission (Coplan *et al.*, 1978), or with a solid-state detector to provide an energy measurement such as the ULECA sensor on ISEE-3 (Hovestadt *et al.*, 1978). Both techniques allowed breaking free from the assumption that all components of the solar wind flow at the

[1] Sometimes these are called "minor" ions, but that term should be avoided considering their important role in modern solar wind research.

same velocity, which had always to be made when interpreting E/q spectra. In addition, the resolution of the m/q spectra was independent of the kinetic temperature, while the usefulness of E/q spectra was seriously limited at times of high temperatures (i.e., in high-speed streams). Thus, second-generation sensors contributed to new and improved determinations of abundances (Bochsler, Geiss, and Kunz, 1986; Schmid, Bochsler, and Geiss, 1988) and charge states (Ipavich *et al.*, 1986). However, the high charge states C^{6+} and O^{8+} remained hidden behind the peak of He^{2+} and the high charge states of Si and Fe remained hidden behind the main charge states of the CNO group.

This shortcoming was overcome with a third generation of sensors using a combination of electrostatic deflection (giving E/q), time-of-flight (giving m/q), and solid-state detectors (measuring E). The SWICS sensor on Ulysses (discussed below) was the first such instrument to provide data from the free-flowing solar wind, but it was not the first such sensor in space. This was the CHEM sensor (Gloeckler *et al.*, 1985) on the terrestrial magnetosphere mission AMPTE/CCE. At times when both the aphelion of the CCE orbit was oriented sunward and the solar wind pressure was exceptionally high the Earth's magnetosphere was sufficiently compressed for the spacecraft to be exposed to the shocked solar wind in the magnetosheath. This allowed the first determination of all charge states of the CNO group in either exceptionally fast or exceptionally dense solar wind (i.e., high-speed streams or interplanetary coronal mass ejections—all discussed below—during brief periods of time, Gloeckler *et al.*, 1986; von Steiger *et al.*, 1992). Furthermore, due to the coincidence technique used for ion registration, SWICS allows for measurement of low-flux particle populations in the <100 keV range that was previously not observed.

Prior to the launch of Ulysses all experimental data had been acquired on spacecraft either in Earth orbit or in interplanetary orbits near the ecliptic plane, which is essentially coincident with the solar equator (up to an angle of 7.25°). Since solar activity occurs mostly at low latitudes the interpretation of the variable solar wind streams was complicated by the structure and evolution of the solar source regions and by solar rotation effects. During its coronal expansion the solar wind relaxes into a structure characterized by a single, but generally highly structured, current sheet. The picture obtained after the heliospheric expansion of that structure was often referred to as the "ballerina skirt" model, a term proposed by Alfvén and pictured in Schwenn (1990). The single most important achievement of Ulysses is to have overcome this limitation by reaching $\pm 80°$ in heliographic latitude, thus adding the third dimension to our picture of the solar wind and the heliosphere by looking at the ballerina's head (and feet), so to say. A second important feature of Ulysses is that it carries the first true mass spectrometer in interplanetary space, as already mentioned. But Ulysses has also its limitations: it represents only the European half of the originally conceived two-spacecraft International Solar Polar Mission (ISPM), as the American spacecraft counter-rotating on the same orbit was canceled in the early 1980s (cf. Bonnet and Manno, 1994, ch. 5, or Fisk, 2003). This complicates separating spatial from temporal variations, specifically asymmetries between the northern and southern hemisphere. Equally unfortunate is the absence of optical instruments.

When Ulysses is at high latitudes the solar wind source regions are ill-observed with telescopes on or near Earth. This leads to one of the main remaining problems for modeling of the high-latitude heliosphere.

Earlier accounts of the solar wind can be found, e.g., in Schwenn (1990) and in Marsch (1991) summarizing the pre-Ulysses picture at the time of its launch, by von Steiger *et al.* (1997), Neugebauer (2001), or Neugebauer and von Steiger (2001), giving the Ulysses picture of the minimum heliosphere, or in Zurbuchen (2007).

In this chapter we attempt to give an account of Ulysses solar wind observations during more than a complete solar cycle. In Section 3.2 we present the three-dimensional morphology of the different solar wind streams and their distribution in the heliosphere; in Section 3.3 we discuss the distribution functions of the solar wind ions and what can be deduced therefrom; in Section 3.4 we concentrate on the solar wind elemental and charge-state composition and its implications; in Section 3.5 we turn to transient phenomena such as corotating interaction regions and interplanetary coronal mass ejections; and in Section 3.6 we summarize the new, four-dimensional picture of the solar wind obtained with Ulysses.

3.2 MORPHOLOGY

The morphology of the solar wind around the solar activity cycle is illustrated in Figure 3.2. It is remarkable how closely solar wind speed profiles reflect the shape of the underlying corona both at low and at high solar activity (left and right panels, respectively).[2]

The top-left panel of Figure 3.2 shows that during Ulysses' first, or minimum, polar orbit the heliosphere was dominated by two high-speed streams at latitudes poleward of $\pm 30^\circ$ or even lower (Hundhausen, 1973). These high-speed streams are magnetically unipolar and emanate from two coronal holes that form at high latitudes during the declining phase of the solar cycle and dominate the polar regions of the corona at solar minimum. Conversely, the low-latitude heliosphere equatorward of $\pm 20^\circ$ is characterized by mixed-polarity, variable, and on average slow solar wind. The association of the slow wind with the coronal streamer belt is evident in this picture. What is less obvious is the association of the fast streams with the coronal holes since these occupy a much smaller solid angle in the corona, usually only poleward of $\pm 60^\circ$ at solar minimum. The dominant view is that the fast streams expand superradially from the coronal holes, an interpretation supported by magnetic (presence of high magnetic variability, $\delta\mathbf{B}/\mathbf{B}$), kinetic (high kinetic proton temperature, T_p), and composition signatures (low freezing-in temperatures from heavy ion charge-state ratios—e.g. O^{7+}/O^{6+}—see Section 3.4). A minority view maintains that the solar wind has a radial flow pattern in the entire heliosphere (Woo and Habbal, 1999), but we find this difficult to accept, particularly in view

[2] Of course, the similarity must be understood only in a qualitative sense since the corona images are momentary snapshots while it took Ulysses 6 years to obtain each of the two speed profiles.

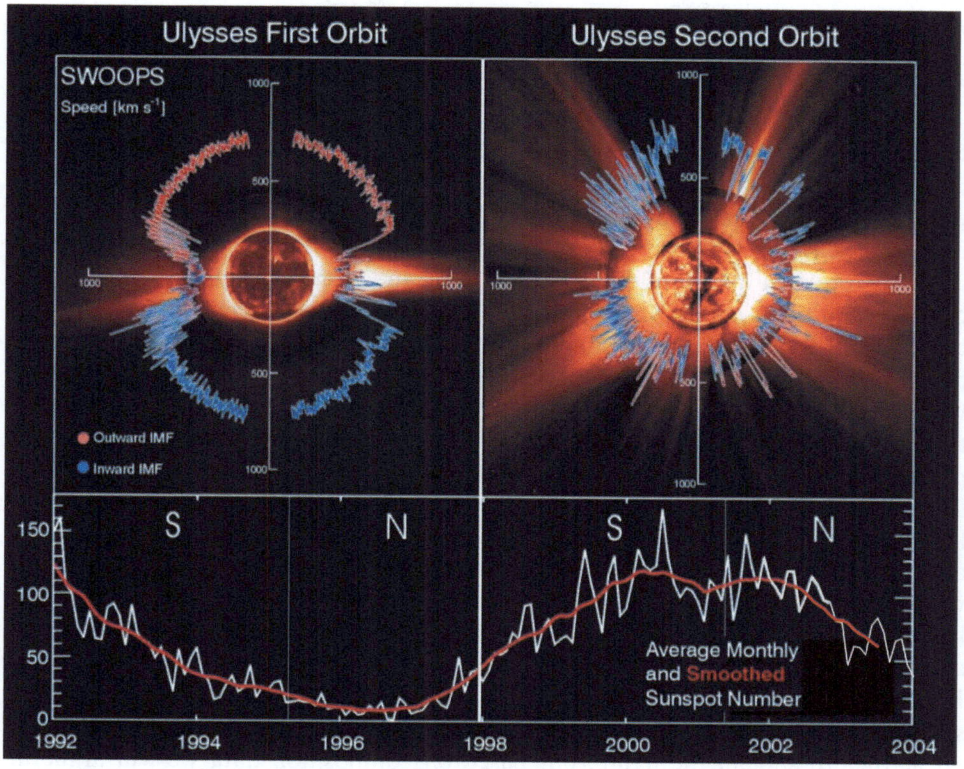

Figure 3.2. Morphology of the solar wind during the solar cycle. Bottom panel: Average and smoothed monthly sunspot number; top panels: polar plots of the solar wind speed as observed with the SWOOPS sensor on Ulysses during the two polar orbits completed so far, with the color of the speed curve indicating the magnetic polarity. The background images are composites of corresponding SoHO LASCO and EIT images illustrating the typical shape of the solar corona at minimum (top left) and at maximum (top right) activity. The difference of the solar wind speed distribution between the minimum and the maximum heliosphere is striking (adapted from McComas *et al.*, 2003).

of the composition signatures discussed below. The degree of superradial expansion has been estimated by Gosling *et al.* (1995a) as $f_g \sim 4.8$ from the first fast latitude scan of Ulysses (the right-hand half of the top-left panel of Figure 3.2). This is larger than, but consistent with, estimates from modeling (Suess *et al.*, 1998). Thus, the fast solar wind streams dominate the solar-minimum heliosphere, filling it to a degree of more than 60%. This is a result that could be obtained only with a mission such as Ulysses going to a significantly higher latitude than 30°, but not from near the ecliptic plane with its small (7.25°) inclination to the heliographic equator. The morphology of the solar wind during Ulysses' first full polar orbit was comprehensively analyzed by McComas *et al.* (2000). After removing the radial gradients the authors find that the solar wind emanating from the coronal holes is very uniform, with none of the high-latitude parameters showing much latitudinal variation. They also find a slight asymmetry between the two hemispheres but caution to interpret it as a true, spatial effect. Most likely the asymmetry was driven by the solar wind source evolving with time, with more energy going into the wind during the declining phase of the cycle (when Ulysses was in the southern hemisphere) than around solar minimum (when Ulysses was in the northern hemisphere). The parameter that showed the least latitudinal variability over all latitudes (i.e., including the band of slow solar wind) was the scaled momentum flux density, $M_p = n_p m_p v_p^2$, consistent with the Helios results of Schwenn (1990).

The top-right panel of Figure 3.2 shows an altogether different, much less ordered, and more complex picture of the heliosphere at solar maximum. The fact that Ulysses found the simple, minimum picture first and the complex, maximum picture second is a somewhat fortuitous consequence of the Challenger catastrophe, which caused its launch to be delayed from Spring 1986 to the Fall of 1990, well after the maximum of cycle 22. Apparently both the solar wind speed and the magnetic polarity have lost their simple ordered structure during the second polar orbit, much like the underlying corona has. Slow solar wind can be observed up to the highest latitudes reached by Ulysses, reflecting the large tilt (about 50°) of the band of solar activity and current sheet caused by the reversing global magnetic field. Fast streams with either magnetic polarity are also present at all latitudes. At the highest northern latitude Ulysses already caught a glimpse of the newly forming polar coronal hole for about three solar rotations. It had the expected (reversed) polarity for cycle 23 (cf. Figure 3.2), but the hole had not yet developed sufficiently to cause the same degree of superradial expansion found at solar minimum (McComas, 2003). Other fast streams are harder to spot in the figure, but based on composition data we will argue in Section 3.4 that they are present all the same with very similar properties as at solar minimum. What changes is the latitudinal distribution of the two quasi-stationary solar wind types and, of course, the mixing-in of many more transient streams from coronal mass ejections (Section 3.5.2). Given all this variability, McComas *et al.* (2003) find that the momentum flux density

$$M_{\text{tot}} = m_p n_p v_p^2 + m_\alpha n_\alpha v_\alpha^2 \simeq m_p n_p v_p^2 (1 + 4A),$$

where A is the alpha-to-proton ratio, nevertheless remains constant at a value of about 3 nPa to within a factor of 1.5 throughout the entire mission. The momentum flux density is the physical quantity responsible for "inflating the heliosphere" (i.e., what determines the standoff distance to the termination shock). It is found to be minimal at low latitudes (<10°), which (at least to some extent) can be attributed to the missing partial pressure from the alpha particles—see Figure 3.3 (from McComas *et al.*, 2003). This may cause the heliosphere as a whole to be girded at low latitudes and assume an hourglass shape, a conclusion supported by global simulations by Pauls and Zank (1996).

Ulysses is now on its third polar orbit; given the orbital period of $T = 6.19$ yr this is is a little more than a complete solar cycle after the first polar orbit. Thus, Ulysses currently encounters the heliosphere again in the declining to minimum configuration, just a little later than back then, albeit with the magnetic polarity reversed. We

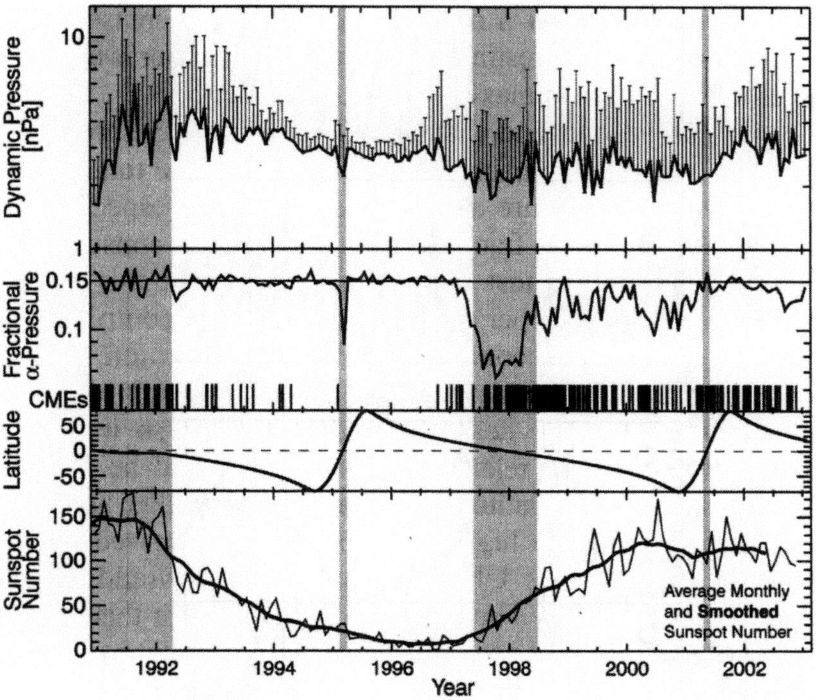

Figure 3.3. From top to bottom: total dynamic pressure of the solar wind, fractional pressure carried by the alpha particles, ICME observations, Ulysses latitude, and sunspot number. Pressures are plotted as solar rotation average values. At low latitudes (<10°, shaded bands) the total pressure tends to assume a mimimum value, partially caused by a lack of contribution from alpha particles (from McComas *et al.*, 2003).

therefore may have expected the current (2004–2007) transition from low latitudes to the south polar region to look very similar to what it looked on the first orbit (1992–1994). Back then a very regular corotating structure of slow solar wind alternating with a high-speed stream was encountered for more than a full year, or some 15 solar rotations. This was easily interpreted as the signature of a relatively flat but tilted streamer belt region of slow solar wind that Ulysses was progressively leaving behind as it climbed from ~13°S to ~33°S, where it became fully immersed in the polar fast stream (panel A of Figure 3.4; see also the lower left quadrant of the top-left panel of Figure 3.2). But the current transition looks very different and much less regular! This is all the more surprising as we are now in an even later phase of the solar cycle with the sunspot number only about half as much as what it was back then. Since the polar coronal holes generally become more and more centered around the heliographic poles towards solar minimum we would also expect the streamer belt to become more aligned with the heliographic equator. Yet this was not observed: panel B of Figure 3.4 shows that Ulysses found a very different heliospheric structure during the

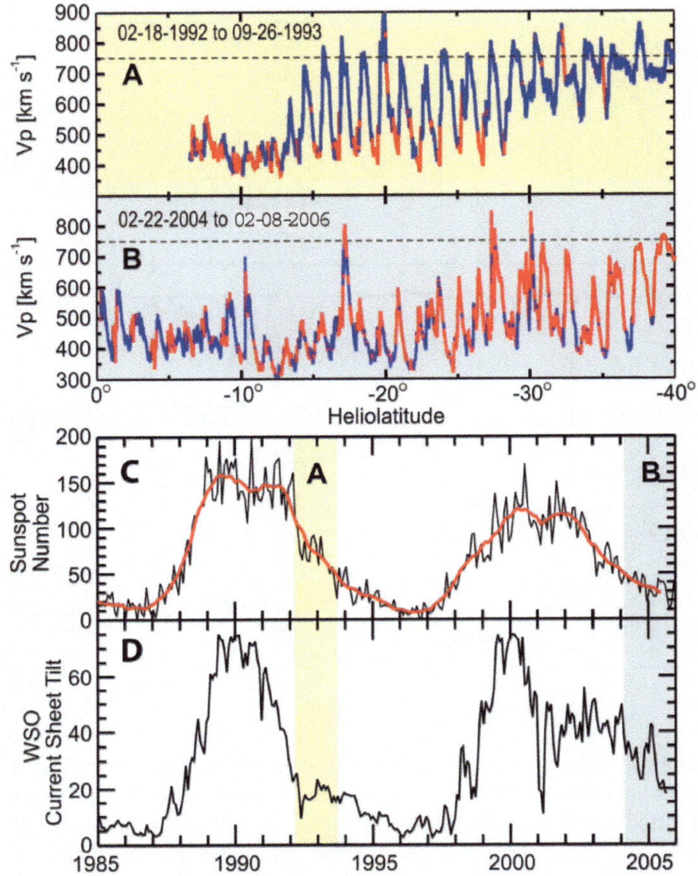

Figure 3.4. Transitions of Ulysses from the slow solar wind at low latitudes to the fast stream from the south polar coronal hole on its first polar orbit (panel A) and third polar orbit (panel B). Plotted is the solar wind speed from SWOOPS with the magnetic field polarity coded as the color of the speed curve (red: inward; blue: outward). The two bottom panels indicate the sunspot number (C) and the tilt of the current sheet (D) as calculated from Wilcox Solar Observatory images (adapted from McComas *et al.*, 2006).

declining phase of solar activity cycle 23 than during the previous cycle 22 (McComas *et al.*, 2006). Clearly the pattern of alternating slow wind with fast streams is much less regular, if at all present, and it took Ulysses to get almost to 40°S until it became fully immersed in the polar fast stream. The reason for this difference becomes apparent from the two lower panels in Figure 3.4: even though solar activity was significantly lower (panel C), the warp of the current sheet (panel D), and with it of the streamer belt, was still larger in cycle 23 as compared with the same phase in the Ulysses mission in cycle 22. We presume that this additional warp was caused by the remaining activity occurring at higher heliographic latitudes than it did in the previous cycle.

3.3 DISTRIBUTION FUNCTIONS

The three-dimensional velocity distribution functions of solar wind protons and alpha particles have been studied quite extensively prior to Ulysses (Marsch, 1991). By the time the ions reach the range of heliographic distances covered by the Ulysses orbit, much of the evolution of their three-dimensional distribution functions has already occurred. We therefore concentrate, in this section, on the aspects of the distribution functions which are specific to the realm of Ulysses (e.g., the addition of pickup ions), and to a comparison of the distribution functions of different heavy ion species.

3.3.1 H and He distribution functions

Figure 3.5 shows sample distribution functions (phase space density *versus* ion speed normalized with the solar wind bulk speed, $W = v/V_{SW}$) of H^+, He^+, and He^{++} obtained with SWICS in the slow solar wind (Gloeckler, 1999). These functions have three distinct parts: a thermal core of solar wind ions around $W = 1$, a flat part up to $W = 2$ that is made up of interstellar pickup ions, and a suprathermal tail at $W > 2$.

The cores of the distribution functions are best represented by kappa functions with a parameter that is found to be fairly small: $\kappa = 4.5$ for H^+ and $\kappa = 3.0$ for He^{++} (Gloeckler and Geiss, 1998a, b). These values are characteristic for the slow solar

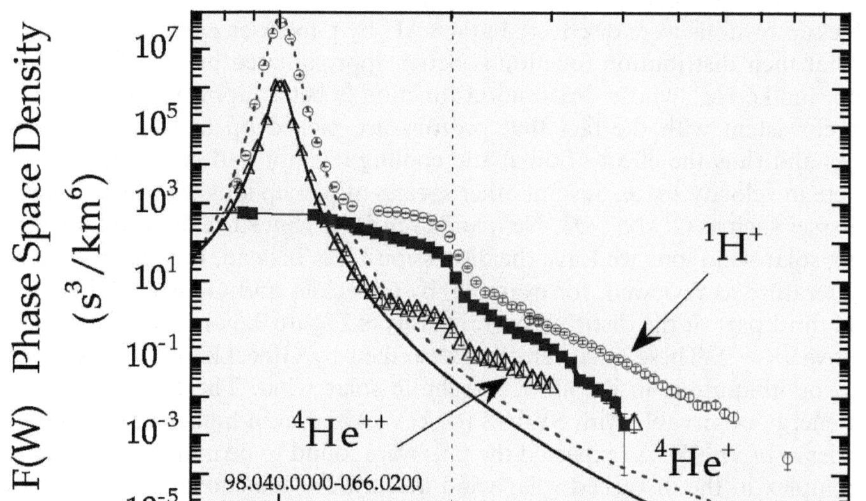

Figure 3.5. Sample distribution functions of H and He observed in the slow solar wind with Ulysses-SWICS. The functions have three distinct parts: a solar wind core around $W = 1$, pickup ions out to $W = 2$, and a suprathermal tail at $W > 2$ (from Gloeckler, 1999).

wind in general, while they are larger (i.e., the wings are less pronounced) in high-speed streams, but it seems that the solar wind distribution functions always have significant wings on top of the purely thermal Maxwellian. In the case of He^{++} it could also be shown that there are kappa wings in all directions except in the magnetic field direction pointing back towards the Sun, which indicates that they may be created in a statistical process by outward-propagating waves along the magnetic field (von Steiger and Zurbuchen, 2003).

Pickup ions stem from interstellar neutrals penetrating deep into the heliosphere, up to the point where they become ionized by the solar EUV radiation and subsequently picked up by the $\mathbf{E} \times \mathbf{B}$ field carried with the solar wind. Their expected flux and energy spectrum were first calculated by Vasyliunas and Siscoe (1976). Since the speed of the interstellar neutrals of $\sim$25 km/s is very low compared with the typical solar wind speed the ions are essentially picked up from rest (i.e., with $-V_{SW}$ in the solar wind frame). Thus, they gyrate with solar wind speed around the magnetic field, which in turn is convected outward also at the solar wind speed. Ions picked up at a particular distance are thus forming a shell distribution function between 0 and $2V_{SW}$, which is progressively shrinking by adiabatic cooling as it is convected outwards. The integral distribution function of ions picked up at all distances thus forms a full sphere with radius V_{SW} in velocity space. Pickup ions were first discovered as He^+ with AMPTE near 1 AU by Möbius *et al.* (1985), but all other interstellar neutrals are essentially fully ionized before reaching 1 AU and therefore cannot be detected there any more. With Ulysses going out to several AU, and carrying the appropriate instrumentation, it became possible to observe other interstellar species for the first time. Pickup protons were discovered at 4.8 AU by Gloeckler *et al.* (1993), who also found that their distribution function is better approximated by a shell distribution function, unlike He^+ whose distribution function is better approximated by a solid sphere, consistent with the fact that protons are picked up at larger heliocentric distances and thus the effect of adiabatic cooling has not had sufficient time to fill the sphere in velocity space. Several other species of pickup ions were also discovered with Ulysses such as C^+, N^+, O^+, Ne^+, and even $^3He^+$, but since these are not strictly speaking solar wind ions we leave the discussion here. Instead, the reader is referred to the literature as reviewed, for example, by Gloeckler and Geiss (1998b).

The third part of the distribution functions in Figure 3.5 are their suprathermal tails above $W = 2$. These tails were first described by Gloeckler (1999), who found them to be ubiquitous in the slow, in-ecliptic solar wind. The tails extend to the highest energy observable with SWICS (60 keV/e) and even higher with HI-SCALE (Gloeckler *et al.*, 1995). As expected the tails were found to be more pronounced and more complex in the disturbed solar wind inside CIRs and downstream of shocks. But it was quite a surprise to find that the tails never disappeared, even in the absence of any shocks, but remained present with a tail strength S_t of a few percent (defined as the integrated phase space density in the tail, $2 < W < 3$, divided by the integrated phase space density in pickup ions, $1.5 < W < 2$) at all times. This suggests that there is a pre-acceleration of ions in the slow, in-ecliptic solar wind that is not directly related to shocks. Since the tail in He^+ is stronger than the one in He^{++} (cf. Figure 3.5) it may be concluded that the bulk solar wind (which contains virtually no He^+) is at best a minor source of pre-accelerated tail particles. Further observations revealed that under many different circumstances the spectrum of suprathermal particles is always a power-law with a spectral index of $-3/2$ when expressed as differential intensity, or -5 when expressed as a distribution function in velocity space (Gloeckler *et al.*, 2000). None of the existing theories for particle acceleration, stochastic acceleration, or diffusive shock acceleration predicted a power-law spectrum with a unique index of -5. Diffusive shock acceleration yields power-law spectra, but with indices that depend sensitively on the shock jump conditions. From these facts Fisk and Gloeckler (2006) concluded that there must be some additional process at work that forces the spectra of the accelerated particles into a unique shape, a process that occurs commonly in many different plasma conditions. This new explanation relies solely on thermodynamic constraints, which is attractive because anything that occurs so commonly must be due to fundamental properties of the system. The theory is based on compressional turbulence, in which stochastic acceleration naturally yields power-law spectra. Starting out with two distinct sets of particles in compressional turbulence, core particles that are not very mobile and simply get compressed and expanded alternately, and very mobile tail particles that diffuse upward in energy to form the suprathermal tail, Fisk and Gloeckler (2006) arrive at the conclusion, only using basic thermodynamic arguments, that a power-law spectrum of -5 in the distribution function is unique because it is the only spectrum where the pressure in the tail undergoes isentropic compressions and expansions (see Fisk and Gloeckler, 2007, for a concise summary of the chain of argument).

3.3.2 Heavy ion distribution functions

It has been long known that alpha particles in the solar wind basically flow at the same bulk speed as the protons, and that they have the same thermal speed as well (Neugebauer and Snyder, 1966). The same behavior was later found to be true for other heavy ions under most solar wind conditions (Ogilvie *et al.*, 1980), which were found to depart from thermodynamic equilibrium having mass-proportional kinetic temperatures. These temperatures are presumably established by wave–particle interactions in the solar wind; only at times of very slow, dense solar wind were departures from the mass-proportional scaling law found, owing to the fact that Coulomb collisions had sufficient time to work towards thermal equilibrium (Hefti *et al.*, 1998). A comprehensive study of the kinetic properties of ~30 heavy ion species in the solar wind was performed by von Steiger and Zurbuchen (2002, 2006) using data from Ulysses-SWICS. They find that all ion species flow at the same bulk speed and have the same thermal speed (i.e., mass-proportional kinetic temperatures) to a remarkable degree of accuracy (see Figure 3.6). This was interpreted as the result of an interplay between the effect of Coulomb collisions, on the one hand, and of wave–particle interactions, on the other, with the latter becoming dominant at the heliographic distances covered by Ulysses. Coulomb collisions constantly push the ions towards equilibrium in the solar wind frame (i.e., equal bulk speed and equal kinetic temperature), with a collision frequency scaling as q^2/m. Wave–particle interactions at nondispersive waves scatter the ions in the rest frame of these waves; the waves propagate along the magnetic field with the Alfvén speed, v_A, so the interactions naturally result in a positive differential speed of $\sim v_A$ relative to the protons and in a thermal speed also of $\sim v_A$ of all heavy ion species. Such differential speeds are often observed in the inner solar system, but at the distance of Ulysses they

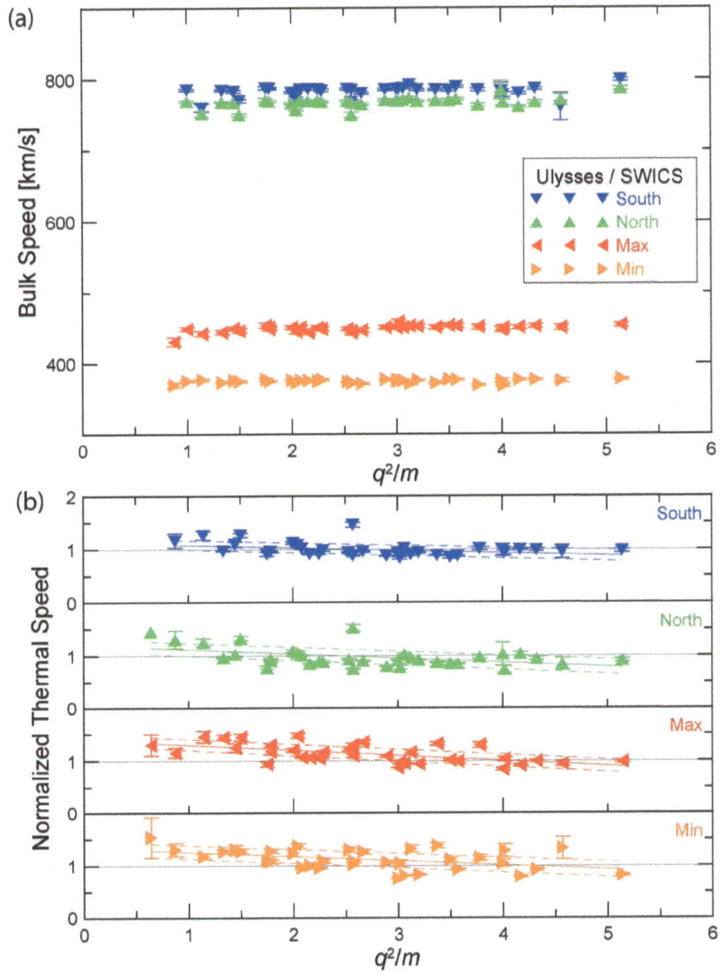

Figure 3.6. Average radial solar wind bulk speed (top) and thermal speed (bottom) of 32 heavy ion species as a function of their q^2/m observed with Ulysses-SWICS during four different ~300-day time periods in the fast (North, South) and in the slow solar wind (Max, Min). All bulk speeds were found to be equal within experimental uncertainty; all thermal speeds (normalized to that of alpha particles) were also uniform in the fast wind but showed a small trend with q^2/m in the slow wind (from von Steiger and Zurbuchen, 2006).

have essentially vanished because the magnetic field direction is predominantly transverse there. Interestingly, there is still a small systematic trend to higher thermal speeds with decreasing q^2/m observed in the slow solar wind. This is the residual imprint of Coulomb collisions that is very visible at 1 AU (Hefti *et al.*, 1998) and decreases with increasing heliocentric distance, but is still discernible just barely at the distance of Ulysses. No such trend with q^2/m is visible in the fast wind because there the much stronger waves have outperformed the collisions to fully equalize the thermal speeds.

3.4 COMPOSITION

As already mentioned in Section 3.1, Ulysses carries SWICS, which is the first fully resolving mass spectrometer flown in the free-flowing solar wind (Gloeckler *et al.*, 1992). Together with its unique orbit this has contributed to a significant improvement of where on the Sun the different solar wind types originate. SWICS routinely measures distribution functions of more than 30 different solar wind ion species of at least 10 different elements (H, He, C, N, O, Ne, Mg, Si, S, and Fe) under all solar wind conditions. Energy spectra of all ion species are obtained simultaneously with a time resolution of 13 minutes. However, for most ion species (except protons and alpha particles) multiple spectra have to be accumulated in order to obtain sufficient statistics, thus reducing the useful time resolution to between 3 hours and a day, or even longer for very rare species. Taking the first three moments of the distribution functions finally yields the density, speed, and thermal speed of each of the ion species. For a more detailed description of the SWICS data analysis procedure see the appendix of von Steiger *et al.* (2000).

The principal motivation for solar wind composition studies is twofold: on the one hand, we seek to determine the composition of the outer convective zone of the Sun (as represented by the photosphere) in order to infer the composition of the protosolar nebula, as this represents the baseline from which the entire solar system was formed some 4.6 Gyr ago (cf. von Steiger *et al.*, 2001, and references therein). On the other hand, composition differences between different solar wind types (or other reservoirs) are indicative for the conditions and processes where these reservoirs originate. Thus, composition studies naturally fall into two different types: charge-state composition and elemental composition. Charge-state composition (i.e., the distribution of the different charge states of a single element) probes the conditions and processes in the corona at a temperature of the order of 10^6 K, whereas elemental composition (i.e., the abundances of the elements summed over all charge states) probe the conditions and processes in the chromosphere and lower transition region at a temperature of the order of 10^4 K. In Figure 3.7 an overview of three solar wind parameters obtained with SWICS is given: the speed of alpha particles, v_α, the freezing-in temperature obtained from oxygen charge states, T_O, and the Mg/O abundance ratio. The figure illustrates that both charge-state composition and elemental composition are somehow related to the solar wind speed, the most obvious feature being an anticorrelation of T_O and v_α. In the next three subsections we will first turn to charge states, then to elements, and finally to some new ideas about the cause of this anticorrelation.

With the exception of helium, Ulysses-SWICS cannot determine the isotopic composition of the observed elements due to its relatively limited mass and mass-per-charge resolution (which is typically 30% and 3%, respectively). Isotopic composition can now be measured routinely with a different sensor type, the isochronous mass spectrometer, such as the MTOF sensor that is part of the CELIAS package flown on SoHO (Hovestadt *et al.*, 1995). We therefore do not discuss isotopic composition here but instead refer to the literature (e.g., Wimmer-Schweingruber, Bochsler, and Wurz, 1999).

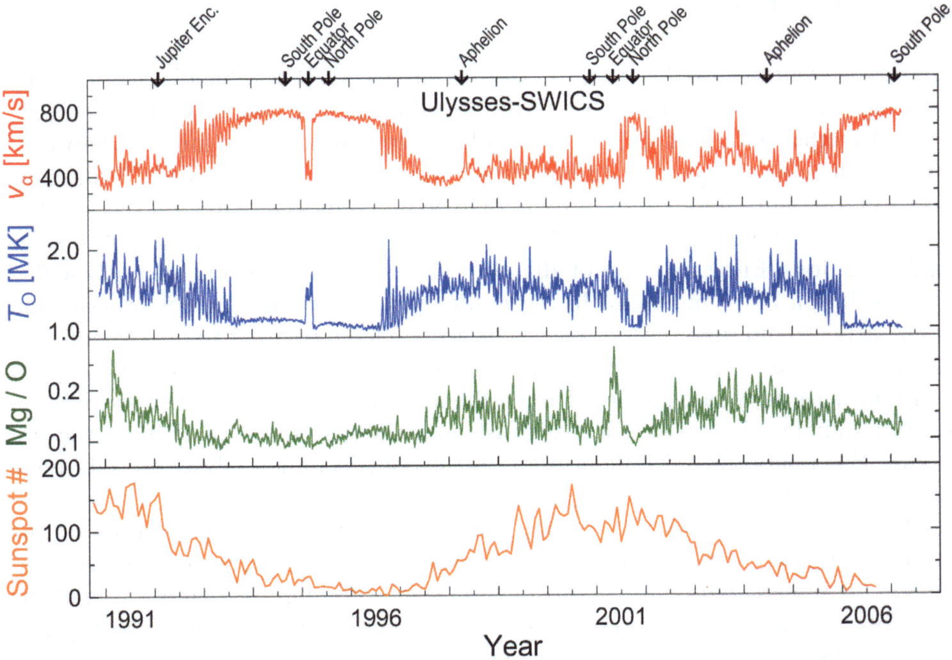

Figure 3.7. Solar wind composition parameters obtained with SWICS over the entire Ulysses mission so far: bulk speed of alpha particles, v_α, freezing-in temperature obtained from oxygen charge states, T_O, and the Mg/O abundance ratio. Plotted are running averages of daily values with a 7-day window for v_α and T_O or a 13-day window for Mg/O. The bottom panel gives the mean monthly sunspot number for comparison.

3.4.1 Charge-state composition

At the base of the corona all ion species are in collisional equilibrium with the ambient hot electrons. As the coronal plasma rises up to become the solar wind, it encounters a decreasing density profile at the same time as it is picking up speed. This means that the collision timescale between ions and electrons is becoming longer and the expansion timescale is becoming shorter, up to the point where the two intersect. Above that altitude the solar wind is expanding so rapidly and the electron density has become so low that the ion charge states are not changed any further; this is the freezing-in concept of Hundhausen, Gilbert, and Bame (1968). Thus, the charge states of heavy ions observed in the solar wind are indicative of the coronal temperature at the altitude where the collision timescale equals the expansion timescale. It is customary to convert the ratio between any two charge states to the corresponding temperature an ambient electron gas would have if it were in equilibrium (Arnaud and Rothenflug, 1985; Arnaud and Raymond, 1992; Mazzotta *et al.*, 1998). This procedure assumes thermal equilibrium up to the freezing-in altitude and complete freezing-in thereafter, whereas in reality the process is more gradual as modeled, for example, by Bürgi and Geiss (1986). Moreover, the freezing-in process may further be

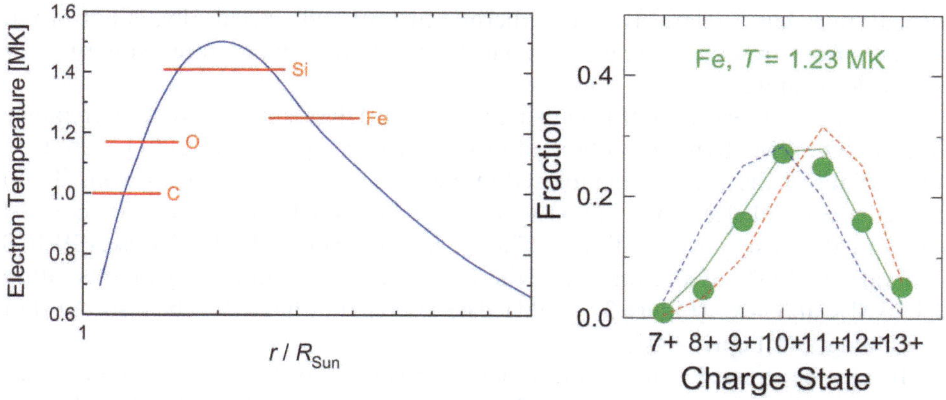

Figure 3.8. Left: Temperature profile in the south polar coronal hole as inferred from solar wind charge-state ratios observed with Ulysses-SWICS. The horizontal bars indicate the altitude range in the corona where the charge states of the indicated elements freeze-in. The charge states indicate that the corona has a temperature maximum of about 1.5 MK at 2–3$R_\odot$. Right: Observed charge-state distribution of iron ions yielding the temperature indicated in the left panel (green dots). The distribution is represented well with a single freezing-in temperature (green line). Any significant admixture of plasma with a temperature 0.1 MK higher or lower (red and blue dashed lines) would broaden the observed charge-state distribution (after Geiss *et al.*, 1995).

modified by the presence of suprathermal tails in the electron distribution function and/or of differential streaming between different ion species.

Since the ionization/recombination rates with hot electrons are temperature-dependent each ion pair freezes-in at a different altitude in the corona. This fact was used by Geiss *et al.* (1995) to obtain a temperature profile in the south polar coronal hole from charge-state observations in the southern high-speed stream, see Figure 3.8. The charge states of each element are well represented by a single temperature. This is a consequence of the fact that the ionization/recombination rates vary strongly from one element to another (by two orders of magnitude in the case at hand), but not so much between the relevant charge states of one element. The typical rates of the four elements were combined with a profile of the coronal electron density to roughly obtain the freezing-in altitude. This together with the freezing-in temperatures of the four elements results in the profile depicted in Figure 3.8. Apparently C and O freeze-in only little above the base of the corona. Si freezes-in at a higher temperature and a lower altitude than Fe, indicating a coronal temperature maximum of at least 1.4 MK between 2 and 3$R_\odot$. The single freezing-in temperature per element can also be used to infer that the coronal hole is thermally homogeneous to less than ±100,000 K. Consider the charge-state distribution of iron ions as depicted in the right panel of Figure 3.8: it is so well represented by a single temperature that any significant admixture of plasma with a higher or lower temperature would necessarily broaden the observed distribution. Since this is not observed we may conclude that the solar wind plasma in fast streams coming from coronal holes has a relatively simple thermal history and, specifically, is not mixed together from reservoirs (e.g., coronal loops) with temperatures differing by 100,000 K or more.

The direct observation of iron charge states on Ulysses has led to a discrepancy with SoHO-SUMER remote observations (cf. von Steiger *et al.*, 2001): based on observations of the 1,242 Å line formed by Fe XII (i.e., Fe^{11+}), Wilhelm *et al.* (1998) inferred a coronal electron temperature of just barely 1 MK at the base of a coronal hole and decreasing with altitude. This is inconsistent with the Ulysses-SWICS observation of 25% of all iron ions in the 11+ charge state, indicating a temperature of 1.23 MK at 3–4$R_\odot$ (Figure 3.8, right). The discrepancy has yet to be resolved (but see Laming and Lepri, 2007).

It must be stressed that the above paragraphs exclusively apply to the fast solar wind streams from coronal holes, which are thought to stream freely from the base of the corona into interplanetary space. The situation outside such fast streams, in the slow solar wind or in interplanetary coronal mass ejections, is much less simple, to the point that it is not even clear if the freezing-in concept applies in the simple form described above. Of course, charge-state distributions may be obtained all the same, it is only their interpretation that becomes less straightforward. Von Steiger *et al.* (2000) have reported average charge-state distributions of four elements obtained in four time periods of close to 1 year each; these are reproduced in Figure 3.9. As already mentioned, the charge-state distributions obtained in fast streams from coronal holes (periods North and South) are well represented by a single temperature for each element. The two other spectra (periods Max and Min) were both obtained in the slow solar wind, the first in 1991–1992 during a period of relatively high solar activity after the maximum of cycle 22, and the other in 1997–1998 during a period of minimum solar activity between cycles 22 and 23. Despite the difference in solar activity the two charge-state distributions Min and Max are rather similar to each other, and obviously they are quite different from the periods North and South. They not only have a significant excess of higher charge states, but they are also broader, indicating a mixture of sources at different, on average higher, temperatures. This has led Zurbuchen *et al.* (2002) to conclude that the slow solar wind is made up from a continuum of dynamic states. The excess of the highest charge state of each element in period Max is due to interplanetary coronal mass ejections, of which there were several in that period but virtually none in period Min; we will return to this point in Section 3.5.2. In summary, the distributions North and South indicate that the fast streams originate from the cool and thermally uniform corona, the coronal holes, whereas the distributions Max and Min indicate that the slow solar wind originates from a hotter and thermally inhomogeneous environment (i.e., from the streamer belt region).

The difference between the two solar wind domains is so clear that the argument may be reversed and charge states may be used to *define* them: we can be sure that a solar wind stream originates in a coronal hole if the charge-state temperature remains below a threshold value. This is illustrated in Figure 3.10, where almost 6,000 daily average values of T_O obtained from the O^{7+}/O^{6+} ratio are plotted versus T_C from the C^{6+}/C^{5+} ratio. The two charge-state temperatures obviously are very closely related

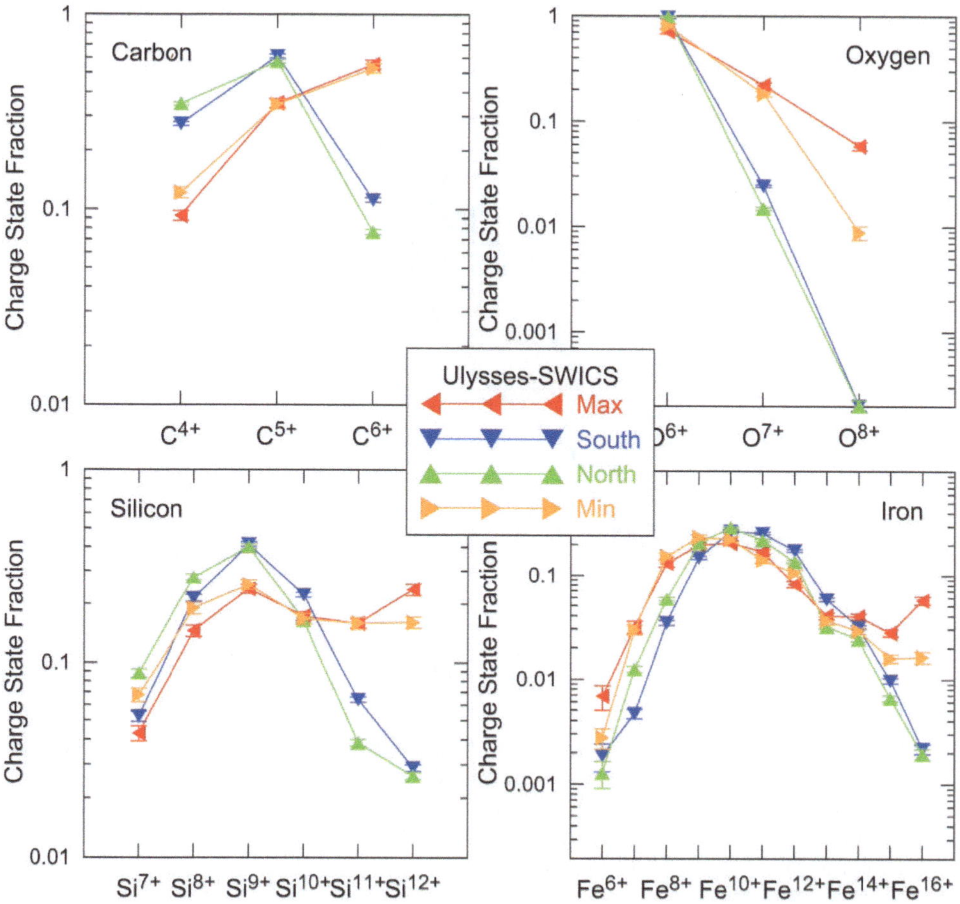

Figure 3.9. Average state distribution functions of C, O, Si, and Fe obtained during four time periods of ~300 days. The spectra marked North and South were obtained in fast streams from coronal holes, while the spectra marked Min and Max were obtained in the slow solar wind. Clearly, the latter show a significant shift to higher charge states as compared with the former (from von Steiger *et al.*, 2000).

by $T_O = 1.18T_C$ with a correlation coefficient of $r^2 = 0.99$. The fast solar wind can easily be spotted near the lower-left corner, well resolved from the extended continuum of slow solar wind and ICME temperatures. To put it simply: we define the coronal hole-associated solar wind as any stream with $T_O < 1.2$ MK (corresponding to $O^{7+}/O^{6+} < 0.01$) and/or with $T_C < 1.0$ MK (corresponding to $C^{6+}/C^{5+} < 0.28$). The values of the threshold ratios make it clear that T_C is an even better coronal thermometer than T_O in coronal holes because the count rate of the O^{7+} charge state may become so low there that a daily value cannot reliably be determined.

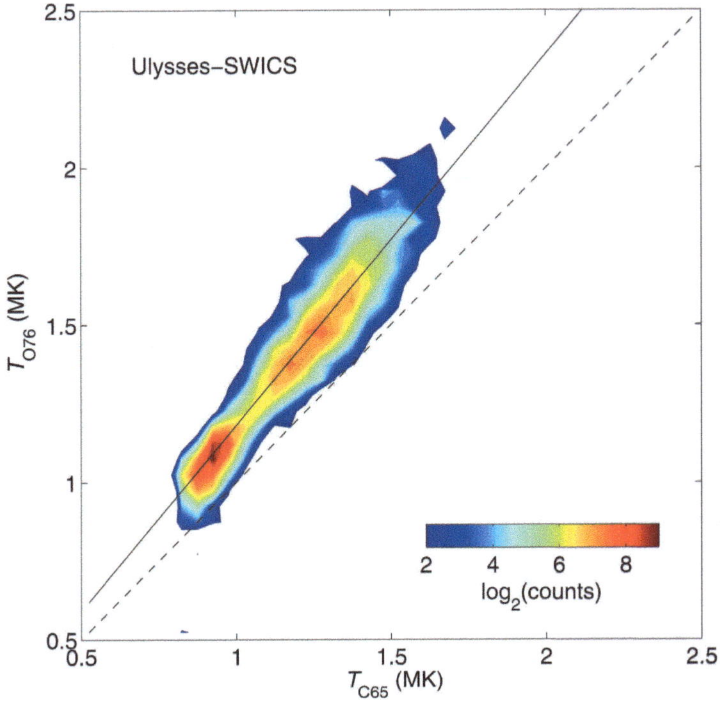

Figure 3.10. Comparison of oxygen to carbon freezing-in temperatures. In this contour diagram of almost 6,000 daily average values the separation of fast streams and slow solar wind is so clear that we propose to use it as the defining parameter for telling the two quasi-stationary solar wind types from each other. Note also that the two temperatures remain closely related over the entire range, $T_O = 1.18 T_C$ (solid line).

3.4.2 Elemental composition

As opposed to charge-state ratios, abundance ratios of elements (summed over all charge states) are generally thought to remain unaffected by coronal heating and solar wind acceleration. Nevertheless, solar wind abundances generally differ from solar abundances, as was realized long before Ulysses (Meyer, 1981): elements with a low first ionization potential (FIP) were found to be overabundant by a factor of 3–5 relative to high-FIP elements in the solar wind as compared with the solar photosphere. This may be seen in Figure 3.7 (third panel) since Mg/O is a low- to high-FIP element ratio. The cause of this difference is the FIP fractionation effect, which Geiss (1982) attributed to a separation mechanism of ions from neutral atoms in the partially ionized portion of the solar atmosphere (i.e., the chromosphere) and lower transition region. Ulysses observations have basically confirmed the presence of the FIP effect, but modified and extended the simple picture in two important respects.

First, it was found that the FIP fractionation factor $f = (X/O)_{SW}/(X/O)_{\odot}$ (with X/O the abundance ration of a low-FIP element X relative to oxygen) was significantly lower than 3–5 in the fast streams from coronal holes. This had been con-

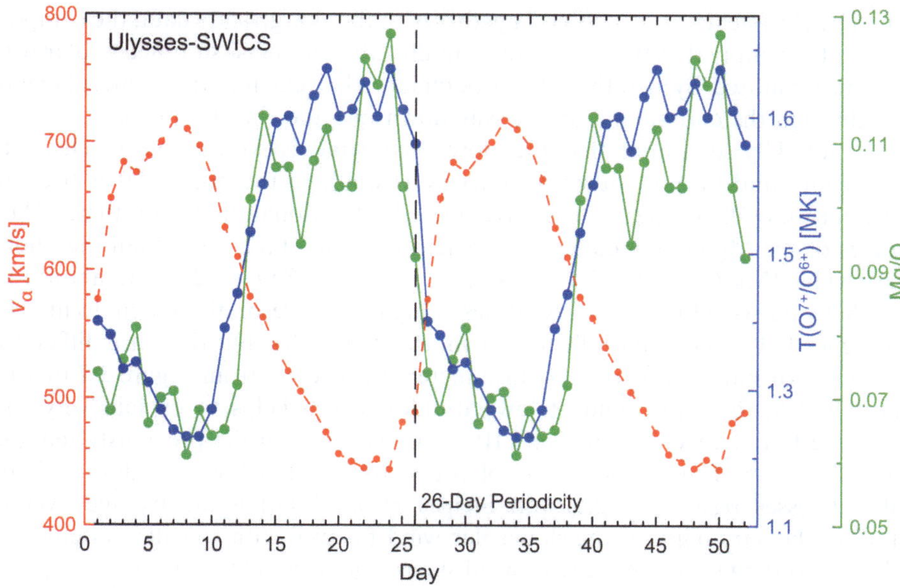

Figure 3.11. Superposed epoch analysis of alternating slow and fast solar wind streams. Plotted are the same three parameters as in Figure 3.7 during the period in 1992–1993 when the streams alternated with a period of 26 days (i.e., once per solar rotation). The data are repeated twice in order to better visualize the periodicity (after Geiss, Gloeckler, and von Steiger, 1995).

jectured shortly before Ulysses from glimpses of such fast streams compared with fast coronal mass ejections, both observed in the shocked solar wind behind the Earth's bow shock, in the magnetosheath (von Steiger *et al.*, 1992). From the massive amount of coronal hole associated SWICS data obtained in the large polar streams at solar minimum in 1993–1996 (and again since 2006) it was found that the average FIP enrichment factor in these streams was just about a factor of 2 or even less (von Steiger *et al.*, 2000). The difference between the two quasi-stationary solar wind types was best illustrated using a superposed epoch analysis by Geiss, Gloeckler, and von Steiger (1995), see Figure 3.11. The figure shows the superposition of the same three parameters as in Figure 3.7 during 15 solar rotations in 1992–1993 when the stream types alternated once every rotation of 26 days. This makes it evident that both the charge-state composition and the element abundances are significantly different in fast streams as compared with slow solar wind, with a sharp boundary between the two stream types. Wimmer-Schweingruber, von Steiger, and Paerli (1997) have shown that the boundaries in all parameters are conlocal with the stream interface at the leading edge of the fast stream to within the time resolution with which they could be determined. A subsequent analysis by McComas, Elliott, and von Steiger (2002) of streams from the smaller coronal holes found at all latitudes during solar maximum revealed that compositional changes were nearly as sharp and clear as in the large solar-minimum polar streams although there is some evidence of a transition layer—in particular, at the trailing edge of a stream—a result also found using ACE-

SWICS by Zurbuchen *et al.* (1999). Together with the charge-state data, the composition data thus imply that the two fundamentally different quasi-stationary solar wind types are separated by a rather sharp boundary that extends from interplanetary space through the corona and all the way down into the chromosphere.

Second, Ulysses found that the slow solar wind is so variable in elemental composition (and in most other parameters as well), to the point that it becomes hardly meaningful to speak of an average FIP fractionation factor there. Daily averages of the Mg/O abundance ratio reach from only little more than the photospheric value ($Mg/O_{\odot} = 0.074$, Grevesse, Asplund, and Sauval, 2007) to about four times that value (see Figure 3.7). Analyses at higher time resolutions seem to indicate even higher FIP fractionation factors at shorter timescales, but these are difficult to ascertain since statistical variability also increases. It seems from Figure 3.7 that the highest FIP factors are found at low latitudes, which helps to explain why with SWICS we found generally smaller FIP fractionation than the previously reported factor of $f = 3$–5. These factors were of course obtained at low latitudes, while the smaller Ulysses result ($f = 2.5$, von Steiger *et al.*, 2000) is an average over all latitudes. The variability of the slow solar wind fits well with so-called interchange models, according to which the slow wind stems from closed loops reconnecting with open field lines, hence interchanging their topological properties as they wander along the solar streamer belt (Fisk, Schwadron, and Zurbuchen, 1998). The natural age variability of these loops together with the fact that the FIP fractionation factor correlates with their ages (Widing and Feldman, 2001) readily accounts for the observed variability.

We have argued above that the composition of the fast polar streams is as close as we can get to the solar composition with *in situ* observations. This is particularly important for elements such as neon that cannot be observed by remote sensing in the photosphere for lack of transitions in the relevant energy range. The "solar" neon abundance given in tables such as that of Grevesse, Asplund, and Sauval (2007) is not really a solar value, but an approximation thereof obtained from other sources such as remote sensing of the corona or solar energetic particles. Bahcall, Basu, and Serenelli (2005) have used this ignorance to argue that the solar neon abundance might be higher by a factor of 2.5–3 (0.4–0.5 dex) than this estimate in order to reconcile the helioseismology results with the latest values of solar abundances—in particular, of CNO—just as if the neon abundance were a freely disposable parameter. But the solar wind value of neon observed with Ulysses-SWICS ought to be taken into account, and it makes such a high neon abundance seem quite unlikely. Von Steiger *et al.* (2000) find $Ne/O = 0.083$ in fast streams but caution that this is a difficult measurement since neon occurs in the single charge state Ne^{8+} that lies close to the most abundant of the heavy ions, O^{6+}. Nevertheless, a recent independent analysis (Gloeckler and Geiss, 2007) seems to confirm this low value of Ne/O. The value is even lower than the solar estimate of Grevesse, Asplund, and Sauval (2007) ($Ne/O = 0.15$), which makes it very difficult to believe that the real solar value could be as high as $Ne/O = 0.4$—as would be needed for the helioseismology results to fit. The solution of this conundrum is as yet outstanding; it may well lie in the abundances of other elements than just that of neon.

3.4.3 Correlation between composition and kinetic parameters

It is readily apparent in Figure 3.7 that there is an anticorrelation between the solar wind speed, v_α, and the oxygen charge-state temperature, T_O. In Figure 3.12 we plot just these parameters in two polar plots much like the ones of SWOOPS (cf. Figure 3.2), whereby the anticorrelation becomes particularly clear. This anticorrelation was studied in some detail in a pair of papers: Gloeckler, Zurbuchen, and Geiss (2003); Fisk (2003). During a 166-day time period around the solar minimum in 1996–1997, Gloeckler, Zurbuchen, and Geiss (2003) first determined a correlation $V = 144/T - 88$, where V is the solar wind speed in km/s and T is the freezing-in temperature from oxygen charge states in MK. The correlation is found to be very tight except at times when an ICME passes by (see Section 3.5.2). On the other hand, Fisk (2003) derives a theoretical relation between coronal electron temperature and terminal solar wind speed squared (i.e., its energy) of the form

$$\frac{V^2}{2} = \frac{C_1}{T} + C_2,$$

where $C_{1,2}$ are constants. The theory is based on the picture of open field lines migrating across the solar surface by successively reconnecting with closed loops and thus displacing themselves by the separation of the loops' footpoints. Each of these reconnection events releases energy and mass onto the open field line (i.e., into the corona and solar wind). In turn, these two quantities determine the final energy of a solar wind parcel, or V^2. The quantities can be determined using solar observations

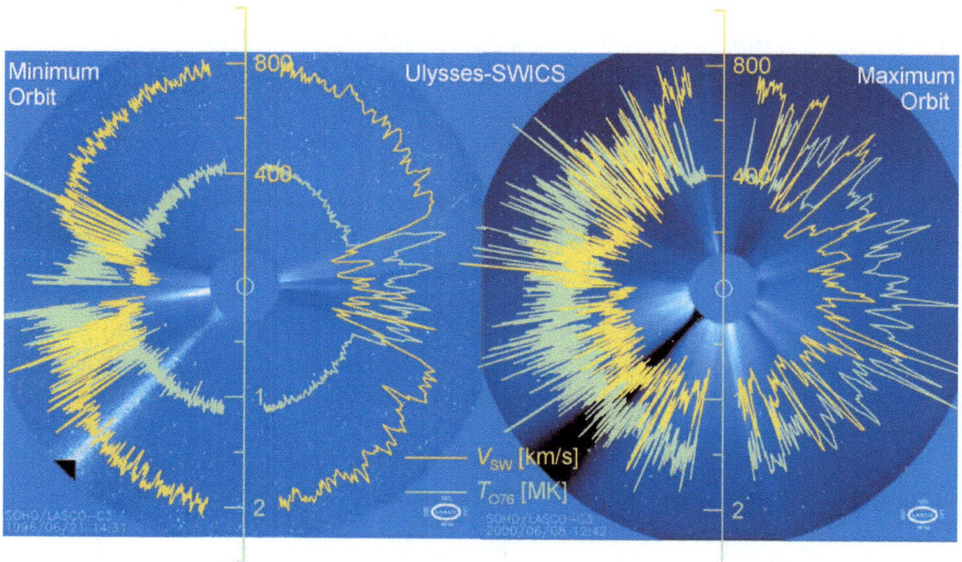

Figure 3.12. Polar plots of the solar wind speed, v_α, and the oxygen charge-state temperature, T_O, in a similar format as in Figure 3.2. There is a clear anticorrelation between these two parameters (from von Steiger and Fröhlich, 2005).

or estimates of typical loop heights and other solar quantities, thus determining the constants $C_{1,2}$. Note that only C_1 involves quantities that are not determined in a straightforward manner, while $C_2 = -GM_\odot/r_\odot = -(437\,\text{km/s})^2$ is simply the gravitational potential at the solar surface and thus unadjustable.[3] Fitting their data to Fisk's $V^2 \propto 1/T$ relation, Gloeckler, Zurbuchen, and Geiss (2003) find an equally satisfying fit (again with the ICME periods removed) as for $V \propto 1/T$. It is noteworthy that this fit yields an intercept value very close to the unadjustable constant C_2 and has the added benefit of a physical underpinning. This can finally be used to reverse the relation and ask about the loop heights with which the migrating field lines reconnect. In the quiet Sun, loop heights were found to show a strong dependence on latitude, reaching up to ~100,000 km at low latitudes; conversely, in polar coronal holes the lowest heights of ~15,000–30,000 km were observed with minimum fluctuation and no dependence on latitude.

We conclude this section by noting that a significant deviation from the $V^2 \propto 1/T$ relation might be used as a sensitive ICME detector. ICMEs are discussed in the next section, but to our knowledge no such study has yet been conducted.

3.5 TRANSIENTS

The continuous (or quasi-continuous) solar wind described so far is often permeated by transient phenomena. The two principal kinds of transients are corotating interaction regions (CIRs) occurring primarily at declining to minimum solar activity, and interplanetary coronal mass ejections (ICMEs) occurring all around the solar cycle but mainly around solar maximum. Both were well-known phenomena before Ulysses, yet Ulysses observations have significantly enhanced and improved our understanding of them.

3.5.1 Corotating interaction regions

CIRs are the result of a fast stream running into a previously ejected stream of slow solar wind. Initially, the two radially emitted streams merely shear along each other, but as the magnetic field gradually bends into the Parker spiral the fast stream begins to ram into the previously (i.e., at a more easterly longitude) emitted slow stream. This leads to the build-up of a pressure wave that eventually steepens to form a forward–reverse shock pair. Under normal conditions these interplanetary shocks develop only outside the Earth's orbit and thus can be observed only with deep-space missions such as Ulysses.

Gosling *et al.* (1993) and Pizzo and Gosling (1994) were the first to describe the formation and evolution of CIRs in three dimensions based on Ulysses-SWOOPS observations during the first polar orbit. These authors found that, as Ulysses ascended to high latitudes, the strong, CIR-related forward shocks were observed only to a mid-latitude corresponding roughly to, but slightly less than, the tilt of the

[3] Coincidentally, this is almost precisely the speed of a typical 1 keV/amu solar wind particle.

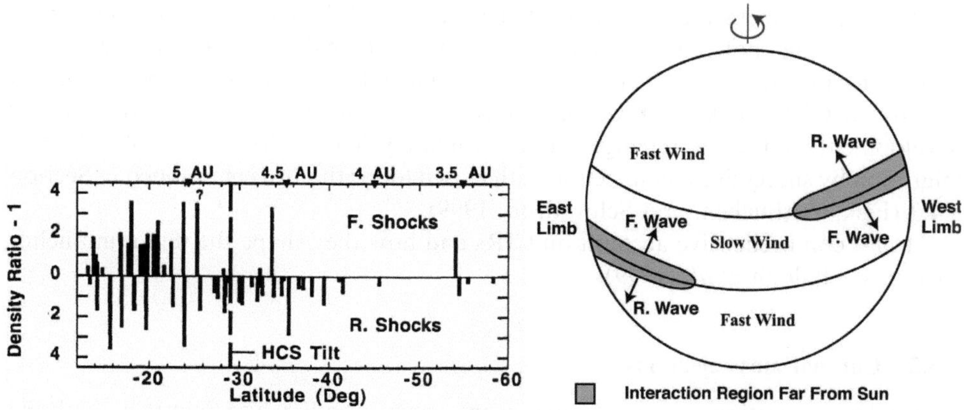

Figure 3.13. Left: Ulysses observations of forward and reverse shocks as a function of heliographic latitude, showing that reverse shocks were observed to much higher latitudes than forward shocks. Right: Sketch of the basic flow geometry at the Sun: CIRs with forward–reverse wave (shock) pairs form at the edge of the streamer belt wherever fast wind follows (i.e., ejected at a more easterly longitude than slow wind). The tilt of the edge causes the reverse shocks to travel to higher latitudes and the forward shocks to travel to lower latitudes from where they are formed (from Gosling *et al.*, 1993, 1995b).

coronal streamer belt region and heliospheric current sheet. At higher latitudes the forward shocks disappeared almost entirely; however, the reverse shocks persisted to a latitude some 10° higher than the streamer belt tilt angle (Figure 3.13, left). This was interpreted as the result of the forward waves propagating to lower heliographic latitudes and the reverse waves to higher latitudes with increasing heliocentric distance, as illustrated in Figure 3.13 (right). Note that the flow pattern is basically established at the onset of the streams near the solar surface although the shocks develop only at a much larger distance. The interpretation is further supported by observations of the corresponding meridional velocity components, which are in excellent agreement with numerical models that describe the origin and evolution of the three-dimensional structure of CIRs (Gosling and Pizzo, 1999).

CIR-related shocks are important sites for the acceleration of particles in the heliosphere. Low-energy (few MeV/amu) charged particles were therefore observed to recur with the solar rotation period in 1992–1993 when Ulysses first traversed to high southern latitudes (see Figure 3.4). But then it came as a considerable surprise that this recurrence pattern persisted almost all the way down to −80° (Lanzerotti *et al.*, 1996), way past the latitude where first the forward shocks and then the reverse shocks had disappeared. In a strictly Parker-type field configuration every field line lies on a cone with a fixed latitude, so it was a mystery where these particles got energized since there is no magnetic connection between high-latitude field lines and the accelerating shocks at lower latitudes. This paradox motivated Fisk (1996) to rethink the heliospheric magnetic field configuration and to come up with a new field model. It is based on the interplay of magnetic field lines rooted in the differentially rotating photosphere and the superradial expansion of these field lines from a rigidly rotating coronal hole. As a result it is found that field lines can have extensive excursions in latitude and, specifically, high-latitude field lines can be connected directly to CIR shocks at lower latitudes farther out in the heliosphere. Further development of the interchange model ultimately led to the picture of field line migration by successive reconnection with closed loops that was mentioned in Section 3.4.3 (Fisk, Zurbuchen, and Schwadron, 1999).

For a comprehensive account on CIRs and how they shape the minimum heliosphere, see Balogh *et al.* (1999).

3.5.2 Coronal mass ejections

Coronal mass ejections are spectacular events when seen in white light (e.g., with the LASCO coronagraph on SoHO). But the identification of their interplanetary counterparts, termed ICMEs, is less than a trivial matter and "is still something of an art" (Gosling, 1997). Zurbuchen and Richardson (2006) have compiled a comprehensive table of 23 ICME signatures subdivided into 5 classes: magnetic field, plasma dynamics, plasma composition, plasma waves, and energetic particle signatures; an image of a typical ICME is given in Figure 3.14.

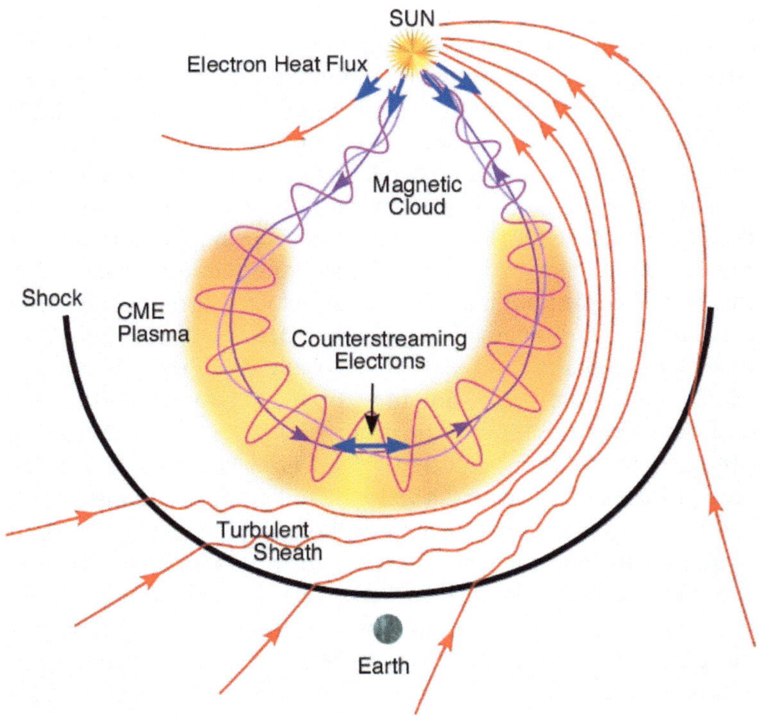

Figure 3.14. Basic geometry of an ICME indicating and relating its principal signatures (from Zurbuchen and Richardson, 2006).

Ulysses' contributions to ICME studies are twofold: mapping ICME occurrence rates as a function of latitude, on the one hand, and adding new plasma composition parameters for ICME identification, on the other.

Plasma composition parameters as ICME signatures

It was long known that a high alpha-to-proton ratio ($>8\%$, say) is a signature of an ICME (Hirshberg, Bame, and Robbins, 1972, what we call ICME today was called driver plasma back then). This signature is easy to spot and never occurs outside ICMEs, but it is only present in a fraction of very roughly 50% of all events. Ulysses (and later missions carrying composition instrumentation) has now added new composition signatures that come fairly close to ideal ICME identifiers (an ideal signature would detect all ICMEs with no false identifications).

One such signature is the average charge state of iron, $\langle Q_{\mathrm{Fe}} \rangle$. Under quasi-stationary solar wind conditions Fe is found distributed over several charge states with a broad maximum around Fe^{10+} (Ipavich *et al.*, 1992; von Steiger, Geiss, and Gloeckler, 1997); the distribution is not very different between fast streams and slow solar wind (see also Figure 3.9). But this changes, sometimes drastically, during the passage of an ICME. Lepri *et al.* (2001) found that $\langle Q_{\mathrm{Fe}} \rangle$ is strongly enhanced, so Fe^{16+} becomes the dominant charge state (even higher charge states are rarely observed because Fe^{16+} has an Ne-like configuration that is hard to ionize further). In a follow-up paper, Lepri and Zurbuchen (2004) showed that $\langle Q_{\mathrm{Fe}} \rangle > 12$ is a very strong ICME identifier that is present in a large fraction of (but not all) ICMEs, and has a negligible probability for false positive ICME identification. It is further shown that the presence of high Fe charge states is correlated with the magnetic connectivity to the flare site from where the CME has originated in the corona (Reinard, 2005), thus putting into perspective "the solar flare myth" (Gosling, 1993) by showing that flares and ICMEs are not entirely unrelated after all. The latter conclusion is made from the observation that ICMEs with a high $\langle Q_{\mathrm{Fe}} \rangle$ are becoming rarer at high latitudes, which have less magnetic connectivity to the active regions at low to mid-latitudes. We will return to the latitude distribution of ICMEs below.

Another ICME signature is the O^{7+}/O^{6+} charge-state ratio, which has already been used for the separation of the two quasi-stationary solar wind types (although, as argued above, the C^{6+}/C^{5+} ratio would be superior for that purpose). Of course, in the light of the above paragraph it comes as no surprise that O^{7+}/O^{6+} is often enhanced in ICMEs. This was already found by Neukomm (1998) and by Henke *et al.* (2001), who found a good positive correlation of the presence of a magnetic cloud, a sure ICME identifier, with high O^{7+}/O^{6+}. However, the definition of a clear-cut threshold value is much less evident in this case than it is for $\langle Q_{\mathrm{Fe}} \rangle$. Richardson and Cane (2004) overcame this difficulty by defining a solar wind speed-dependent threshold value instead of a constant one. From observations with ACE-SWICS they determined a correlation between O^{7+}/O^{6+} and the solar wind speed (in units of km/s) $(O^{7+}/O^{6+})_{\mathrm{ACE03}} = 3.004 \exp(-V/173)$ to hold in the ambient solar wind away from all ICMEs, and define as ICME threshold if O^{7+}/O^{6+} exceeds twice that value. Although this definition is still somewhat arbitrary it does a significantly better job at picking ICMEs than a constant threshold value. Kilchenmann (2007) has performed the same analysis with data from Ulysses-SWICS and finds an ambient (non-ICME) solar wind correlation of $(O^{7+}/O^{6+})_{Uly} = 3.776 \exp(-V/128)$, which is about a factor of 2 lower than the relation of Richardson and Cane (2004) at typical solar wind speeds. However, this apparent discrepancy is not physical because ACE-SWICS was recalibrated after 2004. Using current ACE Level 2 data of the same time period Kilchenmann (2007) finds a relation of $(O^{7+}/O^{6+})_{ACE07} = 1.210 \exp(-V/200)$, which agrees with the Ulysses relation to within 15% at typical solar wind speeds. Note that this discrepancy does in no way invalidate the work of Richardson and Cane (2004) because it is internally consistent, but when comparisons are made the recalibrated, not the published, ACE-SWICS data must be used.

ICMEs at high latitudes

ICMEs are obviously associated with active regions on the Sun, which are found at mid- to low latitudes in the solar corona but never within a coronal hole. It was therefore not a small surprise when Gosling *et al.* (1994) discovered a new class of ICMEs that are fully embedded in the fast solar wind stream from the polar coronal hole at solar minimum. These ICMEs are characterized by a forward–reverse shock pair driven into the ambient fast solar wind by virtue of their high internal pressure and are therefore termed overexpanding ICMEs (see Figure 3.15). To be sure, such ICMEs are rare events, with only six of them observed with Ulysses during its entire solar-minimum orbit. One of them was even observed simultaneously both at low and at high latitudes (Gosling *et al.*, 1995c). Interestingly, these ICMEs do not show any of the compositional signatures discussed above, but are indistinguishable from the ambient fast solar wind regarding their composition (Neukomm, 1998). It is therefore conceivable that overexpanding ICMEs are not strictly speaking ICMEs, but rather the wake of a solar ejection ICME passing by at lower latitudes, as recently modeled by Manchester and Zurbuchen (2006).

As with the quasi-stationary solar wind the rate of CMEs changes drastically from solar minimum to maximum. CMEs, which at solar minimum are confined to low latitudes almost exclusively, are distributed nearly uniformly over all position angles at solar maximum (Gopalswamy *et al.*, 2006). Likewise, we might expect to observe ICMEs equally uniformly at all heliolatitudes, and indeed we can find ICMEs even at the highest latitudes reached by Ulysses (e.g., von Steiger, Zurbuchen, and Kilchenmann, 2005). However, their rate of occurrence surprisingly seems to decrease with increasing heliolatitude even at a time of increasing and high solar activity, as was already apparent from Figure 3.3. This has been demonstrated quantitatively by Lepri and Zurbuchen (2004) by comparing simultaneous observations of the ICME rate on ACE and Ulysses. Von Steiger, Zurbuchen, and Kilchenmann (2005) have compiled the latitude distribution of all ICMEs observed on Ulysses in 1998–2001 (i.e., during the rise and maximum phases of cycle 23, see Figure 3.16). Evidently, there is an anisotropy of the monthly ICME rate with a strong preference for ICMEs near the equator. Note that the plotted ICME rate has

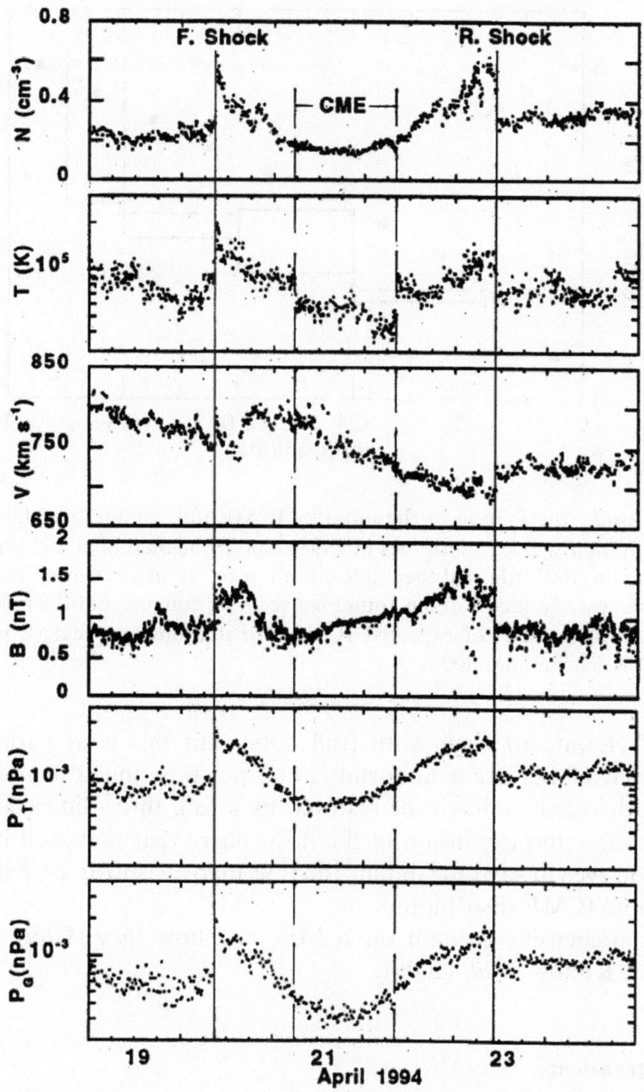

Figure 3.15. Overexpanding ICME observed with Ulysses at 61°S heliographic latitude. Its main feature is a forward–reverse shock pair driven into the ambient fast solar wind stream (from Gosling *et al.*, 1994).

been obtained using all, not only compositional, signatures. This makes it unlikely that the anisotropy is biased by the fact that high-latitude ICMEs are less likely to have a composition signature as discussed above. Von Steiger *et al.* (2005) also discussed what latitude distribution of CMEs might underlie the observed ICME distribution, and find that an isotropic CME model distribution (open dots in Figure 3.16) is a very bad fit to the data. A better fit can be obtained by assuming an

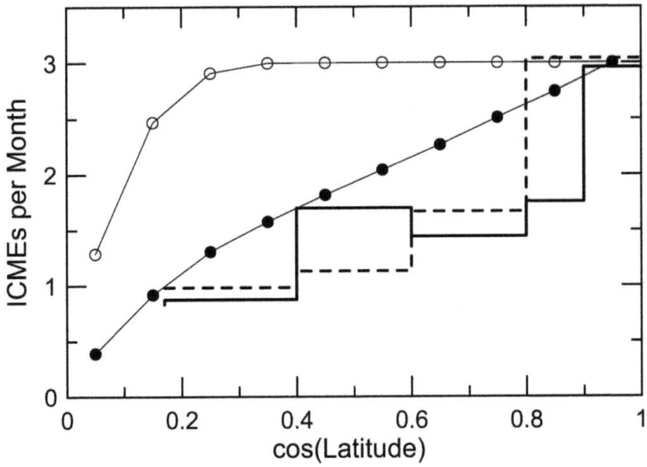

Figure 3.16. Latitude distribution of the monthly ICME rate obtained by Ulysses during its second descent to high southern latitudes in 1998–2000 (solid step line) and during its second fast-latitude scan in 2001 (dashed step line), both near or at maximum solar activity. A significant anisotropy of a factor of 3 is found between the equator and the pole. The symbols represent very simple models of expected ICME distributions, see text (from von Steiger, Zurbuchen, and Kilchenmann, 2005).

anisotropic CME rate to begin with (full dots), but this is at variance with the observations that CMEs occur uniformly at all position angles at solar maximum. This apparent discrepancy might be resolved by a full, three-dimensional model of ICME propagation and expansion in the heliosphere that may well involve super-radial expansion even at solar maximum and thus map an isotropic CME distribution to an anisotropic ICME distribution.

For a comprehensive account on ICMEs and how they shape the maximum heliosphere, see Kunow *et al.* (2006).

3.5.3 Other transients

We conclude this section by briefly mentioning three other types of transient events: magnetic holes, microstreams, and reconnection events.

Magnetic holes are brief (~10–15 s), isolated intervals where the magnetic field strength drops to a value of less than 50% of the ambient field strength with no significant field rotation ($<5°$, say). In rare cases the holes may last up to ~30 minutes (Zurbuchen *et al.*, 2001). During the early, in-ecliptic phase of the Ulysses mission Winterhalter *et al.* (1994) found numerous such magnetic holes at a rate of ~50 per month. These holes, which are interpreted as relic structures of the mirror instability, were found to be preferentially associated with interaction regions. But when Ulysses went on its first high-inclination orbit at solar minimum, leaving all interaction regions behind poleward of ~30°, the rate of magnetic holes did not drop to zero but to a small nonzero constant (Winterhalter *et al.*, 2000). This indicates that magnetic holes may be formed by more than one process: one operating near the ecliptic plane associated with large-scale dynamic solar wind features, and another operating fairly uniformly at all latitudes. The second one of these processes is much less well understood since there are no large velocity gradients to work with at these latitudes. Two possibilities are relics of mirror waves generated in the corona, or relics of an anisotropic ion population generated by pickup ions, but both of them have their shortcomings (Winterhalter *et al.*, 2000).

Microstreams were first observed by Thieme, Marsch, and Schwenn (1990) on Helios and later by Neugebauer *et al.* (1995) when Ulysses was inside the south polar fast stream in 1994. They are defined as localized velocity peaks or dips within a fast solar wind stream with an amplitude of ~40 km/s and a mean half-width of ~10 hours. Using a superposed epoch analysis it could be shown that the density and the temperature profiles of the fast microstreams had the expected compression and pile up on their leading edges, although without any forward or reverse shocks. The corresponding profiles of the slow microstreams were found to be inverted images of those of the fast microstreams. The recurrence rate was found to be on timescales of 2–3 days, with no apparent latitude variation. The cause of microstreams remains largely unclear, even to the point as to whether they are spatial or temporal structures. The only spatial structures inside coronal holes that might be associated with microstreams are polar plumes. Since these are known to have composition signatures different from the surrounding coronal hole (Widing and Feldman, 1992), von Steiger *et al.* (1999) conducted a study to look for such signatures in microstreams but failed to find any. It is concluded that microstreams are not associated with polar plumes, but more likely are dynamic structures generated by processes that do not affect composition.

Reconnection "events" are not strictly speaking transient events, but quasi-stationary, localized phenomena with an exhaust that is traversed by a spacecraft such as Ulysses in a matter of minutes, thus giving the impression of isolated events. Reconnection occurs at thin current sheets separating plasmas having nearly oppositely directed magnetic fields. At the reconnection site magnetic energy is converted to bulk flow energy in a pair of oppositely directed exhausts, one of which (usually the one in the anti-sunward direction) may be detected by a spacecraft as a characteristic signature: a brief (few minutes, sometimes longer) interval of accelerated or decelerated plasma flow within a bifurcated current sheet in which changes in magnetic field and flow velocity are correlated at one edge and anticorrelated at the other. Gosling *et al.* (2006a, b) have identified 91 such events during the entire Ulysses mission; the events are found to be distributed over the entire range in heliocentric distance and almost the entire range in heliolatitude covered by its orbit. Many reconnection events are associated with ICMEs, but not all of them: the first such event detected with Ulysses was associated with a flux rope that was both too small and had an increased, not decreased, proton temperature to be called an ICME (Moldwin *et al.*, 1995). The other events occurred in low-speed solar wind, whereas none were found, of course, inside the unipolar, coronal hole associated fast streams. The main difficulty in recognizing these events is their relatively short duration, putting spacecraft with a higher cadence of plasma measurements such as ACE (Gosling *et al.*, 2005) or Helios (Gosling, Eriksson, and Schwenn, 2006) at an advantage over Ulysses for such studies.

3.6 THE ULYSSES PICTURE: THE SOLAR WIND IN FOUR DIMENSIONS

When Ulysses was launched some 17 years ago our picture of the solar wind in the heliosphere was that of a ballerina skirt: frilled, flapping up and down, complex, yet limited to the vicinity of the ecliptic and the solar equatorial planes. We had very little understanding of the solar wind from polar regions. With Ulysses now on its third polar orbit (and still going strong) this has changed profoundly, and in a way it has become simpler.

The structure of the heliosphere, shaped by the solar wind, has found to be dipolar near solar minimum. It is dominated by two uniform polar fast streams separated by a band of slow and variable solar wind. At solar maximum the picture superficially looks more complicated, but solar wind observations—in particular, those of composition—indicate that it is composed of the same two quasi-stationary solar wind types, but that their source regions in the corona are distributed in a less orderly manner. Magnetic observations even indicate that the simple, bipolar structure of the heliosphere applies throughout the solar cycle, but that the single current sheet separating the two unipolar regions is highly inclined (and more strongly warped) at solar maximum as it flips over to settle down (and flatten) at low latitudes again at the next solar minimum. It is difficult to see how that simple picture could ever have been obtained from an in-ecliptic perspective. Yet it is the global structure of the heliosphere that matters, for example, for cosmic ray modulation. Ulysses has truly added the third dimension to our understanding of the heliosphere by traveling to the regions poleward of 30° in heliolatitude.

But Ulysses has added yet another dimension to our picture of the heliosphere: time. Thanks to its long mission duration, far in excess of the originally planned end of mission after the first set of polar passes in 1995, Ulysses has now mapped the heliosphere for more than a complete solar activity cycle. If we are fortunate the agencies running the mission will have the insight to extend it still further, to its technical limitations (mainly the decay of its RTG power supply). Ulysses could live long enough to complete the 22-year Hale cycle. This is of potentially paramount importance as it would allow us to establish whether the observed north–south asymmetries depend on the 11-year solar cycle or on the 22-year magnetic cycle. For example, during the first polar orbit Ulysses observed the electron temperature in the south polar coronal hole (measured as a charge-state ratio of heavy ions) to be hotter than the northern one. Now, on the third polar orbit, the southern coronal hole is found to be cooler (see the second panel of Figure 3.7), and it remains to be seen whether the northern coronal hole will be hotter this time around. If so, the coronal electron temperature would appear to track the magnetic polarity, not the location in the corona. The coronal electron temperature is important because this is the quantity that determines the dynamic processes accelerating the solar wind, and only Ulysses is in the right location and has the instrumentation to measure it.

Clearly, Ulysses is not the only spacecraft that contributes to our understanding of the Sun and the heliosphere. Many other missions, in Earth orbit or in interplanetary space, have made equally important contributions. The Voyagers are currently exploring the outer boundary of the heliosphere; the Helios spacecraft have mapped the inner heliosphere; at 1 AU IMP-8, Wind, and ACE are continuously monitoring the solar wind impinging on Earth; SOHO has watched the Sun continuously since 1995 with hardly a blink; and Stereo and Hinode, after their recent successful launches, are about to add even more observations, and hopefully a better understanding. Our picture of the Sun and heliosphere has thus been created with the Great Heliospheric Observatory (i.e., the combination of all these and many more missions). Clearly, our understanding of the heliosphere can only be advanced through interdisciplinary studies using the results from as many missions as possible.

Yet Ulysses is the only spacecraft that has ever traveled poleward of 35°, which is just about the boundary to the high-latitude heliosphere. There is no mission firmly planned on anything like a high-inclination orbit (although several such mission proposals are around). Once Ulysses ceases to work due to lack of power (or of political will) this could seriously limit the usefulness of new missions such as Stereo or Hinode. A new mission on a polar orbit around the Sun ought to be very high on the priority list of any agency that has the relevant capabilities. Everything else could be seen like a self-imposed retreat to a flat, two-dimensional heliosphere.

3.7 ACKNOWLEDGMENTS

I thank all experiment teams on Ulysses, in particular the SWOOPS and SWICS teams for building and running their fine instruments without so much of a glitch. I further thank the Ulysses Mission Operations Team for doing such an outstanding job of running that unique mission, and the Ulysses Data Management Team for never failing to a timely delivery of the acquired data. Special thanks are due to Len Fisk, Johannes Geiss, George Gloeckler, Jack Gosling, Dave McComas, Marcia Neugebauer, and Thomas Zurbuchen for many discussions.

3.8 REFERENCES

Arnaud, M., and J. Raymond (1992), Iron Ionization and Recombination Rates and Ionization Equilibrium, *Astrophys. J.*, **398**, 394–406.

Arnaud, M., and R. Rothenflug (1985), An Updated Evaluation of Recombination and Ionization Rates, *Astron. Astrophys. Suppl.*, **60**, 425–457.

Bahcall, J. N., S. Basu, and A. M. Serenelli (2005), What Is the Neon Abundance of the Sun?, *Astrophys. J.*, **631**, 1281–1285. doi:10.1086/431926.

Balogh, A., J. T. Gosling, J. R. Jokipii, R. Kallenbach, and H. Kunow (eds.) (1999), *Corotating Interaction Regions*, Vol. 7 of Space Sciences Series of ISSI, Kluwer Academic, Dordrecht.

Bame, S. J., J. R. Asbridge, A. J. Hundhausen, and M. D. Montgomery (1970), Solar Wind Ions: $^{56}Fe^{8+}$ to $^{56}Fe^{12+}$, $^{28}Si^{7+}$, $^{28}Si^{8+}$, $^{28}Si^{9+}$, and $^{16}O^{6+}$, *J. Geophys. Res.*, **75**, 6360–6365.

Bame, S. J., J. R. Asbridge, W. C. Feldman, and J. T. Gosling (1977), Evidence for a Structure-free State at High Solar Wind Speeds, *J. Geophys. Res.*, **82**, 1487–1492.

Biermann, L. (1951), Kometenschweife und solare Korpuskularstrahlung, *Zeit. Astrophys.*, **29**, 274–286.

Bochsler, P., J. Geiss, and S. Kunz (1986), Abundances of Carbon, Oxygen and Neon in the Solar Wind during the Period from August 1978 to June 1982, *Sol. Phys.*, **102**, 177–201.

Bonnet, R. M., and V. Manno (1994), *International Cooperation in Space*, Harvard University Press, Cambridge, MA.

Bürgi, A., and J. Geiss (1986), Helium and Minor Ions in the Corona and Solar Wind: Dynamics and Charge States, *Sol. Phys.*, **103**, 347–383.

Coplan, M. A., K. W. Ogilvie, P. Bochsler, and J. Geiss (1978), Ion Composition Experiment, *IEEE Trans. Geosci. Electron.*, **GE16**, 185–191.

Fisk, L. A. (1996), Motion of the Footpoints of Heliospheric Magnetic Field Lines at the Sun: Implications for Recurrent Energetic Particle Events at High Heliographic Latitudes, *J. Geophys. Res.*, **101**, 15547–15553.

Fisk, L. A. (2003), Acceleration of the solar wind as a result of the reconnection of open magnetic flux with coronal loops, *Journal of Geophysical Research (Space Physics)*, **108**(A4), 1157, doi:10.1029/2002JA009284.

Fisk, L. A., and G. Gloeckler (2006), The Common Spectrum for Accelerated Ions in the Quiet-Time Solar Wind, *Astrophys. J. Lett.*, **640**, L79–L82, doi:10.1086/503293.

Fisk, L. A., and G. Gloeckler (2007), Acceleration and Composition of Solar Wind Suprathermal Tails, *Space Science Reviews*, **108**, doi:10.1007/s11214-007-9180-8.

Fisk, L. A., N. Schwadron, and T. H. Zurbuchen (1998), On the Slow Solar Wind, *Space Science Reviews*, **86**, 51–60.

Fisk, L. A., T. H. Zurbuchen, and N. A. Schwadron (1999), On the Coronal Magnetic Field: Consequences of Large-scale Motions, *Astrophys. J.*, **521**, 868–877.

Geiss, J. (1982), Processes Affecting Abundances in the Solar Wind, *Space Science Reviews*, **33**, 201–217.

Geiss, J., G. Gloeckler, and R. von Steiger (1995), Origin of the Solar Wind from Composition Data, *Space Science Reviews*, **72**, 49–60.

Geiss, J., G. Gloeckler, R. von Steiger, H. Balsiger, L. A. Fisk, A. B. Galvin, F. M. Ipavich, S. Livi, J. F. McKenzie, K. W. Ogilvie, and B. Wilken (1995), The Southern High Speed Stream: Results from the SWICS Instrument on Ulysses, *Science*, **268**, 1033–1036.

Geiss, J., F. Bühler, H. Cerutti, P. Eberhardt, C. Filleux, J. Meister, and P. Signer (2004), The Apollo SWC Experiment: Results, Conclusions, Consequences, *Space Science Reviews*, **110**, 307–335, doi:10.1023/B:SPAC.0000023409.54469.40.

Gloeckler, G. (1999), Observation of Injection and Pre-Acceleration Processes in the Slow Solar Wind, *Space Science Reviews*, **89**, 91–104, doi:10.1023/A:1005272601422.

Gloeckler, G., and J. Geiss (1998a), Interstellar and Inner Source Pickup Ions Observed with SWICS on Ulysses, *Space Science Reviews*, **86**, 127–159.

Gloeckler, G., and J. Geiss (1998b), Measurement of the Abundance of Helium-3 in the Sun and in the Local Interstellar Cloud with SWICS on Ulysses, *Space Science Reviews*, **84**, 275–284.

Gloeckler, G., and J. Geiss (2007), The Composition of the Solar Wind in Polar Coronal Holes, *Space Science Reviews*, **108**, doi:10.1007/s11214-007-9189-Z.

Gloeckler, G., T. H. Zurbuchen, and J. Geiss (2003), Implications of the observed anticorrelation between solar wind speed and coronal electron temperature, *Journal of Geophysical Research (Space Physics)*, **108**(A4), 1158, doi:10.1029/2002JA009286.

Gloeckler, G., F. Ipavich, W. Stüdemann, B. Wilken, D. Hamilton, G. Kremser, D. Hovestadt, F. Gliem, R. A. Lundgren, W. Rieck *et al.* (1985), The Charge-Energy-Mass Spectrometer for 0.3300 keV/e Ions on the AMPTE/CCE, *IEEE Trans. Geosc. and Rem. Sens.*, **GE23**, 234–240.

Gloeckler, G., F. M. Ipavich, D. C. Hamilton, B. Wilken, W. Stüdemann, G. Kremser, and D. Hovestadt (1986), Solar Wind Carbon, Nitrogen and Oxygen Abundances Measured in the Earth's Magnetosheath with AMPTE/CCE, *Geophys. Res. Lett.*, **13**, 793.

Gloeckler, G., J. Geiss, H. Balsiger, P. Bedini, J. C. Cain, J. Fischer, L. A. Fisk, A. B. Galvin, F. Gliem, D. C. Hamilton *et al.* (1992), The Solar Wind Ion Composition Spectrometer, *Astron. Astrophys. Suppl.*, **92**, 267–289.

Gloeckler, G., J. Geiss, H. Balsiger, L. A. Fisk, A. B. Galvin, F. M. Ipavich, K. W. Ogilvie, R. von Steiger, and B. Wilken (1993), Detection of Interstellar Pick-Up Hydrogen in the Solar System, *Science*, **261**, 70–73.

Gloeckler, G., E. C. Roelof, K. W. Ogilvie, and D. B. Berdichevsky (1995), Proton Phase-Space Densities ($0.5\,\mathrm{keV} < E_p < 5\,\mathrm{MeV}$) at Mid-Latitudes from Ulysses SWICS/HI-SCALE Measurements, *Space Science Reviews*, **72**, 321–326.

Gloeckler, G., L. A. Fisk, T. H. Zurbuchen, and N. A. Schwadron (2000), Sources, Injection and Acceleration of Heliospheric Ion Populations, in *Acceleration and Transport of Energetic Particles Observed in the Heliosphere* (R. A. Mewaldt, J. R. Jokipii, M. A. Lee, E. Möbius, and T. H. Zurbuchen,eds.), Vol. 528 of American Institute of Physics Conference Series, pp. 221–228.

Gopalswamy, N., Z. Mikić, D. Maia, D. Alexander, H. Cremades, P. Kaufmann, D. Tripathi, and Y.-M. Wang (2006), The Pre-CME Sun, *Space Science Reviews*, **123**, 303–339, doi:10.1007/s11214-006-9020-2.

Gosling, J. T. (1993), The Solar Flare Myth, *J. Geophys. Res.*, **98**, 18937–18949.

Gosling, J. T. (1997), Coronal Mass Ejections: An Overview, in *Coronal Mass Ejections* (N. Crooker, J. A. Joselyn, and J. Feynman, eds.), Vol. 99 of Geophysical Monograph, pp. 916, American Geophysical Union, Washington, DC.

Gosling, J. T., and V. J. Pizzo (1999), Formation and Evolution of Corotating Interaction Regions and Their Three Dimensional Structure, *Space Science Reviews*, **89**, 21–52, doi:10.1023/A:1005291711900.

Gosling, J. T., S. Eriksson, and R. Schwenn (2006), Petschek-type magnetic reconnection exhausts in the solar wind well inside 1 AU: Helios, *Journal of Geophysical Research (Space Physics)*, **111**(A10), 10102, doi:10.1029/2006JA011863.

Gosling, J. T., S. J. Bame, D. J. McComas, J. L. Phillips, V. J. Pizzo, B. E. Goldstein, and M. Neugebauer (1993), Latitudinal Variation of Solar Wind Corotating Stream Interaction Regions: Ulysses, *Geophys. Res. Lett.*, **20**, 2789–2792.

Gosling, J. T., D. J. McComas, J. L. Phillips, L. A. Weiss, V. J. Pizzo, B. E. Goldstein, and R. Forsyth (1994), A New Class of Forward–Reverse Shock Pairs in the Solar Wind, *Geophys. Res. Lett.*, **21**, 2271–2274.

Gosling, J. T., S. J. Bame, W. C. Feldman, D. J. McComas, J. L. Phillips, B. Goldstein, M. Neugebauer, J. Burkepile, A. J. Hundhausen, and L. Acton (1995a), The band of solar wind variability at low heliographic latitudes near solar activity minimum: Plasma results from the Ulysses rapid latitude scan, *Geophys. Res. Lett.*, **22**, 3329–3332, doi:10.1029/95GL02163.

Gosling, J. T., S. J. Bame, D. J. McComas, J. L. Phillips, V. J. Pizzo, B. E. Goldstein, and M. Neugebauer (1995b), Solar Wind Corotating Stream Interaction Regions out of the Ecliptic Plane: ULYSSES, *Space Science Reviews*, **72**, 99–104.

Gosling, J. T., D. J. McComas, J. L. Phillips, V. J. Pizzo, B. E. Goldstein, R. J. Forsyth, and R. P. Lepping (1995c), A CME-Driven Solar Wind Disturbance Observed at Both Low and High Heliographic Latitudes, *Geophys. Res. Lett.*, **22**, 1753–1756.

Gosling, J. T., R. M. Skoug, D. J. McComas, and C. W. Smith (2005), Direct evidence for magnetic reconnection in the solar wind near 1 AU, *Journal of Geophysical Research (Space Physics)*, **110**(A1), 1107, doi:10.1029/2004JA010809.

Gosling, J. T., S. Eriksson, R. M. Skoug, D. J. McComas, and R. J. Forsyth (2006), Petschek-Type Reconnection Exhausts in the Solar Wind Well beyond 1 AU: Ulysses, *Astrophys. J.*, **644**, 613–621, doi:10.1086/503544.

Grevesse, N., M. Asplund, and A. J. Sauval (2007), The solar chemical composi- tion, *Space Science Reviews doi: 10.1007/s11214-007-9173-7. in press.*

Gringauz, K. I., V. G. Kurt, V. I. Moroz, and I. S. Shklovskii (1961), Results of Observations of Charged Particles Observed Out to $R = 100{,}000$ km, with the Aid of Charged-Particle Traps on Soviet Space Rockets, *Soviet Astronomy*, **4**, 680–695.

Hefti, S., H. Grünwaldt, F. M. Ipavich, P. Bochsler, D. Hovestadt, M. R. Aellig, M. Hilchenbach, R. Kallenbach, A. B. Galvin, J. Geiss *et al.* (1998), Kinetic properties of solar wind minor ions and protons measured with SOHO/CELIAS, *J. Geophys. Res.*, **103**, 29697–29704.

Henke, T., J. Woch, R. Schwenn, U. Mall, G. Gloeckler, R. von Steiger, R. J. Forsyth, and A. Balogh (2001), Ionization State and Magnetic Topology of Coronal Mass Ejections, *J. Geophys. Res.*, **106**, 10597–10613.

Hirshberg, J., S. J. Bame, and D. E. Robbins (1972), Solar Flares and Solar Wind Helium Enrichments: July 1965–July 1967, *Sol. Phys.*, **23**, 467.

Hovestadt, D., G. Gloeckler, C. Y. Fan, L. A. Fisk, F. M. Ipavich, B. Klecker, J. J. O'Gallagher, M. Scholer, H. Arbinger, J. Cain *et al.* (1978), The Nuclear and ionic Charge Distribution Particle Experiments on the ISEE-1 and ISEE-C Spacecraft, *IEEE Trans. Geosci. Electron.*, **GE16**, 166.

Hovestadt, D., M. Hilchenbach, A. Bürgi, B. Klecker, P. Laeverenz, M. Scholer, H. Grünwaldt, W. I. Axford, S. Livi, E. Marsch *et al.* (1995), CELIAS Charge, Element and Isotope Analysis System for SOHO, *Sol. Phys.*, **162**, 441–481.

Hundhausen, A. J. (1973), Nonlinear model of high-speed solar wind streams, *J. Geophys. Res.*, **78**, 1528–1542.

Hundhausen, A. J., H. Gilbert, and S. Bame (1968), Ionisation state of the interplanetary plasma, *J. Geophys. Res.*, **73**, 5485–5493.

Ipavich, F. M., A. B. Galvin, G. Gloeckler, D. Hovestadt, S. J. Bame, B. Klecker, M. Scholer, L. A. Fisk, and C. Y. Fan (1986), Solar Wind Fe and CNO Measurements in High-Speed Flows, *J. Geophys. Res.*, **91**, 4133–4141.

Ipavich, F. M., A. B. Galvin, J. Geiss, K. W. Ogilvie, and F. Gliem (1992), Solar Wind Iron and Oxygen Charge States and Relative Abundances Measured by SWICS on Ulysses, in *Solar Wind Seven* (E. Marsch and R. Schwenn, eds.), Vol. 3 of COSPAR Colloquia Series, pp. 369–374, Pergamon Press.

Kilchenmann, A. (2007), *Interplanetary Coronal Mass Ejections Observed with SWICS/Ulysses*, Ph.D. thesis, Universität Bern.

Krieger, A. S., A. F. Timothy, and E. C. Roelof (1973), A Coronal Hole and its Identification as the Source of a High Velocity Solar Wind Stream, *Sol. Phys.*, **29**, 505–525.

Kunow, H., N. U. Crooker, J. A. Linker, R. Schwenn, and R. von Steiger (eds.) (2006), *Coronal Mass Ejections*, Vol. 21 of Space Sciences Series of ISSI, Springer-Verlag, Dordrecht.

Laming, J. M., and S. T. Lepri (2007), Ion Charge States in the Fast Solar Wind: New Data Analysis and Theoretical Refinements, *Astrophys. J.*, **660**, 1642–1652, doi:10.1086/513505.

Lanzerotti, L. J., C. G. Maclennan, T. P. Armstrong, E. C. Roelof, R. E. Gold, and R. B. Decker (1996), Low energy charged particles in the high latitude heliosphere., *Astron. Astrophys.*, **316**, 457–463.

Lepri, S. T., and T. H. Zurbuchen (2004), Iron charge state distributions as an indicator of hot ICMEs: Possible sources and temporal and spatial variations during solar maximum, *Journal of Geophysical Research (Space Physics)*, **109**(A18), A01112, doi:10.1029/2003/JA009954.

Lepri, S. T., T. H. Zurbuchen, L. A. Fisk, I. G. Richardson, H. V. Cane, and G. Gloeckler (2001), Iron charge distribution as an identifier of interplanetary coronal mass ejections, *J. Geophys. Res.*, **106**, 29231–29238.

Manchester, W. B., and T. H. Zurbuchen (2006), Are high-latitude forward–reverse shock pairs driven by CME overexpansion?, *Journal of Geophysical Research (Space Physics)*, **111**(A10), 5101, doi:10.1029/2005JA011461.

Marsch, E. (1991), Kinetic Physics of the Solar Wind Plasma, in *Physics of the Inner Heliosphere* (R. Schwenn and E. Marsch, eds.), Vol. 2, pp. 45–133, Springer-Verlag, Berlin.

Mazzotta, P., G. Mazzitelli, S. Colafrancesco, and N. Vittorio (1998), Ionization balance for optically thin plasmas: Rate coefficients for all atoms and ions of the elements H to N I, *Astron. Astrophys. Suppl.*, **133**, 403–409.

McComas, D. J. (2003), The Three-Dimensional Structure of the Solar Wind over the Solar Cycle, in *Solar Wind Ten* (M. Velli, R. Bruno, and F. Malara, eds.), AIP Conf. Proc. 679, pp. 33–38, American Institute of Physics, Melville, New York.

McComas, D. J., H. A. Elliott, and R. von Steiger (2002), Solar wind from high-latitude coronal holes at solar maximum, *Geophys. Res. Lett.*, **29**(9), 1314, doi:10.1029/2001GL013940.

McComas, D. J., B. L. Barraclough, H. O. Funsten, J. T. Gosling, E. Santiago-Muñoz, R. M. Skoug, B. E. Goldstein, M. Neugebauer, P. Riley, and A. Balogh (2000), Solar wind observations over Ulysses' first full polar orbit, *J. Geophys. Res.*, **105**, 10419–10434.

McComas, D. J., H. A. Elliott, N. A. Schwadron, J. T. Gosling, R. M. Skoug, and B. E. Goldstein (2003), The three-dimensional solar wind around solar maximum, *Geophys. Res. Lett.*, **30**, 1517, doi: 10.1029/2003GL017136.

McComas, D. J., H. A. Elliott, J. T. Gosling, and R. M. Skoug (2006), Ulysses observations of very different heliospheric structure during the declining phase of solar activity cycle 23, *Geophys. Res. Lett.*, **33**, 9102, doi:10.1029/2006GL025915.

Meyer, J.-P. (1981), A Tentative Ordering of all Available Solar Energetic Particles Abundance Observations, in *Proc. 17th Int. Cosmic Ray Conf., Paris*, Vol. 3. pp. 145–152.

Möbius, E., D. Hovestadt, B. Klecker, M. Scholer, G. Gloeckler, and F. M. Ipavich (1985), Direct Observation of He^+ Pickup Ions of Interstellar Origin in the Solar Wind, *Nature*, **318**, 426–429.

Moldwin, M. B., J. L. Phillips, J. T. Gosling, E. E. Scime, D. J. McComas, S. J. Bame, A. Balogh, and R. J. Forsyth (1995), Ulysses observation of a noncoronal mass ejection flux rope: Evidence of interplanetary magnetic reconnection, *J. Geophys. Res.*, **100**, 19903–19910, doi: 10.1029/95JA01123.

Neugebauer, M. (2001), The solar-wind and heliospheric magnetic field in three dimensions, in *The Heliosphere near Solar Minimum. The Ulysses Perspective* (A. Balogh, R. G. Marsden, and E. J. Smith, eds.), pp. 43–106, Springer-Praxis, Chichester, UK, ISBN 1-85233-204-2.

Neugebauer, M., and C. W. Snyder (1962), Solar Plasma Experiment, *Science*, **138**, 1095–1097.

Neugebauer, M., and C. W. Snyder (1966), Mariner 2 Observations of the Solar Wind, *J. Geophys. Res.*, **71**, 4469–4484.

Neugebauer, M., and R. von Steiger (2001), The Solar Wind, in *The Century in Space Science* (J. Bleeker, J. Geiss, and M. C. E. Huber, eds.), Chap. 47, pp. 1115–1140, Kluwer Academic, Dordrecht.

Neugebauer, M., B. E. Goldstein, D. J. McComas, S. T. Suess, and A. Balogh (1995), Ulysses Observations of Microstreams in the Solar Wind from Coronal Holes, *J. Geophys. Res.*, **100**, 23389–23395.

Neukomm, R. O. (1998), *Composition of Coronal Mass Ejections Derived with SWICS/Ulysses*, Ph.D. thesis, Universität Bern.

Ogilvie, K. W., P. Bochsler, J. Geiss, and M. A. Coplan (1980), Observations of the Velocity Distribution of Solar Wind Ions, *J. Geophys. Res.*, **85**, 6069–6074.

Parker, E. N. (1958), Dynamics of the Interplanetary Gas and Magnetic Fields, *Astrophys. J.*, **128**, 664–676.

Parker, E. N. (2001), A History of the Solar Wind Concept, in *The Century in Space Science* (J. Bleeker, J. Geiss, and M. C. E. Huber, eds.), Chap. 9, pp. 225–255, Kluwer Academic, Dordrecht.

Pauls, H. L., and G. P. Zank (1996), Interaction of a nonuniform solar wind with the local interstellar medium, *J. Geophys. Res.*, **101**, 17081–17092, doi:10.1029/96JA01298.

Pizzo, V. J., and J. T. Gosling (1994), 3-D simulation of high-latitude interaction regions: Comparison with ULYSSES results, *Geophys. Res. Lett.*, **21**, 2063–2066.

Reinard, A. (2005), Comparison of Interplanetary CME Charge State Composition with CME-associated Flare Magnitude, *Astrophys. J.*, **620**, 501–505, doi:10.1086/426109.

Richardson, I. G., and H. V. Cane (2004), Identification of interplanetary coronal mass ejections at 1 AU using multiple solar wind plasma composition anomalies, *Journal of Geophysical Research (Space Physics)*, **109**(A18), 9104, doi:10.1029/2004JA010598.

Schmid, J., P. Bochsler, and J. Geiss (1988), Abundance of Iron Ions in the Solar Wind, *Astrophys. J.*, **329**, 956–966.

Schwenn, R. (1990), Large-Scale Structure of the Interplanetary Medium, in *Physics of the Inner Heliosphere* (R. Schwenn and E. Marsch, eds.), Vol. 1, Chap. 3, pp. 99–181, Springer-Verlag, Berlin.

Suess, S. T., G. Poletto, A.-H. Wang, S. T. Wu, and I. Cuseri (1998), The Geometric Spreading of Coronal Plumes and Coronal Holes, *Sol. Phys.*, **180**, 231–246.

Thieme, K. M., E. Marsch, and R. Schwenn (1990), Spatial structures in high-speed streams as signatures of fine structures in coronal holes, *Ann. Geophys.*, **8**, 713–723.

Vasyliunas, V. M., and G. L. Siscoe (1976), On the Flux and the Energy Spectrum of Interstellar Ions in the Solar System, *J. Geophys. Res.*, **81**, 1247–1252.

von Steiger, R., and C. Fröhlich (2005), The Sun, from core to corona and solar wind, in *The Solar System and Beyond: Ten Years of ISSI* (J. Geiss and B. Hultqvist, eds.), Vol. SR-003, pp. 99–112, ISSI Scientific Report Series, ESA, Noordwijk, The Netherlands.

von Steiger, R., J. Geiss, and G. Gloeckler (1997), Composition of the Solar Wind, in *Cosmic Winds and the Heliosphere* (J. R. Jokipii, C. P. Sonett, and M. S. Giampapa, eds.), pp. 581–616, University of Arizona Press, Tucson.

von Steiger, R., and T. H. Zurbuchen (2002), Kinetic properties of heavy solar wind ions from Ulysses-SWICS, *Adv. Space Res.*, **30**, 73–78.

von Steiger, R., and T. H. Zurbuchen (2003), Temperature Anisotropies of Heavy Solar Wind Ions fromUlysses-SWICS, in *Solar Wind Ten* (M. Velli, R. Bruno, F. Malara, and B. Bucci, eds.), Vol. 679, pp. 526–529, American Institute of Physics Conference Series.

von Steiger, R., and T. H. Zurbuchen (2006), Kinetic properties of heavy solar wind ions from Ulysses-SWICS, *Geophys. Res. Lett.*, **33**, 9103, doi:10.1029/2005GL024998.

von Steiger, R., T. H. Zurbuchen, and A. Kilchenmann (2005), Latitude Distribution of Interplanetary Coronal Mass Ejections during Solar Maximum, in *Connecting Sun and Heliosphere*, Vol. 592, ESA, Noordwijk, The Netherlands.

von Steiger, R., S. P. Christon, G. Gloeckler, and F. M. Ipavich (1992), Variable Carbon and Oxygen Abundances in the Solar Wind as Observed in Earth's Magnetosheath by AMPTE/CCE, *Astrophys. J.*, **389**, 791–799.

von Steiger, R., L. A. Fisk, G. Gloeckler, N. A. Schwadron, and T. H. Zurbuchen (1999), Composition Variations in Fast Solar Wind Streams, in *Solar Wind Nine* (S. R. Habbal, R. Esser, J. V. Hollweg, and P. A. Isenberg, eds.), Vol. 471 of AIP Conf. Proc., Woodbury, NY, pp. 143–146, American Institute of Physics, Melville, New York.

von Steiger, R., N. A. Schwadron, L. A. Fisk, J. Geiss, G. Gloeckler, S. Hefti, B. Wilken, R. F. Wimmer-Schweingruber, and T. H. Zurbuchen (2000), Composition of Quasi-stationary Solar Wind Flows from Ulysses/SWICS, *J. Geophys. Res.*, **105**, 27217–27236.

von Steiger, R., J.-C. Vial, P. Bochsler, M. Chaussidon, C. M. S. Cohen, B. Fleck, V. S. Heber, H. Holweger, K. Issautier, A. J. Lazarus *et al.* (2001), Measuring Solar Abundances, in *Solar and Galactic Composition* (R. F. Wimmer-Schweingruber, ed.), Vol. 598, pp. 13–22, AIP Conf. Proc., Woodbury, NY, American Institute of Physics, Melville, New York.

Widing, K. G., and U. Feldman (1992), Elemental Abundances and their Variations in the Upper Solar Atmosphere, in *Solar Wind Seven* (E. Marsch and R. Schwenn, eds.), Vol. 3, pp. 405–410, COSPAR Colloquia Series, Pergamon Press.

Widing, K. G., and U. Feldman (2001), On the Rate of Abundance Modifications versus Time in Active Region Plasmas, *Astrophys. J.*, **555**, 426–434, doi:10.1086/321482.

Wilhelm, K., E. Marsch, B. N. Dwivedi, D. M. Hassler, P. Lemaire, A. Gabriel, and M. C. E. Huber (1998), The Solar Corona above Polar Coronal Holes as Seen by SUMER on SOHO, *Astrophys. J.*, **500**, 1023–1038.

Wimmer-Schweingruber, R. F., P. Bochsler, and P. Wurz (1999), Isotopes in the Solar Wind: New Results from ACE, SOHO, and Wind, in *Solar Wind Nine* (S. R. Habbal, R. Esser, J. V. Hollweg, and P. A. Isenberg, eds.), Vol. 471, pp. 147–152, AIP Conf. Proc., Woodbury, NY, American Institute of Physics, Melville, New York.

Wimmer-Schweingruber, R. F., R. von Steiger, and R. Paerli (1997), Solar Wind Stream Interfaces in Corotating Interaction Regions: SWICS/Ulysses Results, *J. Geophys. Res.*, **102**, 17407–17417.

Winterhalter, D., M. Neugebauer, B. E. Goldstein, E. J. Smith, S. J. Bame, and A. Balogh (1994), Ulysses Field and Plasma Observations of Magnetic Holes in the Solar Wind and their Relation to Mirror-Mode Structures, *J. Geophys. Res.*, **99**, 23371–23381.

Winterhalter, D., E. J. Smith, M. Neugebauer, B. E. Goldstein, and B. T. Tsurutani (2000), The latitudinal distribution of solar wind magnetic holes, *Geophys. Res. Lett.*, **27**, 1615–1618.

Woo, R., and S. R. Habbal (1999), Imprint of the Sun on the Solar Wind, *Astrophys. J.*, **510**, L69–L72.

Zurbuchen, T. H. (2007), A New View of the Coupling of the Sun and the Heliosphere, *Ann. Rev. Astron. Astrophys.*, **45**, doi:10.1146/annurev.astro.45.010807.154030.

Zurbuchen, T. H., and I. G. Richardson (2006), In-Situ Solar Wind and Magnetic Field Signatures of Interplanetary Coronal Mass Ejections, *Space Science Reviews*, **123**, 31–43, doi: 10.1007/s11214-006-9010-4.

Zurbuchen, T. H., S. Hefti, L. A. Fisk, G. Gloeckler, and R. von Steiger (1999), The Transition between Fast and Slow Solar Wind from Composition Data, *Space Science Reviews*, **87**, 353–356.

Zurbuchen, T. H., S. Hefti, L. A. Fisk, G. Gloeckler, N. A. Schwadron, C. W. Smith, N. F. Ness, R. M. Skoug, D. J. McComas, and L. F. Burlaga (2001), On the origin of microscale magnetic holes in the solar wind, *J. Geophys. Res.*, **106**, 16001–16010, doi: 10.1029/2000JA000119.

Zurbuchen, T. H., L. A. Fisk, G. Gloeckler, and R. von Steiger (2002), The solar wind composition throughout the solar cycle: A continuum of dynamic states, *Geophys. Res. Lett.*, **29**, 1352, doi:10.1029/2001GL013946.

4

The global heliospheric magnetic field

Edward J. Smith

4.1 INTRODUCTION

The heliospheric magnetic field originates on the Sun. Portions of the solar field extend up into the corona, the Sun's outermost atmosphere, where the solar wind originates. The perfectly electrically conducting solar wind plasma carries the magnetic field along with it to completely fill the heliosphere. Spacecraft observations show that the magnetic field is present at all radial distances and heliographic latitudes. The solar wind and magnetic field are time-varying because of changes on the Sun and dynamic changes intrinsic to the expanding magnetized plasma. However, the heliospheric magnetic field has a global structure that is revealed by averaging the measurements. The time interval of the averaging depends on the relevant length scale of interest. Averages from minutes to hours to the solar rotation period are commonly used. This chapter addresses the global structure of the heliospheric magnetic field and relates it to the Sun's magnetic field. Time variations are also discussed with emphasis on slow, large-scale variations. Specifically, the large topic corresponding to short, smaller scale variations that includes waves, turbulence, discontinuities, etc., although they may be mentioned, are not the focus of this chapter.

This chapter is not a general treatise on the heliospheric magnetic field. The point of view is almost always related to the many contributions made by the Ulysses mission. The scientific background of each major topic is characteristically discussed as needed to understand the Ulysses achievements. One aspect of the major influence Ulysses has had is that the term, interplanetary magnetic field (abbreviated IMF), is rapidly falling out of usage. We now speak of the heliospheric magnetic field or HMF, a fact that reflects the significant change in perspective from earlier observations that were restricted to the low latitudes containing the planets and were truly interplanetary to the three-dimensional perspective provided by Ulysses.

Of course, a lot of observations were made before Ulysses was launched in 1990. Even before it became possible to observe the solar wind and HMF, a model had been developed that has survived the test of the observations with slight modification. The Parker model (1963) provides a theoretical description of the HMF that has continued to serve as the standard against which to compare the observations. Accordingly, we begin with a rather detailed description of the Parker magnetic field model since it provides the context in which to view the observations to be discussed and contains the essential physics necessary to understand them.

4.2 THE HELIOSPHERIC MAGNETIC FIELD: A GLOBAL PERSPECTIVE

Understanding a subject is aided by proceeding from its simplest to its more complex aspects. The easiest view of the HMF to grasp is a global perspective supported by the underlying theoretical considerations. This approach relates the HMF to the global properties of the Sun's magnetic field and to the solar wind that transports it into space. A three-dimensional view is adopted based on the revealing observations now available as a result of the Ulysses mission.

In describing the HMF, it has become standard to use a preferred set of coordinates called solar–heliospheric or RTN coordinates. The primary vectors that define this system are **R**, radially outward from the center of the Sun, and **H**, along the Sun's axis of rotation. The component, **T**, is defined by $\mathbf{T} = \mathbf{H} \times \mathbf{R}$ and is positive in the sense of rotation of the Sun (positive or counterclockwise looking down from above). The component **N** is then $\mathbf{R} \times \mathbf{T}$ and positive is northward. This convention is closely related to the standard spherical–polar coordinates (radial distance, co-latitude and longitude or (r, θ, ϕ) except that **N** is northward whereas the θ component points in the opposite direction). The field components are then (B_R, B_T, B_N) or (B_R, B_θ, B_ϕ). The field is also often expressed in terms of the magnitude, B, and two angles, ϕ_B, the azimuthal or longitudinal angle, and a polar angle, θ_B, or latitude angle, δ_B. Care is necessary because the literature frequently involves other symbols than those above (e.g., B_θ or even B_ϕ for B_N). Fortunately, the context usually makes it clear which component is actually being discussed.

4.2.1 The Parker field model

E. N. Parker has a specific theoretical point of view. He avoids electric fields, **E**, and currents, **j**, preferring to work with only the plasma velocity, **V**, and the magnetic field, **B**. The rationale is that **E** can always be derived afterward, if necessary, from $\mathbf{E} = -\mathbf{V} \times \mathbf{B}$ and **j** can be obtained from $\mu_0 \mathbf{j} = \nabla \times \mathbf{B}$. This approach is basically magnetohydrodynamic (MHD) theory and is widely used by other plasma theorists. It is sometimes referred to as the VB paradigm (Parker, 1996). This approach will be used in the following derivations because it is elegant and simple. Readers may note that in the book about the solar wind written by Parker there is little, if any, mention of electric fields and currents.

Admittedly, this approach is still not the one that is most familiar to readers nor, surprisingly, most workers in magnetospheric plasma physics. Because of the usual classical introduction to electromagnetic theory and experiment, people tend to think in terms of currents and electric fields as fundamental and as the causes of the respective phenomena. However, in plasma physics, especially hydromagnetic (or magnetohydrodynamic or magnetofluid) theory, currents are actually caused by stresses in the plasma and electric fields are caused by relative motion between the plasma and the magnetic field. Thus, the difference is not simply a matter of taste but involves fundamental distinctions between cause and effect.

The Parker model (Parker, 1963) is basically a solar wind model—in fact, the first such model. It is a strictly hydrodynamic model that ignores the magnetic field in so far as it might affect the acceleration of the hot coronal plasma to supersonic speeds followed by its escape from the Sun's strong gravitational field. The cause of the solar wind is solely the internal pressure of the plasma and the gradient in pressure that exerts an outward force able to overcome solar gravity.

The magnetic field is added more or less as a "tracer" in the solar wind flow. Parker knew that the solar wind would be magnetized and that **B** would play a significant role once the solar wind left the Sun in determining the properties of hydromagnetic waves and in various other perturbations to the steady flow. He specifically investigated eruptive phenomena at the Sun that might cause "blast waves"—that is, shock waves not driven by the injection of fresh solar plasma but able to propagate as large-amplitude waves into the heliosphere. The conclusions of the model regarding the character of the magnetic field are basically correct, since, at large distances, it does not represent a significant energy density or pressure compared with the convective or "ram" pressure of the solar wind. Another useful source of information about the solar wind model and HMF is Hundhausen (1972) that contains early solar wind and magnetic field measurements including attempts to supplement the Parker model in various ways.

It is customary to treat the magnetic field as "frozen-into" collisionless plasma because of the high electrical conductivity. By eliminating any relative motion between the field and plasma with **V** parallel to **B**, the electric field vanishes and extremely large currents are avoided. The field travels along with the solar wind (**V** and **B** remain parallel) and is transported into space to form the heliospheric magnetic field. In the frame of reference that corotates with the Sun, the solar wind follows a streamline given by $r\,d\phi/dr = v_\phi/v_r = -\Omega r/v_r$ where Ω is the angular rotation rate of the Sun. Integration produces $\phi = -\Omega r/v_r$—recognizable as the expression for an Archimedes spiral. Since **B** is parallel to **V**, $B_\phi/B_R = v_\phi/v_r = -\Omega r/v_r = \tan\phi_P$, the Parker spiral. When the plasma and field vectors are transformed into the inertial/non-corotating frame, the field line is unchanged (according to the special theory of relativity when $V \ll c$) but the solar wind streamline is radial.

Alternatively, in the inertial frame, a radially flowing solar wind parcel reaches a distance, $r = v_r t$, at time t after leaving the Sun. During that interval, the Sun has rotated counterclockwise as viewed from above the north pole and the (sub-solar) longitude of the solar wind parcel is $\phi = -\Omega t = -\Omega r/v_r$. Since one end of **B** is attached to the rotating Sun, the locus of the field line is given by the same equation or $B_\phi/B_R = -\Omega r/v_r = \tan\phi_P$, as above.

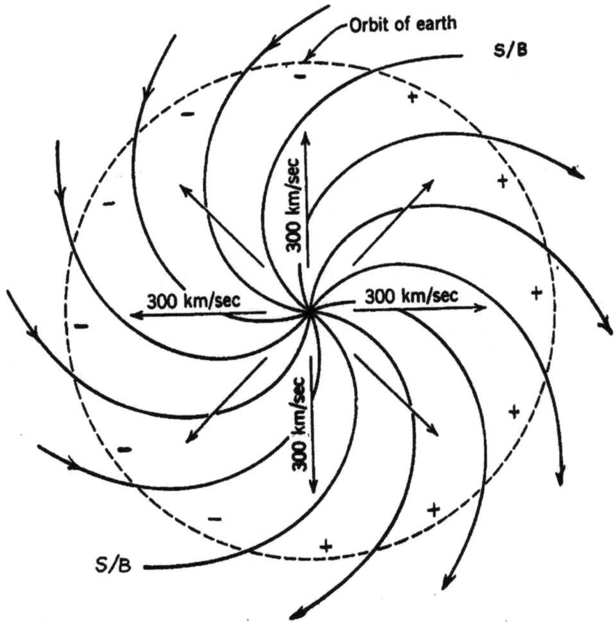

Figure 4.1. The Parker model in the solar equatorial or ecliptic plane. The straight lines emanating from the Sun at the center are radial solar wind velocity vectors with speeds of 300 km/s (considered slow wind today). The spirals are magnetic field lines that start out radially and make a large angle to the radial direction by the time they reach 1 AU (the dotted cycle). Arrows added to the field directions indicate their polarity at the Sun. The pluses designate outward-directed (positive) fields and the minuses inward-directed (negative) fields. The field lines divide the circle into two magnetic "sectors". Two of the spirals are the boundaries between the sectors (designated S/B for sector boundary). Adapted from Parker (1963).

Figure 4.1 shows the radial solar wind velocity vectors (assumed to have a constant speed of 300 km/s) and the spiral magnetic field. Additional features have been added that are discussed later: the polarity of the magnetic field (sunward or outward), the resulting division into magnetic "sectors", and the sector boundary (SB) between them.

The above equations can be converted to latitudes other than the equator by substituting $\Omega r \sin\theta$ for Ωr, where θ is the co-latitude measured from the polar axis. Away from the equator, with $\theta \neq \pi/2$, the field lines form helices lying on the surface of a cone of half-angle, θ (Figure 4.2). Their locus is given by $\theta = \theta_0 =$ constant, $\phi = -\Omega r \sin\theta_0 / V_r$. The height above the equatorial plane is $z = r\cos\theta_0$. The distance from the polar axis, ρ, is simply $\rho = r\sin\theta_0$.

The conservation of magnetic flux requires that $\int B_R \cdot dA =$ constant where A is the element of area on the surface of a sphere enclosing the Sun. In terms of the solid angle (ω), $\int B_R \cdot dA = \int B_R r^2 d\omega$ so that $B_R r^2 =$ constant or $B_R \sim 1/r^2$. It follows from the above equation for B_ϕ / B_R that $B_\phi \sim 1/r$. Hence, the magnetic field magnitude is given by $B = B_R(1 + (\Omega r \sin\theta / V_r)^2)^{1/2}$.

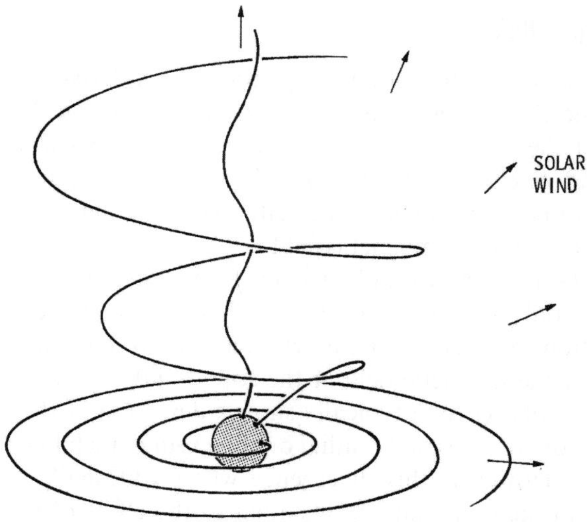

Figure 4.2. The Parker model in the solar meridional plane. The view is perpendicular to the solar equator and rotation axis. Several field lines are shown originating at different solar latitudes. The solar wind velocity vectors are radial. The field lines are helices lying on cones with half-angles equal to the source latitudes. At a given radial distance, the fields are tightly spiraled in the equator and radial over the pole.

The remaining field component, B_N, is necessarily zero because $\mathbf{B}||\mathbf{V}$ in the corotating frame and, hence, in the inertial frame. It is occasionally incorrectly stated that Parker assumed $B_N = 0$; however, it actually follows from the basic assumption that **V** is radial at the Sun.

The Parker model is a steady-state model; however, it has also proven useful in many applications involving time variations in **V** and **B**. For example, there has been recent scientific interest in time variations in B_N at the Sun associated with motions of the footpoints of the magnetic field lines that accompany convective motions of the solar plasma such as in granules or super-granules. This possibility was discussed first by Jokipii and Parker (1969).

With this information as background, we are ready to describe the global properties of the HMF during minimum solar activity. In reviewing the field measurements, the only feature of the solar field that is needed is the largest scale dipole component that can be conveniently described in terms of two opposed magnetic poles. As with other planetary and stellar dipoles, the solar dipole is tilted relative to the Sun's axis of rotation. As the Sun rotates, the magnetic equator wobbles up and down in an inertial frame and fields with opposite polarities (outward or inward) are customarily observed each solar rotation and in opposite hemispheres. This approach provides an opportunity to compare the Parker model with observations. Section 4.3 will introduce complexity associated with magnetic field and solar wind structures that are also present during solar minimum.

4.2.2 B_R and open flux

In the Parker model the heliospheric magnetic field is derived from the radial field component at the Sun. B_R is caused by currents inside the Sun and between the photosphere and the corona but not by currents in the solar wind. Currents in a steady solar wind give rise instead to B_ϕ. Therefore, it is reasonable to begin our discussion with observations of B_R, especially since that component contains implicit information about the solar magnetic field.

In-ecliptic observations extending over many missions and years have documented the essential correctness of the Parker magnetic field model. However, Ulysses observations have clarified important aspects of the Parker model, especially the importance of the magnetic field at the solar wind source.

Images of the solar corona typically show a deviation of the magnetic field lines in the polar cap from being strictly radial but diverging much like those from the pole of a bar magnet. However, this divergence was traditionally attributed to excess plasma pressure in the polar caps rather than to the effect of the magnetic field.

In spite of the presumed dominance of the plasma pressure in the corona, early in-ecliptic measurements showed that the magnetic field energy density, and consequently the magnetic pressure, equaled or exceeded the plasma pressure when both were extrapolated back to the corona (Davis, 1966). This possibility was considered by Parker (1963) but not enough was known when he formulated his theory to justify any but the simplest assumptions. The early in-ecliptic measurements near the orbit of Earth showed that the energy density of the solar wind ($nMV^2/2$) exceeded the magnetic energy density (B^2/μ) and internal plasma energy density ($3nkT/2$, where k is the Boltzmann constant and T is temperature) by a factor of approximately 100. However, the magnetic field at the Sun is radial and with $B_R \sim r^{-2}$, $B_R = 3.5\,\mathrm{nT}$ at 1 AU (216 solar radii) grows by $(216)^2$ and becomes $1.6 \times 10^5\,\mathrm{nT} = 1.6$ gauss at the Sun. B^2/μ is then increased to $0.21\,\mathrm{erg/cm^3}$. On the other hand, conservation of mass implies that $n \sim r^{-2}$ so that $nMV_R^2/2 = 64 \times 10^{-10}\,\mathrm{erg/cm^3}$ at 1 AU ($n = 5\,\mathrm{cm}^3$, $V_R = 420\,\mathrm{km/s}$) and becomes $3 \times 10^{-4}\,\mathrm{erg/cm^3}$ at the Sun. The magnetic energy density is 700 times larger and dominates the energy density of the solar wind. This conclusion led to models that included the effect of the magnetic field at the solar wind source. Ulysses observations have provided convincing evidence that clearly show the vital role of the magnetic field in the source region.

Ulysses provided the first opportunity to study the dependence of B_R on heliographic latitude. Prior studies of the polarity (inward/outward) of the interplanetary magnetic field by in-ecliptic spacecraft clearly indicated that the fields were associated with the Sun's global magnetic field (i.e., with the solar magnetic dipole and the polar cap fields). That suggested that the fields should be stronger at high latitudes just as for a dipole field.

However, the initial Ulysses measurements near solar minimum unexpectedly showed that $r^2 B_R$ was $\approx 3\,\mathrm{nT}\ (\mathrm{AU})^2$ independent of latitude without increasing toward the poles (Smith and Balogh, 1995). Panel (a) of Figure 4.3 shows this parameter was constant in the south and north hemispheres above latitudes of $\pm 20°$. Panel (b) of the figure compares the Ulysses measurements with simultaneous

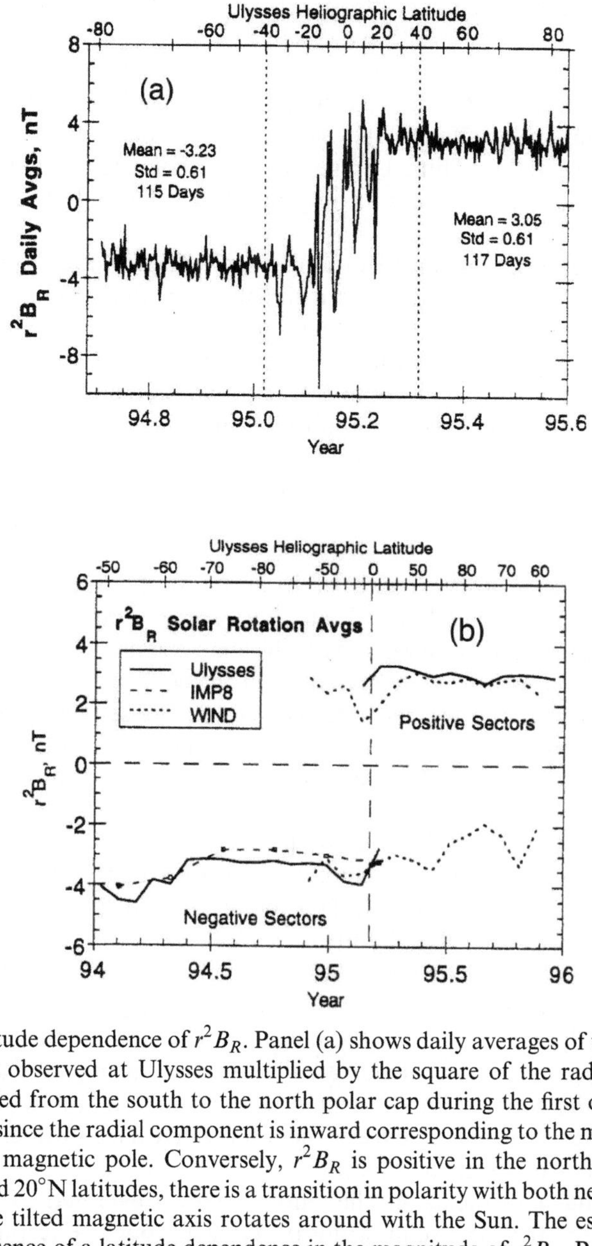

Figure 4.3. Latitude dependence of r^2B_R. Panel (a) shows daily averages of the radial magnetic field component observed at Ulysses multiplied by the square of the radial distance as the spacecraft traveled from the south to the north polar cap during the first orbit. In the south, r^2B_R is negative since the radial component is inward corresponding to the magnetic polarity of the Sun's south magnetic pole. Conversely, r^2B_R is positive in the north solar hemisphere. Between 20°S and 20°N latitudes, there is a transition in polarity with both negative and positive fields seen as the tilted magnetic axis rotates around with the Sun. The essential unexpected feature is the absence of a latitude dependence in the magnitude of r^2B_R. Panel (b) repeats the Ulysses r^2B_R averaged over successive solar rotations of 25 days. The dashed data are averages of measurements made simultaneously by two spacecraft in the ecliptic plane (WIND, IMP 8). The IMP 8 data in the negative magnetic sectors agrees closely with the Ulysses data in the southern hemisphere. The WIND data in positive sectors agrees with the Ulysses data in the north hemisphere. These comparisons show that the Ulysses measurements are not affected by temporal changes during the 1-year excursion. (Smith *et al.*, 1997a)

observations by in-ecliptic spacecraft and shows that time variations were not a significant factor.

The interpretation is that the enhanced magnetic pressure over the polar caps was being relieved by a non-radial expansion of the magnetic field and solar wind to produce equilibrium (i.e., uniform B_R). Since B is dominant, the solar wind is also deflected to lower latitudes. Thus, Ulysses provided convincing evidence that non-radial or "super-radial" expansion was not caused by enhanced plasma pressure but by magnetic pressure.

Subsequent theoretical reconsideration of the effect of the magnetic pressure confirmed this conclusion and showed that the pressure equilibrium is likely reached within 5 solar radii (Suess *et al.*, 1996). Pressure equilibrium results in both the field and solar wind velocity becoming radial. The radial evolution of the field and solar wind begins not at the Sun but at about 5 $R_{\odot}$ and would then be consistent with the Parker model at greater distances. An important implication is that extrapolating the solar wind and magnetic field inward using spacecraft observations assuming radial expansion is only valid down to the pressure equilibrium surface. Below that distance, a model is needed to take account of the non-radial flow from that surface inward to the corona.

Ulysses has provided other evidence of the importance of the magnetic field in the solar wind source region. The Solar Wind Ion Composition Spectrometer/SWICS investigation made the first measurements of the charge state of heavy solar wind ions including oxygen from which the temperature in the corona could be inferred using coronal models. The derived temperature showed an inverse relation with the measured solar wind speed—that is, higher coronal temperatures were correlated with lower speeds (Gloeckler, Zurbuchen, and Geiss, 2003). This finding is contrary to the Parker hydrodynamic model that implies higher temperatures produce faster, not slower, wind because of the higher plasma pressure.

Previous evidence of magnetic field control was also suggested by an observed correlation between solar wind speed and the modeled expansion of the magnetic field at the source (Wang and Sheeley, 1990).

The important role of the magnetic field has led to the development of solar wind models in which the magnetic field plays a crucial role (e.g., Fisk, 2003).

According to the Parker model, the magnetic field lines at the Sun are "open" meaning that they continue to extend radially outward throughout the heliosphere without crossing the equator and returning to the Sun (Figures 4.1 and 4.2). This distinction separates them from "closed" field lines that have both ends on the Sun like dipole field lines or magnetic "loops".

Most solar magnetic fields in the photosphere are closed. Thus, the fields observed by spacecraft in the solar wind are typically open. Exceptions do occur, transient phenomena that involve eruptions of large portions of the corona (coronal mass ejections or CMEs) that carry off magnetic loops (e.g., Crooker, Joselyn, and Feyman, 1997). CMEs transport fields that either close back at the Sun or are self-contained "flux ropes", both of which are configurations very unlike the open field lines in the surrounding solar wind.

In order to assess the relative occurrence of the open and closed fields, an accurate measure of the total open magnetic flux over the entire Sun is needed. Magnetograph observations of the photospheric field cannot distinguish between open and closed fields so those measurements represent the total magnetic flux. Past estimates of the total open flux were based on computational models that extrapolated magnetograph measurements to a "source surface" (Altschuler and Newkirk, 1969; Schatten, Wilcox, and Ness, 1969) typically located at about 2 solar radii at which the fields were required to become radial (i.e., open). Attempts were then made to compare such estimates with B_R observed in the ecliptic by spacecraft (Wang, Lean, and Sheeley, 2000).

The absence of a latitude dependence of B_R has an important consequence for determining the open magnetic flux. The total open flux on the Sun can be easily derived using the value of B_R measured at any latitude since

$$\phi(\text{open}) = \int B_R \, dA = 4\pi r^2 B_R$$

Therefore, $r^2 B_R$ is equivalent to open magnetic flux. The value of $r^2 B_R = 3.0$ nT $(\text{AU})^2 = 3 \times 10^{-9} \times (1.49 \times 10^{11})^2 = 6.66 \times 10^{13}$ webers or 6.66×10^{21} maxwells. Compared with the total flux (e.g., Harvey and Receley, 2002), this estimate shows that at solar minimum about one-half of the flux is open.

4.2.3 B_T and the Parker spiral angle

Qualitatively, the Parker spiral results from having one end of the open field line being attached to the rotating Sun while the other end is carried off in the solar wind. In a frame of reference that corotates with the Sun, the solar wind streamlines form Archimedes spirals and the magnetic field lines are parallel to the streamlines. In a non-rotating or inertial frame of reference, the solar wind streamlines become radial but the field lines continue to follow an Archimedes spiral. Observations of the spiral angle are important because they represent a test of the Parker model and provide quantitative information about the angular rotation rate of the Sun and the heliolatitude of the field at the solar source.

The formula for the spiral angle as a function of distance and co-latitude, θ, is $\tan\phi_P = -\Omega r \sin\theta / V_r$. In general, the field lines are helices lying on the surface of a cone with a half-angle, θ (Figure 4.2). In the solar equator, the field lines are confined to a plane and are similar to a wound-up watch spring.

The earliest indications that the Parker model was basically correct was the observed spiral field, $\tan\phi_B = B_T/B_R$, at the orbit of Earth and beyond and its correspondence with ϕ_P, based on the angular velocity of the Sun, Ωr, and the measured solar wind speed (Thomas and Smith, 1980). Because of ever-present large-amplitude fluctuations in the solar wind and magnetic field (e.g., Tu and Marsch, 1995), the Parker spiral angle is usually only observable on average rather than instantaneously.

For many years, measurements of the spiral angle were restricted to a narrow range of latitudes near the ecliptic plane. However, Ulysses overcame that limitation and allowed observations of the spiral angle from the equator to the poles. Since

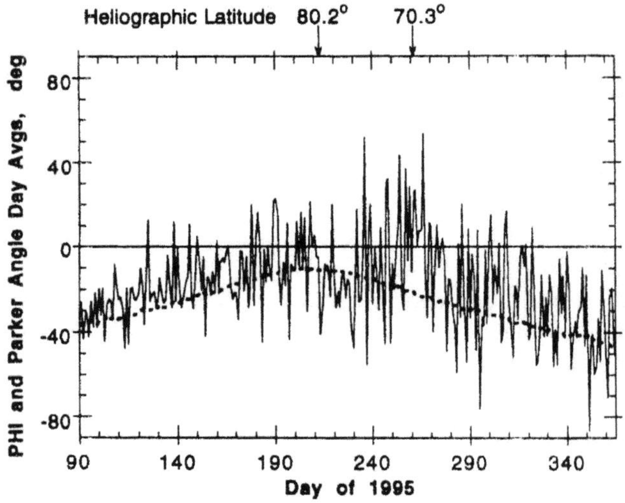

Figure 4.4. The spiral angle in the north polar cap. Daily averages of the measured spiral angle are connected by solid lines. The dashed curve is the Parker spiral based on the observed solar wind speed at Ulysses. The notable features are the large excursions in the measurements, associated with large-amplitude Alfvén waves that cause changes in the field direction, and the tendency for the fluctuations to lie above the dashed curve. The latter implies a tendency for the field directions to be more radial (corresponding to zero) because differential rotation was ignored. (Smith *et al.*, 1997a)

Ulysses traveled southward after leaving Jupiter, it reached the Sun's south polar cap first where the magnetic field was pointed inward (negative polarity). Averages of the observed spiral angle appear in Figure 4.4 as a function of latitude along with the Parker spiral angle computed using the observed solar wind speed (Smith *et al.*, 1997a). The large irregular deviations from ϕ_P are caused by ever-present Alfvén waves (Smith *et al.*, 1997b). Although the observed spiral angle generally follows the trend of the Parker spiral with increasing latitude, significant differences of several degrees are evident and indicate that the field is more radial than predicted.

A common method of displaying ϕ_B is in terms of histograms or probability distributions that contains information about the average/mean value, the most probable value (MPV) and reveal any asymmetries. Figure 4.5 (Forsyth, Balogh, and Smith, 2002) contains such histograms in a series of latitude ranges as Ulysses traveled southward from the equator, northward across the equator, over the north polar cap and returned toward the equator (the first Ulysses orbit). The measured field components have been transformed into a coordinate system with one axis aligned with the theoretical value of the Parker spiral based on the measured solar wind speed. Therefore, in the figure, deviations of $\phi_B - \phi_P$ from zero are departures from the Parker model, a way of accommodating the change in the angles with latitude. The field was restricted to a single sector at high latitude and two sectors at low latitudes.

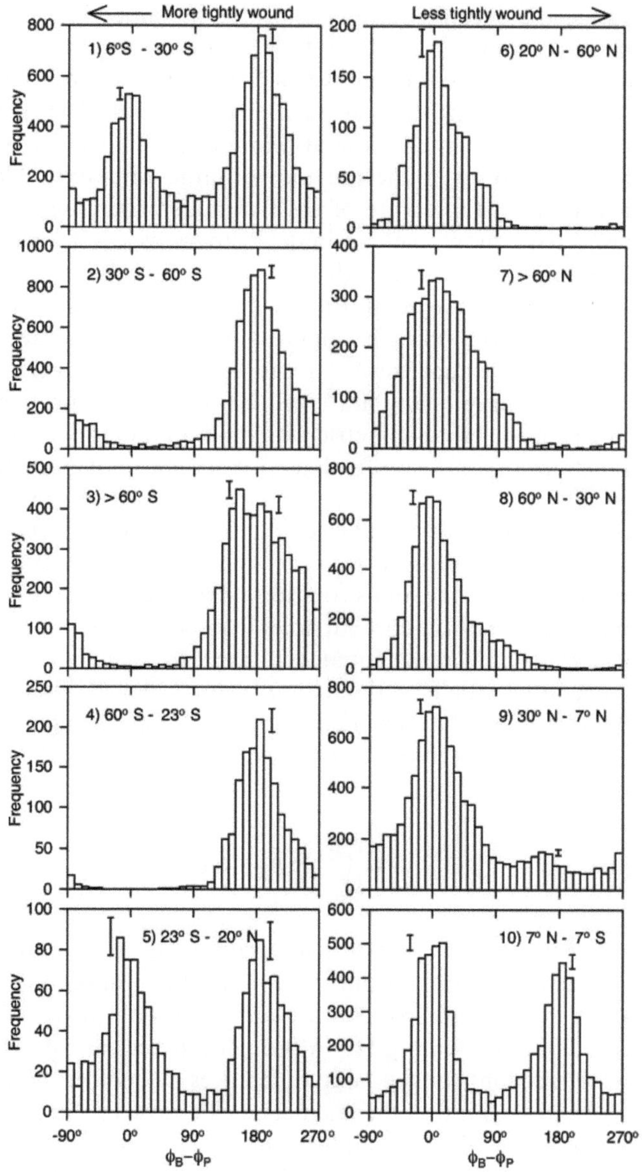

Figure 4.5. Ulysses observations of the spiral angle in the south and north hemispheres. Probability distributions or histograms are shown of the differences between the observed angle, ϕ_B, and the theoretical angle, ϕ_P, based on the measured solar wind speed, the equatorial solar rotation rate and the radial distance of Ulysses. Each panel coincides with a latitude range covered by Ulysses from the equator across both polar caps and back to the equator. At low latitudes, two magnetic sectors appear near 0° (positive) and 180° (negative). The differences are resolved into 10° intervals. Error estimates made for each histogram are represented by vertical bars. A shift in the differences toward more or less tightly wound spirals is indicated. (Forsyth, Balogh, and Smith, 2002)

The histograms, the means, and most probable values reveal a close correspondence between ϕ_B and ϕ_P at all latitudes. The distributions are reasonably well-behaved although the histogram above 60°S is double-peaked. The vertical bars adjacent to the peaks represent the statistical error associated with the number of examples in each bin and show that the appearance of the two peaks may not be statistically significant. However, the presence of large-amplitude Alfvén waves at high latitudes, a characteristic feature of the fast high-latitude wind, contribute to such irregularities. Furthermore, the bins in the histograms are 10° wide so that small but real differences might be buried in the distributions.

Significant differences can arise in using averages as compared with using most probable value (mode). For example, Figure 4.4 averages acquired in the north hemisphere indicate a departure toward more radial field directions, whereas in Figure 4.5 the corresponding histogram (>60°N) has a most probable value near ϕ_P. The histogram is, however, asymmetric with more observations corresponding to less tightly wound (more radial) spirals that shift the mean/average to more radial angles.

An obvious possibility that could account for observed deviations from the Parker spiral is the differential rotation of the Sun. Figures 4.4 and 4.5 are based on a constant period equal to the Sun's rotation period at the equator. The justification for this assumption is that the coronal holes from which the solar wind originates are observed to rotate rigidly at the equatorial rate. However, it is also well-known that the Sun rotates differentially with a slower rate at high latitude.

To investigate this effect, the Parker equation for the spiral angle was recast so as to yield the rotation rate observed at Ulysses (Smith *et al.*, 1997a). The Parker equation can be rewritten as

$$\frac{\Omega}{\Omega_o} = \frac{rV_rB_T}{\Omega_o} r^2 B_R \cos\delta$$

where Ω_o is the rotation rate at the equator. The ratio on the right-hand side is plotted in Figure 4.6 as a function of latitude and compared with a well-known expression for differential rotation. Large discrepancies are apparent with most points lying well below the dotted curve representing Ω_o. Although differential rotation may play a role, it alone cannot account for the large differences evident in the figure.

It has already been shown that the magnetic fields close to the Sun are not radial but are diverted equator-ward. That implies that the latitude at which the spacecraft is located is not necessarily the latitude at which the field line left the Sun and this difference might contribute to the discrepancy.

This possibility was investigated using the above equation and plotting $rV_rB_T/\Omega_o(r^2B_R)$ versus $\cos\delta$ with the results shown in Figure 4.7. Ignoring differential rotation, if the latitude of the field was equal to the latitude of Ulysses and $\Omega = \Omega_o$, the points would lie along the solid line with unit slope. The two cases using data in the south and north hemispheres reveal a systematic discrepancy that is quantified by the dashed line, the linear best fit to the points having a slope of 0.76. The plotted points and straight line are consistent with the fields actually

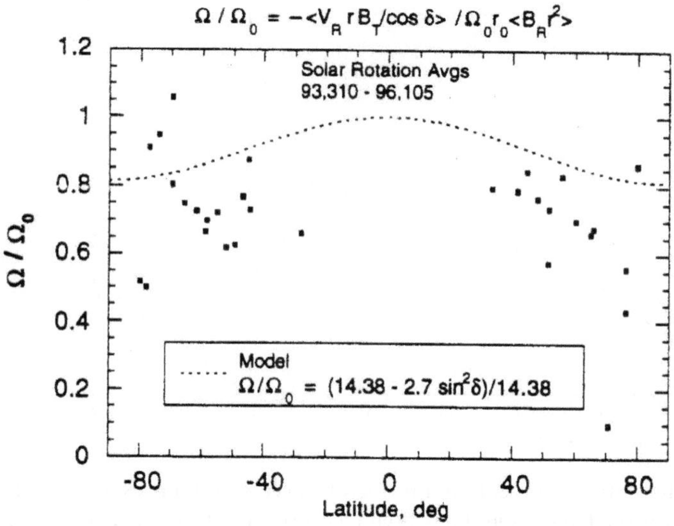

Figure 4.6. Comparison of solar differential rotation with differential rotation inferred from Ulysses measurements. The equation of the Parker spiral angle is reformulated to yield an estimate of the ratio of the Sun's angular rate of rotation, Ω, to the rotation rate at the solar equator, Ω_o. Ulysses measurements of $r^2 B_R$, B_T, and V_R averaged over a solar rotation at Ulysses latitude, δ, are combined to provide an estimate of $\Omega \cos \delta / \Omega_o$. The estimate is shown as a function of δ. The dashed curve is the ratio based on the formula shown at the bottom of the figure that expresses the ratio as a function of δ for comparison with the observations. The agreement is poor with large discrepancies and a systematic displacement to rotation rates that are far too low. (Smith *et al.*, 1997a)

originating at higher latitudes than the spacecraft latitudes at which they were measured.

A quantitative investigation of this possibility was carried out using a model of the solar magnetic field that assumed a dipole plus an equatorial current sheet (Banaszkiewicz, Axford, and McKenzie, 1998). Qualitatively, the model produces an equator-ward displacement of the field lines at the Sun consistent with their origin being above the latitude of the spacecraft.

For the chosen field parameters, the latitude of the field line at the Sun, δS, and the spacecraft latitude are nearly linear-related: $\delta S \approx \delta/3 + \pi/3$. The rotation of the photosphere as given by Howard and Harvey (1970) is

$$\Omega(\delta S) = \Omega_o \left(1 - \sin^2 \frac{\delta S}{8} - \sin^4 \frac{\delta S}{6} \right)$$

For a given δS, Ω and δ can be calculated so that $\Omega/\Omega_o \cos \delta$ is then known and can be compared with the results in Figure 4.7. The model calculations (the solid curve) lead to close agreement with the dotted best fit straight line. Thus, two factors cause the discrepancies from the Parker spiral, the expansion of the field from high to low latitudes near the Sun and the corresponding reduction in the rate of solar and field

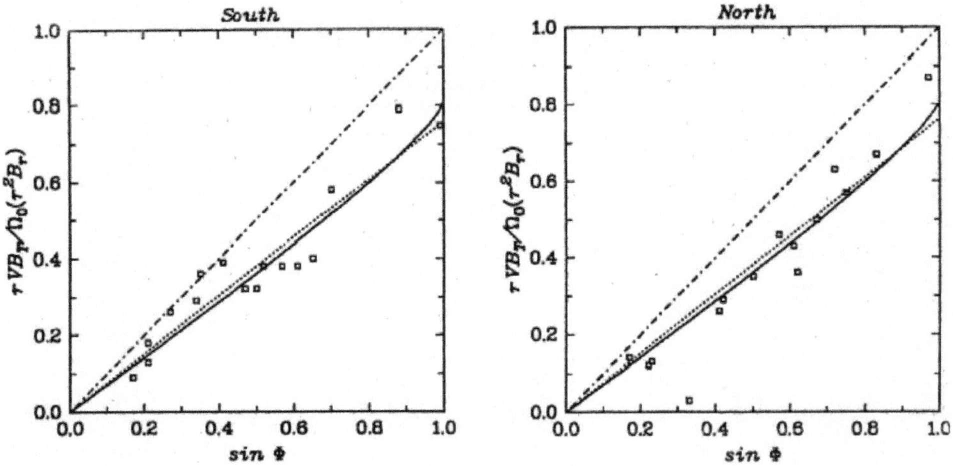

Figure 4.7. Alternative representation of the observed spiral angles as a function of latitude. The equation for the Parker spiral has been recast so that the calculated parameter represents $\Omega/\Omega_o \cos\delta$. Each data point is a solar rotation average and separate panels show the results in the north and south hemispheres. The solid straight line has slope = 1 corresponding to agreement. The inferred values of $\Omega/\Omega_o \cos\delta$ ($\sin\phi$ here) disagree with $\cos\delta$. The dotted line is a least squares straight line through the data assuming it passes through zero. The solid line that agrees well with the observations and with the dotted line is based on a model that includes equator-ward expansion of the field lines and differential solar rotation (Banaszkiewicz, Axford, and McKenzie, 1998).

line rotation. The law of iso-rotation (e.g., Ferraro and Plumpton, 1961) implies that the angular velocity along the field line is constant and is implicit in the above discussion.

This analysis is actually consistent with the histograms in Figure 4.5. Basically, the differences between the Ulysses latitude and the latitude of the field line at the Sun cause relatively small changes in the spiral angle. For example, near the equator in the slow wind with $\delta = 30°$, $r/r_o = 3.8\,\text{AU}$ and $V_r = 600\,\text{km/s}$ or $0.357\,\text{AU/day}$, $\phi_P = \tan^{-1}[(2\pi/26)(3.8)\cos 30°/0.357]$ or $\phi_P = 1.15\,\text{rad} = 65.8°$. If the field line originates at $\delta S = 70°$ and $\Omega/\Omega_o = 0.76$ as implied by the above equations, $\phi_B = 59.5°$, a difference of only 6.3°. At higher Ulysses latitudes, the differences in the two angles decrease and are inside the interval of 10° in the histograms and would probably be unobservable. In addition, Forsyth, Balogh, and Smith (2002) calculated the means in each latitude interval. The departures from the Parker angles vary between 7.4° and 13.6° and are consistent with the observed fields being more radial as expected for $\Omega/\Omega_o < 1$.

This analysis is consistent with the conclusion based on the latitude independence of $r^2 B_R$ that the field lines near the Sun are deflected equator-ward. In addition, the expansion of the field lines derived above is qualitatively consistent with the Wang and Sheeley (1990) expansion factor. The expansion is least at high latitudes where the solar wind speed is greatest.

The Ulysses observations at high latitudes show that the basic concept behind the Parker field model is sound but that account must be taken of details not contemplated in a steady-state model. In Section 4.4.5 it will be shown that large departures from the Parker spiral of several tens of degrees occur in association with a specific solar wind structure. The explanation involves the changing solar wind speed at the base of the rotating field line. Such a complication can be incorporated into a Parker-like model and provides additional information regarding the behavior of the magnetic field and solar wind at the Sun.

4.2.4 The north–south component, B_N

The north–south component of the HMF has been of keen interest since the very first interplanetary field observations by Pioneer V in 1960. The reason was, and is, the role of a southward-directed magnetic field component in causing the geomagnetic tail, storms, etc. Dungey (1961) was first to recognize the importance of merging or reconnection of a southward solar wind field with the northward-directed geomagnetic field in allowing plasma to enter the Earth's magnetosphere. That interest has continued into the present and still accounts for a continuing interest in B_N. In addition, the component provides another test of the Parker model.

The Parker model implies that B_N is zero. This result is not an assumption (as is sometimes stated) but is a consequence of the assumption of radial solar wind flow. The simplest argument is to note that, as described above, the solar wind streamlines in the frame corotating with the Sun have only two components: a radial component, V_r, and an azimuthal component, $V_\phi = -\Omega r \sin\theta$. There is no meridional or B_N component. Since, the magnetic field lines are the same as the streamlines in the non-rotating system, they also have only radial and azimuthal components.

A long history of in-ecliptic observations over a large range of radial distances has confirmed that B_N is indeed zero on average (e.g., Thomas and Smith, 1980). Long intervals of several days with non-zero B_N can, in fact, occur in association with large-scale solar wind structures and coronal mass ejections. Because of persistent fluctuations in B_N, as in the other components, it is common to use a statistical approach to determine average or most probable values from histograms of large numbers of measurements.

Ulysses extended these observations from the equator to the poles. Figure 4.8 contains Ulysses results as a function of latitude in the usual form of histograms (Forsyth, Balogh, and Smith, 2002). Each histogram corresponds to a different range of latitudes during Ulysses first orbit. The meridional angle, $\delta_B = \sin^{-1}(B_N/B)$, obtained from hourly averages, is shown in 5° intervals. Most of the histograms have most probable values of, and are symmetric about, $\delta_B = 0°$. A few histograms are not smooth but contain irregularities such as small double peaks. The error bars for each histogram establish that such features are within statistical error. The means are all less than 1°, typically a few tenths of a degree. Therefore, this analysis is consistent with the Parker model over all magnetic latitudes.

Admittedly, accumulating or averaging the data over long intervals could suppress departures from the model especially if they are periodic or quasi-periodic.

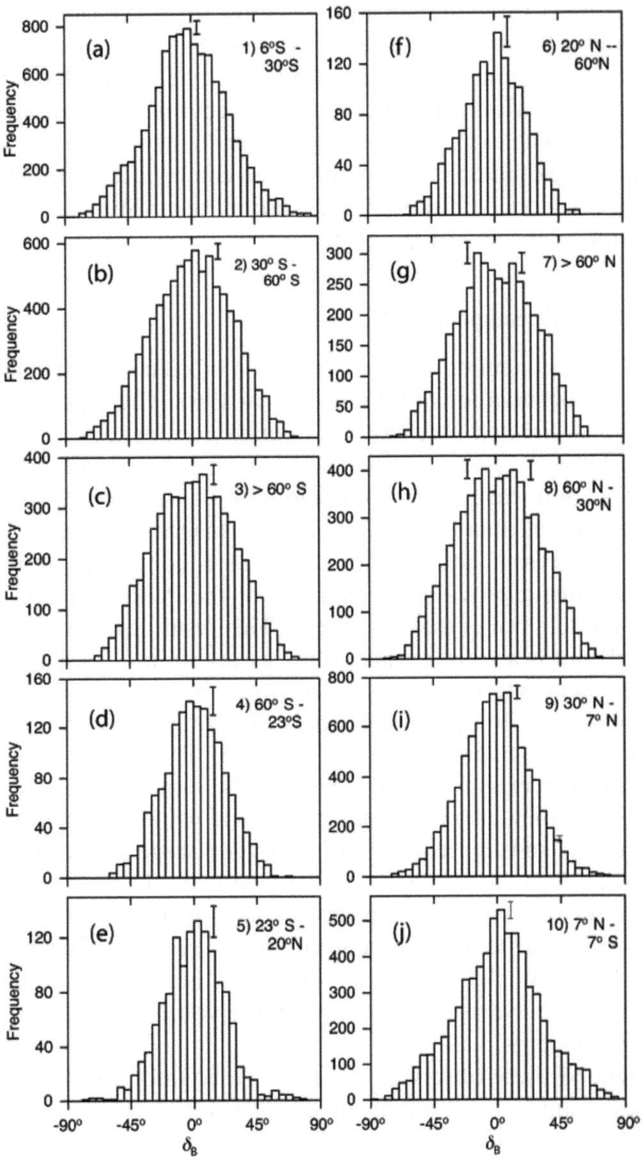

Figure 4.8. The north–south field angle measured at Ulysses as a function of latitude. The measured field component, B_N, and the field magnitude were used to compute the angle, $\delta_B = \sin^{-1}(B_N/B)$. The angles were then assembled into histograms or probability distributions, the number of observations for a given angle within 5° bands. Histograms are shown from the equator to both solar poles and back to the equator. Error estimates are shown as vertical bars. In general, the most probable values and averages are consistent with $\delta_B = 0$ as predicted by the Parker model. There are some discrepancies, however, such as double peaks and asymmetric distributions that might indicate small deviations from theory. (Forsyth, Balogh, and Smith, 2002)

There are valid reasons for anticipating that departures from the Parker model might be evident at higher latitudes. Two possibilities will be discussed in greater detail in a later section because they include other aspects of the magnetic field than implied by the Parker model but are mentioned briefly here. First, there is theoretical (Suess, Thomas, and Nerney, 1985; Pizzo and Goldstein, 1987) and experimental evidence (Winterhalter *et al.*, 1990) of a "flux deficit", a deficit in magnetic flux with radial distance as compared with the Parker model, that has been interpreted as a spreading of field lines away from the equator leading to the development of a B_N component.

The other reason is evidence from Ulysses that energetic particles accelerated in the middle heliosphere near the equator are able to access much higher latitudes than the acceleration sites, suggesting that the particles are following field lines that deviate from the Parker spiral. The statement that $B_N = 0$ is equivalent to stating that the helical Parker fields lie on a cone having a constant half-angle, θ_B. By considering factors not included in the Parker model, departures of the field lines in latitude can be introduced (Fisk, 1996).

4.3 THE HELIOSPHERIC MAGNETIC FIELD AT SOLAR MINIMUM

The following discussion of the HMF at solar minimum adds complexity to the heliosphere in the form of structure associated with the Sun and solar wind. The resulting changes to the global perspective also involve a time-dependent dynamic phenomenon. The effect on the HMF of this structure provides a further test of the Parker model.

At solar minimum, the underlying structure of the solar–heliospheric magnetic field remains relatively simple. It is dominated by the solar magnetic dipole and by stable structures that change slowly over several solar rotations. Consequently, they are referred to as corotating structures. A major influence is the tilt of the magnetic dipole to the Sun's rotation axis by a few tens of degrees. The tilt angle undergoes a slow systematic change with a corresponding change in the orientation of heliospheric structure. The dipole structure is also manifested in the solar wind structure with fast wind originating in the vicinity of the Sun's magnetic poles and slow wind dominating low latitudes. The interaction of the fast and slow wind caused by the dipole tilt strongly influences heliospheric structure.

In spite of the dominance in structure, short-lived intermittent events occur on the Sun that disrupt the prevailing solar wind–magnetic field structure. They result in large plasmoids being injected into the heliosphere. Although these coronal mass ejections are much more common at solar maximum, a brief discussion of their properties and interaction with the preexisting solar wind is included to provide a more comprehensive description of the heliosphere at solar minimum.

4.3.1 Dipole tilt, sector structure, and heliospheric current sheet

Open field lines can point inward or outward and typically exhibit one polarity in one solar hemisphere and the opposite polarity in the other. This property has long been attributed to structure imposed by the solar magnetic dipole. When the magnetic pole in the north is positive (the field points outward), the HMF polarity is positive in the north hemisphere. In the south solar hemisphere, the HMF polarity is then inward (negative), the same as the polarity of the south magnetic pole.

Generally, the magnetic dipole is not aligned with the Sun's rotation axis but, near solar minimum, is tilted to it by tens of degrees. As the Sun rotates, in-ecliptic spacecraft are located in first one, then the other magnetic hemisphere. This change in field structure is manifested observationally in the "sector structure". Although the field is not radial, it points inward or outward along the Parker spiral so that the polarity is still easily determined. If the observed polarities are plotted as pluses and minuses around the periphery of a circle corresponding to solar longitude, the circle appears to be divided into "sectors" (Wilcox and Ness, 1965). An example of this "sector structure" is included in Figure 4.1. Sometimes four or more sectors are observed indicating a departure of the HMF from a simple dipole-like field and the development of a more complex configuration.

The nature of the boundary between sectors, the "sector boundary", was uncertain originally but is now accepted to be a thin current sheet lying between the two opposite polarity fields (Figure 4.9). The current is perpendicular to and separates the two opposing fields. The sector boundaries are actually crossings of the heliospheric current sheet (HCS) as the tilted dipole/current sheet rotates along with the Sun. The HCS is the heliospheric magnetic equator separating open field lines from the two magnetic poles. However, it is not confined to a plane. Since the solar dipole is tilted, the HCS is inclined relative to the solar heliographic equator. Since the HCS sur-

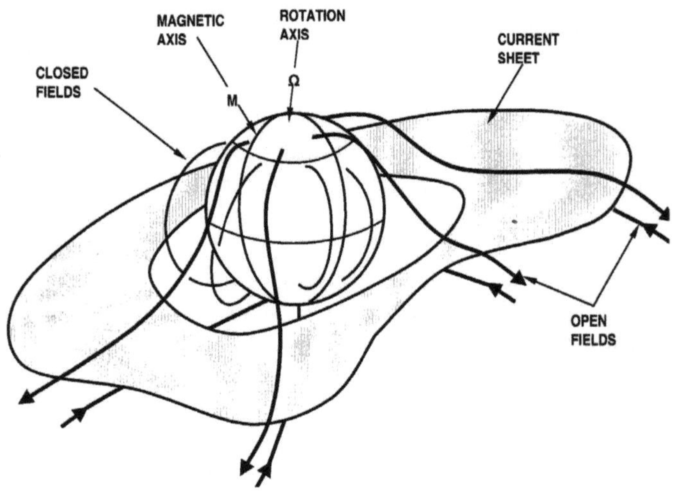

Figure 4.9. Three-dimensional schematic of tilted dipole with open and closed fields. The heliospheric current sheet is shown near the Sun. Magnetic field lines originate from a magnetic dipole whose axis, **M**, is tilted relative to the Sun's rotation axis, Ω. Open field lines from the north and south polar caps lie above and below the current sheet. Some closed field lines begin and end on the Sun. (Smith, 2001)

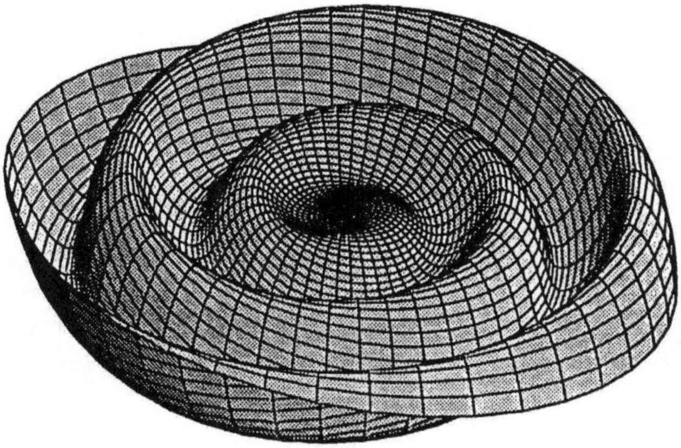

Figure 4.10. The current sheet in the heliosphere. In contrast to Figure 4.9, this three-dimensional figure shows the HCS at large distances from the Sun. Two sectors and two solar rotations are shown and the warped current sheet extends out to about 15 AU. The tilt angle between the HCS and the Sun's rotation axis is about 30° a value characteristic of solar minimum. (Jokipii and Thomas, 1981)

rounds the Sun and extends throughout the heliosphere, it takes a shape that is described as a flying carpet, a hat's brim, or a ballerina's skirt (Figure 4.10). The occurrence of more than two fairly wide sectors is attributed to "warps" or "folds" in the current sheet caused by a more complex solar magnetic field structure.

Multiple crossings of the HCS over fairly short intervals are also a frequent occurrence. Over intervals of minutes to days, two or more crossings may be observed. Proposed explanations are that the HCS has small-scale bumps or warps or that surface waves are propagating along it. An alternative is that the HCS consists of multiple current sheets (Crooker *et al.*, 1993). The basis of this suggestion is that multiple current sheets are known to occur at the Sun's surface.

4.3.2 Sector structure and source surface models

The inclination of the HCS was inferred well before the Ulysses mission, based on the variation in sector structure as the Earth and in-ecliptic spacecraft traveled around the Sun so that their changing heliographic latitude caused an annual variation (Rosenberg and Coleman, 1969). Even at the moderately high northern latitude of 7.25°, during a solar rotation, they spend more time above the HCS than below it and the above polarity is seen more often than the below polarity.

Interest in the sector structure prompted the discovery of a correlation between the HMF polarity and daily variations in the geomagnetic field in the Earth's polar regions (Svalgaard, 1975). Since observations of geomagnetic variations extended back in time over four sunspot cycles, it was possible to study the amplitude of the annual variation of the sector structure (i.e., the HCS inclination) and how it changed with the solar cycle (Svalgaard and Wilcox, 1975). In spite of an inability to determine

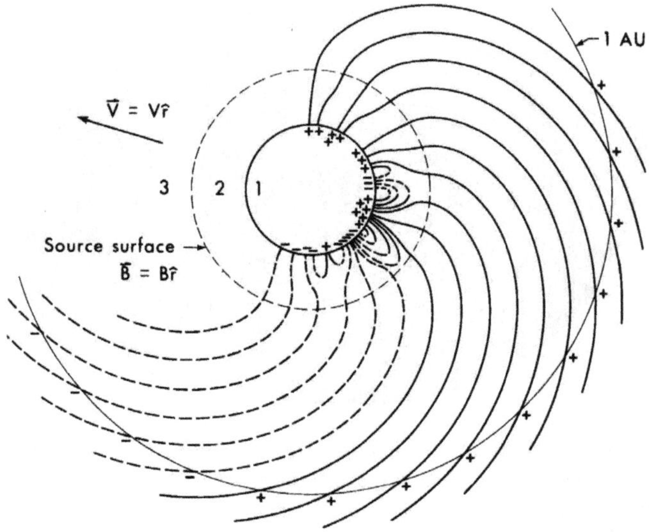

Figure 4.11. Principal features of potential field source surface models. The Sun is shown with various polarity regions (+, −) in the photosphere. The dashed circle is the source surface on which fields reaching it become radial and then spiral as they extend outward into the heliosphere. A few closed fields that do not reach upward to the source surface are also shown. The numbers (1, 2, 3) identify the different field regions and the radial solar wind velocity, *V*, is represented. (Schatten, Wilcox, and Ness, 1969)

the inclination near solar maximum, it was evident that the inclination increased as the solar cycle progressed from minimum to maximum and then decreased again at the next minimum.

Work on the sector structure led to the development of three-dimensional models of the HMF that are used to predict the sector structure and when the Earth or spacecraft will intercept the current sheet. These potential field source surface (PFSS) models extrapolate photospheric magnetic fields observed by ground-based magnetographs to a spherical solar wind "source surface" at which field lines are required to become purely radial (Figure 4.11; Altschuler and Newkirk, 1969; Schatten, Wilcox, and Ness, 1969). The field at the source surface is usually dipole-like consisting of oppositely directed fields in two hemispheres divided by a wavy "neutral line" along which the radial field vanishes. The location of the source surface was systematically varied to obtain the best agreement between in-ecliptic observations of the HCS and the Source Surface Neutral Line (SSNL). The "optimum" location placed the source surface at 2.5 solar radii. These models have continued to be successful in predicting the sector structure in the ecliptic.

4.3.3 Heliospheric current sheet and plasma sheet: properties

Maxwell's equations specify the conditions to be satisfied at a boundary separating two different magnetic fields. The normal component (if it exists) is continuous but the tangential component is not and a current flows along the boundary. At a plane boundary, $\mu_o \mathbf{K} = \mathbf{n} \times \delta\mathbf{B}$, where $\delta\mathbf{B}$ is the vector field change across the boundary, **n** is the normal to the boundary, and **K** is the linear current density. If the current sheet has a finite thickness, δz, the linear current density is related to the current density: $\mathbf{K} = \mathbf{J}\delta z$.

A favorite description of the current sheet is that the field on one side gradually decreases to zero then reappears with the opposite polarity and magnitude on the opposite side of the current sheet. The field, assuming a finite thickness for the current sheet, is often described mathematically using a hyperbolic tangent (a Harris sheet). However, actual current sheets such as the HCS do not follow this scenario. The field is not unidirectional but rotates across the current sheet often with constant magnitude (Figure 4.12).

In hydromagnetic theory, two different structures correspond to a current sheet (e.g., Landau and Lifschitz, 1960). The HCS is expected to be a tangential discontinuity (TD) distinguished by the absence of a field component normal to the

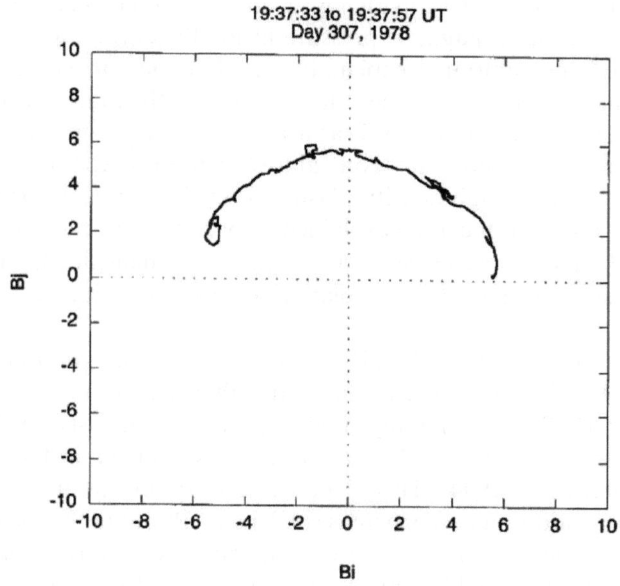

Figure 4.12. Change in the magnetic field on crossing the HCS. The field change during a crossing of the HCS (by ISEE-3 in 1978) is shown looking down on the current sheet. This view is achieved by performing a minimum variance analysis to determine the directions (principal axes) corresponding to maximum, intermediate, and minimum variances. The direction of minimum variance is normal to the HCS and to the plane of the figure. The magnetic field components, B_j and B_i, have been transformed into axes corresponding to the directions of the intermediate and maximum variances. The third component, B_k, is not shown but is invariably small indicating **B** does not cross the current sheet but rotates through it as shown. The field magnitude is conserved. Both the rotation and constant magnitude of the field are common features of HCS crossings. (Smith, 2001)

current sheet without any restriction on the change in field magnitude. The alternative is a rotational discontinuity (RD) for which a normal component exists and the field magnitude is constant. Although the rotating field suggests that the HCS is an RD, rotation of the field across a TD is allowed. The existence of a normal component is then the crucial difference. Searches for a possible normal component have failed to reveal conclusively that the HCS has a normal field component so that although the field rotates like an RD it appears to be a TD (Smith, 2001). In fact, the field magnitude is not always constant. As the field rotates, the magnitude first increases and then decreases (or *vice versa*) without passing through zero so that the field change resembles an "S" superposed on a half-circle.

The existence of a normal component is also important because it is the possible signature of the reconnection of adjacent field lines. In the case of a unidirectional current sheet, adjacent fields would have opposite senses or be anti-parallel. Merging or reconnection of such fields would lead to their being continuous across the current sheet and to a normal component. Actually, the rotation of the field across the HCS avoids oppositely directed fields lying nearby and explains why reconnection and normal components are not observed.

The current sheet model involving a unidirectional field change has to accommodate the reduced magnetic pressure inside the current sheet as a result of B going to zero. The traditional explanation is that the current sheet contains a plasma sheet that provides the pressure needed to exactly compensate the decrease in magnetic pressure. The rotation of $\mathbf{B}$ at nearly constant magnitude eliminates the need for this plasma pressure. However, almost without exception, the HCS does occur in association with high-density plasmas (Gosling *et al.*, 1981; Borrini *et al.*, 1981). The close association can be explained in part by the occurrence of the HCS almost exclusively in low-speed solar wind. High plasma densities are also correlated with slow wind since, in general, the solar wind flux (nV) tends to be a conserved quantity.

The close association with the HCS has led to the designation, Heliospheric Plasma Sheet (HPS). The HPS is essentially an enhancement in plasma density that accompanies the HCS. A commonly used approach in identifying the HPS is to determine the parameter, $\beta = 8\pi nkT/B^2$, from measurements in the vicinity of the HCS (Winterhalter *et al.*, 1994). This identification exploits an observed decrease in B inside the HPS in addition to the increase in density. Presumably, the reduction in B within the plasma sheet results from a diamagnetic decrease (the plasma tends to displace the field). Reasonably abrupt increases and decreases in β define the boundaries of the plasma sheet.

Although the HCS and HPS are referred to as sheets, they have finite thicknesses and are only truly thin sheets on heliospheric scales. The thicknesses are basic parameters and their changes with heliocentric distance are also of interest in attempts to understand the physics of the HCS and HPS. For example, a common theoretical prediction is that current sheets such as the HCS (and, in general) will suffer from the reconnection of the adjacent oppositely directed fields and be disrupted or become "tattered" with time/distance. An alternative view is that instabilities of one kind or another will cause the current sheet to gradually widen.

An accurate measure of the thickness of the HCS is important in part because it permits calculation of the current density. It also influences the speed of energetic particles such as cosmic rays as they drift along the current sheet because of the abrupt change in field direction (Jokipii, Levy, and Hubbard, 1977). Furthermore, current sheets are a common feature of space plasmas, examples being current and plasma sheets in the magnetotails of Earth, Jupiter, and other planets and the draping of magnetic fields around satellites (e.g., the Moon and Titan), and comparative studies are important.

There are many published examples of current sheet crossings and a statistical study of the thickness as observed at 1 AU (Winterhalter *et al.*, 1994). The latter also included the thickness of the plasma sheet identified by the enhancement of the plasma beta. Determining the thickness involves both the orientation of the HCS and the velocity of the solar wind. The normal to the current sheet was obtained from a minimum variance analysis of the magnetic field changes. The median thickness of the HCS was found to be 9,100 km while the thickness of the HPS was 320,000 km.

There are intervals when Ulysses spends relatively long times near the ecliptic plane at significant distances beyond 1 AU. The first occurred when Ulysses was *en route* to Jupiter in 1991. A second interval was in 2003–2004 when the spacecraft descended back toward the ecliptic from its second north polar pass. These cases provide an opportunity to extend the study of HCS/HPS thicknesses well beyond 1 AU. The corresponding distances were approximately 3 and 5 AU and observations near 1 AU by the ACE spacecraft supplemented the Ulysses observations at 5 AU. Thus, it was possible to compare the statistics of the HCS and HPS thicknesses at three separate distances (Zhou *et al.*, 2005).

The median HCS thicknesses are 1,705, 1,638, and 1,452 km at 1, 3, and 5 AU. Surprisingly, there appears to be little, if any, change over a range of 5 AU. The HPS, as expected, is substantially thicker but decreases with distance, the median values being 308, 211, and 138 (in units of 10^4 km). The HCS thicknesses obtained in this study are exceptionally thin compared with other estimates, passing the spacecraft in only seconds not minutes or even hours. However, the thicknesses were derived, not by inspection, but by transforming the current sheet fields into axes in and perpendicular to the current sheet and calculating and plotting the current density. High–time resolution magnetic and plasma measurements were used and discriminate against the much broader density increases related to slow wind in which the HCS/HPS are located that were mentioned above.

The three recent values at 1, 3, and 5 AU are significantly smaller than the median of 9,100 km obtained in the earlier study. Possibly, a solar cycle dependence is responsible since the two studies at 1 AU were carried out in different phases. However, the HPS medians at 1 AU obtained in both studies agree rather closely.

4.3.4 The HMF and testing of source surface models

Spacecraft observations are useful in testing various models and their validity throughout the solar cycle. The source surface models yield estimates of the inclination of the HCS based on the shape of the neutral line. Three-dimensional models also predict the magnetic field strength at all latitudes. Ulysses is uniquely suited to test both parameters since it makes measurements at all latitudes and at solar minimum and maximum.

The inclination is customarily derived simply from the maximum latitudinal extent north and south of the computed neutral line. The HCS inclination is routinely supplied to other investigators for each solar rotation and is widely used in various studies since it influences many aspects of the heliosphere (e.g., solar cycle modulation of galactic cosmic rays, Smith, 2006). The issue has added significance since the inclination is known to change over time including systematic solar cycle variations.

During solar minimum, the inclination of the HCS is restricted to a zone of low latitudes and at higher latitudes the field has only a single polarity and the sector structure disappears. The limited latitude range was confirmed, first by Pioneer 11 which reached a latitude of 16° after the first Jupiter encounter and then by Ulysses at a south latitude of −30° (Smith, Tsurutani, and Rosenberg, 1978). As Ulysses traveled from the equator to the pole, it observed the highest latitude reached by the HCS (its inclination) above which only a single sector was present (Smith *et al.*, 1993). The Ulysses crossing agreed reasonably well with one of the Stanford models (the "classical" model). Ulysses also crossed the highest latitude reached by the HCS three more times during solar minimum and four times during solar maximum. The combined results are presented in Section 4.7.4 where solar cycle variations in HCS inclination are discussed.

Another question is how well the models predict the field strength at latitudes above and below the ecliptic. As Ulysses has discovered, $r^2 B_R$ and, hence, B_R close to the Sun, is independent of latitude whereas the models predict a dipole-like increase from the equator to the poles. Other evidence was available early that indicated the models did not calculate the field strength correctly over the source surface. In particular, the computed magnitude of B_R increases gradually both above and below the neutral line (Figure 4.13). Taken literally, that would indicate that the current would be "thick"—that is, extend over a large fraction of 1 $R_\odot$ near the Sun and increase to much greater widths as the solar wind expands into the heliosphere.

This feature of the model contradicts observations that, even at 1 AU, the current is very thin (compare the discussion of thickness above). The HCS crosses spacecraft in intervals from a few seconds to a few minutes implying the current sheet thickness is a small fraction of a solar radius even at 1 AU.

Testing source surface models against observations has involved many observables other than the HMF such as solar wind density, temperature, and speed. It is also popular to compare the open field structures predicted by models with the coronal fields imaged during solar eclipses.

Furthermore, potential field source surface models have evolved steadily. Early changes included the arbitrary addition of a strong polar cap magnetic field and then a change to treating the photospheric field observations at the inner boundary as strictly radial without including meridional components. The results from Ulysses and other missions have motivated changes to the models such as the inclusion of an equatorial current sheet in addition to the usual currents on the source surface.

In addition to potential field models, MHD models have been developed follow-

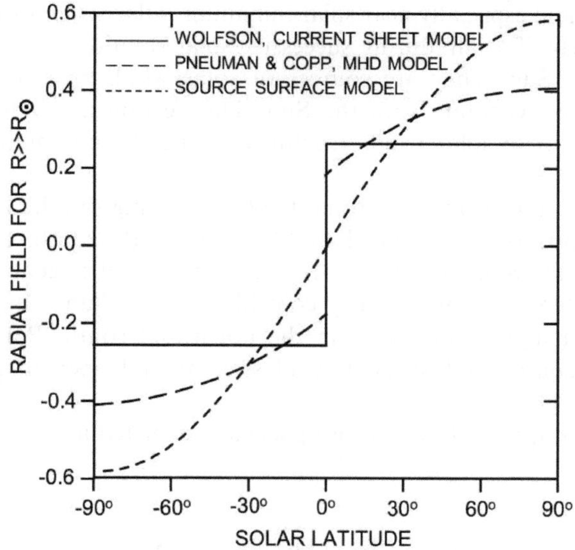

Figure 4.13. Thickness of the HCS according to various models. The radial field strength at large solar distances is plotted as a function of solar latitude to show how the fields and thicknesses vary in three current sheet models. The Wolfson model is an infinitesimally thin current sheet. The field produced by source surface models typically varies over a large range of latitudes (i.e., is very thick). The intermediate case is represented by an MHD model developed by Pneuman and Kopp (1971) that includes currents in the corona unlike source surface models. It changes abruptly at the current sheet, consistent with a thin current sheet and the field strength increases from the equator to higher latitudes. Ulysses data are consistent with the thin current sheet and constant strength as a function of latitude. (Wolfson, 1985)

ing the initial model of Pneuman and Kopp (1971) (e.g., Mikić and Linker, 1996). As with the potential field models, a number of alternatives are now available and the choice between them often depends on the application of the user—for example, in the extrapolation of solar wind and magnetic field observations inward to identify the source region or the extrapolation of solar observations outward to compare with *in situ* data. Altogether, modeling of the solar–heliospheric field continues to be an active area of research and one to which Ulysses can be expected to make important contributions.

4.4 THE HMF AND HELIOSPHERIC STRUCTURE

4.4.1 Solar and solar wind structure

The solar corona exhibits large-scale structure throughout the solar cycle. Visible in coronagraph images, the structures are a manifestation of underlying magnetic field structures. At sunspot maximum, these structures change rapidly, typically within less than a solar rotation, and are a significant aspect of solar activity. As solar activity declines, and especially near solar minimum, solar structure becomes simpler and changes slowly, if at all, during successive solar rotations so that these features corotate with the Sun. They are sources of solar wind, persistent fast and slow streams, which also corotate with the Sun. This section addresses the properties and evolution of these solar wind structures during the minimum phase of solar activity.

Large "holes" appear in the corona, the largest being simultaneously located in the Sun's polar caps, one in the north and the other in the south. They are typically not aligned with the Sun's rotation axis but their geometric center is tilted to the rotation axis by tens of degrees. They coincide with the Sun's magnetic poles and exhibit opposite magnetic polarities in the north and south. Polar coronal holes (PCHs) are the source of fast solar wind streams and open magnetic field lines (Figure 4.14).

Coronal holes appear dark because plasma is depleted as a result of solar wind

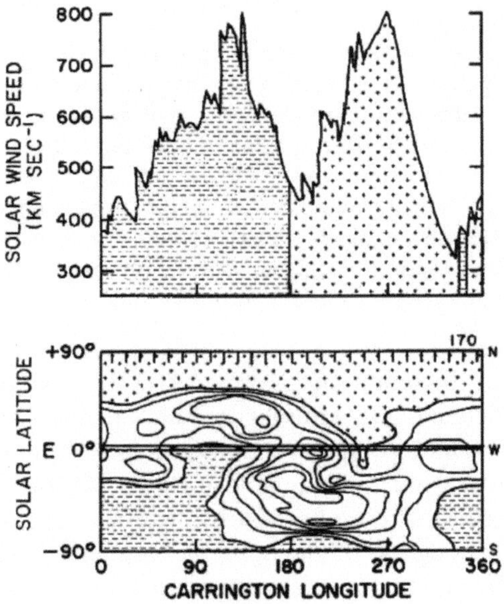

Figure 4.14. Association between the magnetic polarities of polar coronal holes and fast solar wind streams. This figure is one of the earliest to show this relationship. The upper panel contains the solar wind speed over a single solar rotation as a function of solar (Carrington) longitude. The pluses and minuses denote the magnetic field polarity in the two fast streams. Because of the definition of longitude, time proceeds from right to left in the diagram with the speed rising from 300–400 km/s to 800 km/s. The front edges of the fast streams are to the right and the trailing edges to the left. The lower panel shows the measured coronal brightness as a function of latitude and longitude. The brightness contours outline the tilted coronal disk or streamer belt. The two north and south polar coronal holes extend to the equator at longitudes of ~90° and ~270° and the observed magnetic polarities are indicated. The figure shows that the fast streams are correlated with the polar coronal holes. (Hundhausen, 1977)

outflow and the absence of trapped electrons.The edges of the polar holes are marked by polar crown prominences or closed magnetic loops that appear to straddle the coronal hole boundary.

Coronal holes also appear below the polar caps at lower latitudes. They are also dark and sources of solar wind and open fields. Equatorial coronal holes are also persistent and are sources of additional corotating solar wind streams. Generally, coronal holes in the north (south) solar hemisphere have the same sign as the north (south) polar cap magnetic field (Hundhausen, 1977). The relative locations of the equatorial and polar holes cause the magnetic equator (neutral line) to deviate from a simple circle on the photosphere or solar source surface and to adopt a "wavy" shape.

In contrast to coronal holes, streamers are bright structures typically shaped like a helmet, rounded at low altitudes and narrowing to a sharp peak at high altitudes. The contrast with coronal holes reflects a difference in their underlying magnetic structures. Magnetic fields in streamers are closed giving rise to their characteristic shape and their brightness that is caused by electrons trapped in the magnetic loops. Near solar minimum, multiple streamers are located near the solar equator and form a "coronal disk" or "streamer belt" around the Sun (Howard and Koomen, 1974). At solar maximum, streamers appear at essentially all latitudes as a result of the increased complexity and smaller scale structure of the solar magnetic field.

The streamer belt and the heliospheric current sheet/plasma sheet are customarily assumed to be structural counterparts. This association is based on both being high-density regions across which the field reverses direction. However, the field lines adjacent to the HCS are open while the fields throughout most of the streamer are closed. The open field lines passing above and below the HCS may originate immediately adjacent to the streamer but not be actually part of it. Alternatively, the open field lines may originate along the boundary of the polar coronal holes and be diverted equator-ward to pass above and below the current sheet.

The solar wind surrounding the HCS is invariably slow ($\sim$400 km/s) in contrast to the high-speed wind from PCHs ($\sim$800 km/s). This configuration has led to models of the solar wind at the solar surface having a simple structure (Figure 4.15, Pizzo, 1991). A fairly wide equatorial zone is visualized that contains only slow wind with the HCS representing the magnetic equator passing through the middle of the zone. Fast wind occupies the polar caps above and below the equatorial band of slow wind. Typically, the entire structure is tilted relative to the Sun's rotation axis. Close to the Sun, the boundaries between fast and slow wind are assumed to be abrupt or discontinuous. At larger distances, the boundaries have finite widths and have become transition regions.

The figure also shows the location of the HCS inside slow wind and tilted along with the magnetic dipole. The HCS separates slow and fast from the north PCH with one polarity from slow and fast wind from the south PCH with the opposite polarity.

4.4.2 Evolution and interaction of fast and slow wind

The realization that the solar wind consists of two types, fast wind from high latitudes and slow wind from low latitudes, received considerable support from HELIOS

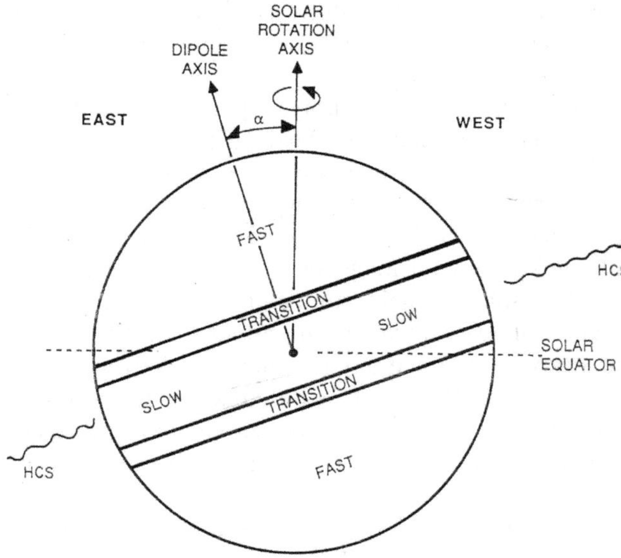

Figure 4.15. Model of the tilted dipole and fast–slow solar wind transition near the Sun. This schematic was developed by Pizzo (1991) to model fast–slow solar wind interactions and has proven to be useful in general. Helios spacecraft observations showed that near the Sun the change from fast to slow wind is abrupt and is represented here by a finite but thin transition. The magnetic dipole is tilted relative to the Sun's rotation axis and the HCS is tilted to the solar equator. As the Sun rotates, an observer/spacecraft at the central meridian encounters slow wind from the solar/magnetic equator, then fast wind arrives from higher magnetic latitudes followed by slow wind, etc.

observations between 1 and 0.3 AU (Schwenn and Marsch, 1990). The boundary between fast and slow wind became steeper with decreasing distance consistent with Figure 4.16. The assumption of an abrupt discontinuous boundary between fast and slow winds at the Sun has proven very useful in understanding their interaction and evolution as the Sun rotates and as they propagate into the heliosphere.

The evolution of the solar wind speed with distance can be understood with reference to Figure 4.15. Consider a spacecraft at constant heliographic latitude (e.g., the equator) as the Sun rotates under it. As the Sun rotates, the solar wind arriving at the spacecraft will change from slow to fast when the longitude of the boundary becomes the same as the heliographic longitude of the spacecraft. Fast wind will then arrive until the opposite side of the Sun rotates into view and the boundary again crosses the longitude of the spacecraft and fast wind is replaced by slow wind. If the tilt angle is sufficiently large, both the north and south boundaries can be crossed twice, so that the pattern becomes slow then fast wind from the north followed by slow then fast wind from the south. The HCS will be crossed twice inside the slow wind.

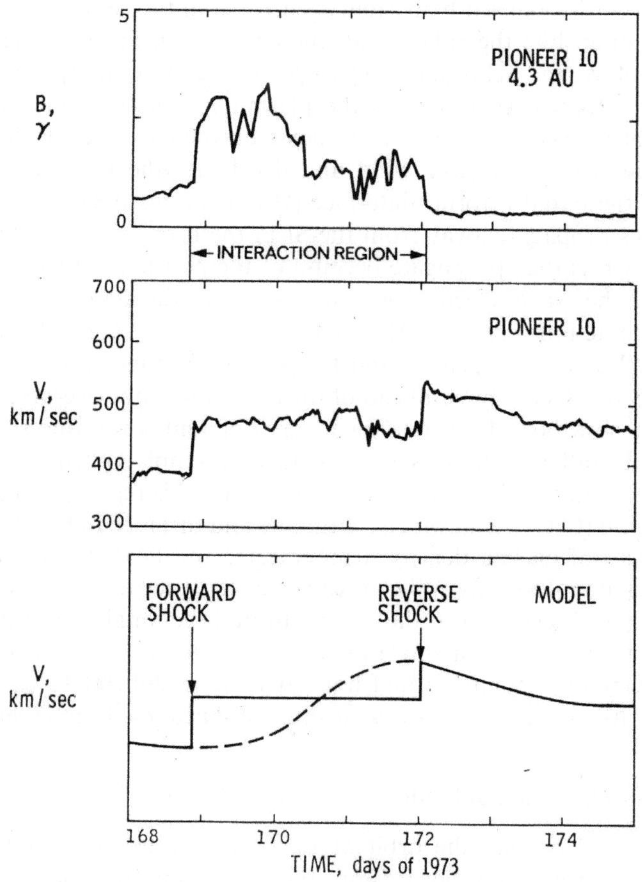

Figure 4.16. Schematic and observations of a CIR at large distances. The magnetic field magnitude and solar wind speed were observed by Pioneer 10 at 4.3 AU and produced an early example of the changes associated with the newly discovered CIR. The schematic in the bottom panel helps identify the forward and reverse shocks seen in the upper panels as abrupt changes in **B** and **V**. The schematic indicates how a gradual change in **V** nearer the Sun has been replaced by a nearly constant **V**. The two shocks have accelerated the initially slow wind and decelerated the fast wind to eliminate high pressure inside the CIR. (Smith and Wolfe, 1976)

However, the change from slow to fast speed takes place over a narrow range in longitude, and before the slow wind can travel very far into the heliosphere it is overtaken by the fast wind. Alternatively, at the crossing from fast to slow wind, the fast wind emitted first simply outruns the slow wind and continues to precede it into space.

As the fast wind overtakes slow wind, it is unable to simply pass through the slow wind because the presence of magnetic fields originating in different plasmas lends an identity to the two plasmas. According to Lenz's law, a change in the magnetic flux inside a conductor is opposed by currents generated in the conductor. In a perfectly conducting medium like the solar wind, the currents ensure that the flux remains constant. Therefore, the fields and the two plasmas cannot interpenetrate and both are compressed. Compression causes the plasma pressures to increase along this "stream interface" (SI). The pressure gradients on the two sides of the SI represent stresses that accelerate the slow wind and decelerate the fast wind and cause a broadening of the initially abrupt interface (Hundhausen, 1985).

The stresses propagate away from the SI in the form of large-amplitude waves that grow steeper as they propagate because of the non-linear relation between the amplitude and the wave speed. These waves define the outer boundaries of the "corotating interaction region" (CIR) that has developed. The field strength inside the CIR as well as plasma density and temperature are increased significantly. At 1 AU, the HCS is sufficiently far ahead of the SI that it is not overtaken by the CIR.

At the fast–slow boundary crossing, a very different corotating structure forms. Fast wind leads rather than lags slow wind and simply outruns it producing a rarefaction region rather than a compression region. This region, called a corotating rarefaction region (CRR), also widens with time and distance but is characterized by decreasing values of plasma density, temperature, and field strength and the wind speed decreases monotonically. When the decreasing speed is extrapolated Sun-ward from the point of observation to the corona to find the longitude of the source, the wind is found to come from a single or narrow range of longitude adjacent to the trailing boundary of a polar coronal hole. It is more difficult to extrapolate CIR plasma back to its source because of the strong non-linear evolution it has undergone.

4.4.3 CIRs, shocks, and dipole tilt

As the CIRs travel beyond the orbit of Earth and reach about 2 AU, the large-amplitude waves propagating away from the stream interface have steepened into a pair of shocks (Smith and Wolfe, 1976; Gosling, Hundhausen, and Bame, 1977). The leading boundary becomes a "forward" shock (FS) propagating away from the Sun (Figure 4.16). The trailing boundary evolves into a "reverse" shock (RS) that is propagating back toward the Sun in the solar wind. However, the speed of the reverse shock is less than the supersonic solar wind speed so it is simultaneously convected outward. The shock pair form sharp inner and outer edges of the CIR as typically observed by spacecraft beyond $\sim$2 AU. The solar wind speed profile has been transformed from a steady increase into two abrupt increases at the forward and reverse shocks with a more or less constant velocity within the CIR. Evidence of the SI has not disappeared as will be discussed below.

The HCS also becomes part of the CIR. Near the Sun, the HCS occurs well upstream of the fast wind. As is seen in Figure 4.15, the HCS is surrounded by slow wind with the increase to the fast wind occurring later. However, as the shocks propagate away from the SI at speeds of $\sim$100 km/s, the CIR widens at a rate of $\sim$1 day/AU or 0.25 AU/AU (Smith and Wolfe, 1979). The leading edge of the CIR approaches and eventually engulfs the HCS so that it appears behind the FS but upstream of the SI.

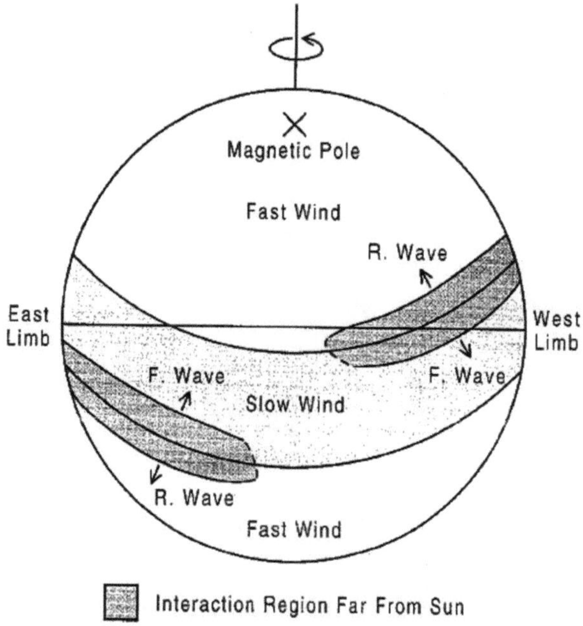

Figure 4.17. Schematic showing the tilted CIRs and the directions of propagation of their forward and reverse shocks. This model is similar to Figure 4.16 but is valid at large distances. The tilts of the CIRs (shaded) in the northern and southern hemispheres agree with modeling by Pizzo (1994). The upward tilt in the north and downward tilt in the south lead to the forward shocks (F. waves) propagating equator-ward and reverse shocks (R. waves) propagating pole-ward. These predictions were confirmed by Ulysses as it traveled to high latitudes. (Gosling and Pizzo, 1999)

Because the solar dipole and fast–slow boundaries are tilted, the CIR structures are also tilted. In Figure 4.15 the slow–fast boundary and the stream interface in the north are tilted upward. On the opposite side of the Sun (at a longitude difference of 180°), the boundary and SI are tilted downward. As the CIR evolves, the forward and reverse shocks will move away from the SI with approximately the same tilt angles relative to the radial solar wind as the stream interfaces. The resulting configuration is shown in Figure 4.17 (Gosling and Pizzo, 1999). In both hemispheres, the FS propagate equator-ward and toward increasing longitude (said to be westward) while the RS propagate pole-ward and toward decreasing longitude (eastward).

As Ulysses traveled to high latitudes for the first time, both forward and reverse shocks accompanying CIRs were observed at low latitudes. However, at higher latitudes, although reverse shocks continued to be seen, forward shocks became less common and eventually were no longer observed. This change to reverse shocks only is what would be expected based on Figure 4.17 and constitutes confirmation of the tilted dipole–CIR geometry (Gosling *et al.*, 1995). The orientations of the forward

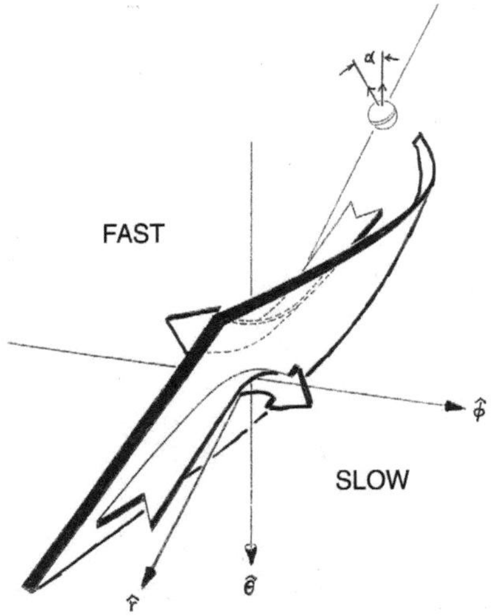

Figure 4.18. Diagram of a stream interface. Fast wind from the left is approaching slow wind on the right. They are separated by the curved surface, the stream interface, only a part of which is shown. The SI is tilted toward the right and spirals outward from the Sun. The arrows show the deflections of the fast and slow wind, neither of which can cross the interface. (Pizzo, 1991)

and reverse shocks observed by Ulysses are also consistent with Figure 4.17 (Burton *et al.*, 1996).

The tilted SI causes three-dimensional changes in the solar wind velocity as shown in Figure 4.18 (Pizzo, 1991). The solar wind cannot cross the SI because the plasmas on opposite sides originate in different magnetic fields and the plasmas are prevented from passing through one another. At the interface, the solar wind flow is deflected both east–west in azimuth and north–south in elevation (Siscoe, 1972). In the reference frame of the SI, the solar wind is approaching from both sides. If the SI is tilted to the radial solar wind flow direction, the stresses acting normal to the SI will deflect the flows so that they are parallel to the interface. The relatively abrupt deflections—with the appearance of V_T and V_N, the east–west and north–south components of the flow speed—are a convenient identifier of the stream interface in addition to the pressure maximum.

Since the fields cannot cross the SI, they are also deflected parallel to it. As the CIR expands, the velocity and field deflections spread out from the interface to occupy the entire region. Clack, Forsyth, and Dunlop (2000) have confirmed that the fields inside CIRs are parallel to the SI. When the two angles of the field direction are displayed by plotting δ_B versus ϕ_B, they lie along a "sinusoid" showing that they are correlated (Figure 4.19). This "angle–angle" display identifies "planar magnetic structures" (PMS), so-called because they are formed by the intersection of a plane with a sphere. Clack, Forsyth, and Dunlop fit the data to a sinusoid and the inferred planes along the sinusoid are parallel to the expected orientation of the stream interface. When this analysis is applied to CIRs in the southern and northern hemi-

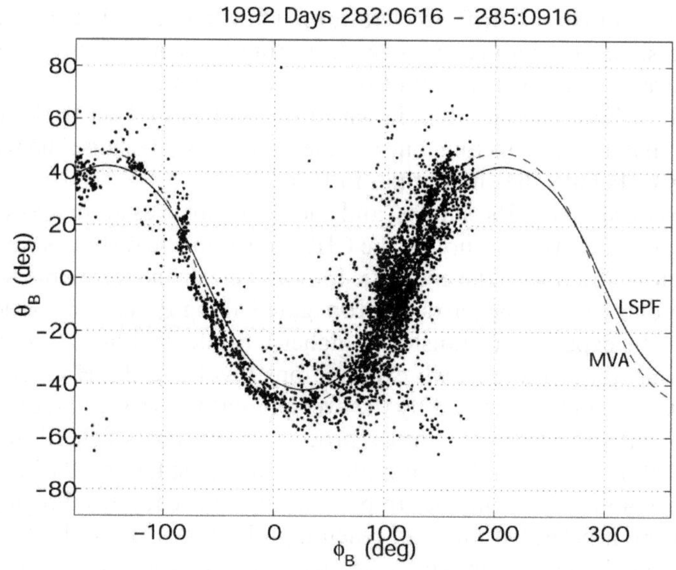

Figure 4.19. Correlated variations in field latitude and longitude angles. The meridional, latitudinal or elevation angle, called θ_B here, and the azimuthal or spiral angle, ϕ_B, were derived from 1-minute averages of the field components measured by Ulysses. The data were obtained over the 3-day interval shown while Ulysses was inside a CIR. The plot reveals a close correspondence between the variations in the two angles that results in them lying along a "sinusoid". The variations lie along the solid and dashed curves (MVA means a minimum variance analysis was used; LSPF is a least squares planar fit). The sinusoidal variation resembles the intersection of a plane with a sphere and variations of this kind are referred to as planar magnetic structures. The variations are, in fact, restricted to planes parallel to the stream interface inside the CIR. Although the field varies along a plane, the average field direction is still consistent with the Parker spiral. The figure does not imply a systematic variation in average field direction along the curve. (Clack, Forsyth, and Dunlop, 2000)

spheres, the planes are tilted in opposite senses in agreement with simulation of the stream interface by Pizzo (1994) and Figure 4.18.

It is important to recognize that the field orientations corresponding to the planar structure are fluctuations and are not a gradual change in the orientation of the average field. The average direction continues to agree with the Parker spiral throughout the CIR. However, the variations in the field direction about the average spiral are restricted to parallel planes.

4.4.4 CIRs, energetic particles, and their access to high latitudes

Shocks are prolific sites of particle acceleration. Charged particles are accelerated to energies up to a thousand times greater than the ~1 keV typical of solar wind protons. Two processes have been proposed as the cause of the acceleration. The particles may gain energy by drifting along the shock surface under the action of a parallel electric field (shock drift acceleration). They may also become energized as they cross the shock numerous times by scattering back and forth between waves generated upstream and downstream of the shock (diffusive shock acceleration). The latter is a form of Fermi acceleration in which the particles reflect back and forth off "boundaries" that are approaching one another in this case because the wind speed is inevitably slower behind the shock than in front of it.

As Ulysses observed CIRs, shocks, and energetic particles at increasing latitudes, it was surprising to find that, although the CIRs and associated shocks were restricted to latitudes below about 45°, the energetic ions and electrons were able to propagate almost to the polar caps. If the particles propagated along Parker magnetic field lines, they would not be expected to transfer to higher latitude field lines (the field lines lie on cones of constant latitude). Two explanations involving the HMF were offered. Energetic particles can be scattered by the ever-present waves (small-scale changes in field direction) onto adjacent field lines, a process called "cross-field diffusion" (Kota and Jokipii, 1995). Alternatively, a modification to the standard Parker model was proposed that would allow field lines to pass through the equatorial regions in which the particles were accelerated and still reach high latitudes (Fisk, 1996).

The Fisk model differs from the Parker model by incorporating three new effects: polar coronal holes, differential solar rotation, and super-radial expansion of the solar wind and magnetic field. There is nothing in the Parker model that prohibits adopting a profile for B_R at the Sun as a function of latitude (e.g., a dipole field tilted at an angle, α, to the rotation axis). Neither is there a restriction on assuming Ω is a function of latitude (i.e., the differential rotation rate, ω), a decrease in rate with increasing latitude established by solar observations of sunspots, and Doppler-shifting of emission lines. A tilted dipole will cause periodic variations in the strength of B_R and B_T with latitude. However, the field lines continue to rotate on a cone with constant θ and the Parker spiral is preserved.

However, since polar coronal holes are persistent, their rate of rotation is known and they are found to rotate rigidly at the same rate as the solar equator rather than at the high-latitude rate customary of differential rotation. Since the holes rotate rigidly, the differentially rotating photospheric magnetic fields rotate into, through, and out of the holes. Presumably, they are closed fields approaching the coronal hole, open upon entering it, and reconnect or close upon leaving (Fisk, 1996).

The other change that characterizes the Fisk model is the introduction of super-radial expansion of the field. Super-radial expansion causes large displacements of field lines in latitude and longitude near the Sun. Super-radial expansion is introduced by way of the Ulysses result that the field lines from the solar dipole expand until B_R is independent of latitude. The location at which this occurs defines the solar wind source surface in this model. (This approach contrasts with the more common potential field source surface that does not involve any assumption regarding super-radial expansion.)

The over-expanding field lines are assumed to originate along with the fast solar wind from a coronal hole surrounding the magnetic dipole, **M**, at its center (Figure 4.20, Fisk 1996). The magnetic pole and the coronal hole are assumed to rotate at the rigid equatorial rate, Ω. The edge of the symmetric coronal hole in the photosphere is

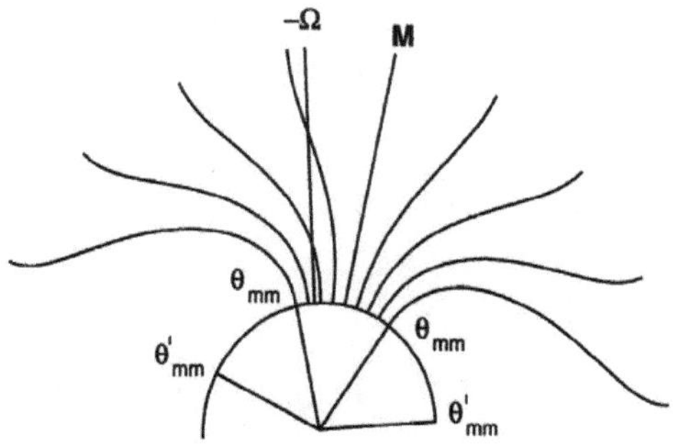

Figure 4.20. Super-radial expansion of polar cap field lines according to the Fisk model. The model takes advantage of Ulysses results that show the fields expand non-radially until reaching a surface (a solar wind source surface different than those derived from potential field models) on which they become radial and are uniformly distributed (B_R is constant). The magnetic dipole, **M**, is tilted relative to the rotation axis, $-\Omega$ (the southern heliographic pole). An open field line originating at the boundary of a coronal hole and heliomagnetic latitude, θ_{mm}, expands to reach latitude, θ'_{mm}. (Fisk, 1996)

defined by the angle it makes with **M** that is specified in the model. The edge of the coronal hole expands to occupy a larger co-latitude angle on the source surface. A field line that leaves the coronal hole at a specific co-latitude angle in magnetic coordinates defined by **M** reaches a larger co-latitude angle on the source surface that can be calculated.

Surprisingly, a symmetry axis exists at the source surface that is neither Ω nor **M** but **P**, the vector from the Sun's center to the point at which the field line originating on the rotation axis in the photosphere arrives at the source surface (Figure 4.21). The angle between **P** and **M** is called β. In the heliographic coordinate system corotating with the coronal hole at the rate, Ω, the over-expanded field lines on the source surface rotate around **P** in circles at rate ω which varies along the trajectories (Fisk, Figure 4.3).

The Fisk model leads to equations for the field components that differ significantly from the Parker equations:

$$B_R = B_o\left(\frac{r_o}{r}\right)^2$$

$$B_\theta = \left(\frac{B_o r_o^2}{rV_R}\right)\omega \sin\beta \sin\left(\phi + \frac{\Omega r}{V_R} - \phi_o\right)$$

$$B_\phi = \left(\frac{B_o r_o^2}{rV_R}\right)\left[\omega\left(\cos\beta\sin\theta + \sin\beta\cos\theta\left(\phi + \frac{\Omega r}{V_R} - \phi_o\right)\right) - \Omega\sin\theta\right]$$

The field strength at the source surface located at r_0 is B_0. The co-latitude and longitude in heliographic coordinates are θ and ϕ, and ϕ_o is the longitude of the magnetic pole. The differential rate of rotation, $\omega = \Omega - \Omega(\theta)$—that is, the difference

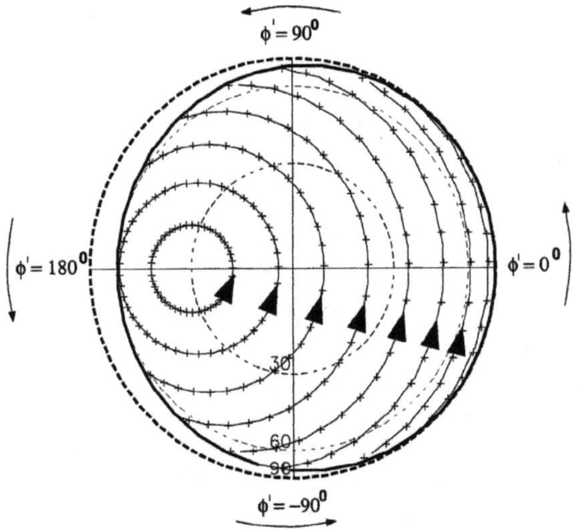

Figure 4.21. Rotation of field lines in the Fisk model. The solar wind source surface is viewed from the south polar axis and is tilted 15° with respect to the heliographic pole that lies along the intersection of the two orthogonal straight lines. The dashed circle is the heliographic equator with four longitudes, ϕ', indicated. The semi-circular arcs are the footpoints of field lines moving along the source surface projected onto the heliographic equator. The heavy solid curve is the maximum heliographic co-latitude reached by fields from the polar coronal hole ($\theta_{mm} = 24°$, θ'_{mm}). Differential rotation carries the field lines around a common point, the point at which the non-rotating field line from the heliographic pole reaches the source surface. The fields rotate counterclockwise because the view is from the south. As the field lines rotate, they cover a wide range of heliographic longitudes and latitudes. (Fisk, 1996)

between the angular velocity at the equator and at high latitudes (not the angular velocity at high latitudes).

When $\beta = 0$, the equations reduce to the Parker equations:

$$B_\theta = 0$$

$$B_\phi = \left(\frac{B_o r_o^2}{r V_R}\right)[(\omega - \Omega)\sin\theta]$$

The effect on the spiral angle can be seen by rewriting the equation for B_ϕ as $\tan\phi_B - \tan\phi_P = (r/V_R)[\omega(\cos\beta\sin\theta + \sin\beta\cos\theta\cos\phi^*)]$. The term in brackets is simply the differential rotation modified to include the changes that take place at the foot of the field line as it rotates around the trajectory in Figure 4.21 and both θ and ϕ vary.

The origin of the B_θ component can also be seen in Figure 4.21. As the field line moves along the trajectory with changing ϕ, θ varies between a minimum and maximum value. This variation produces the desired excursion of the field lines in latitude and allows energetic particles to migrate from low to high latitudes. As the

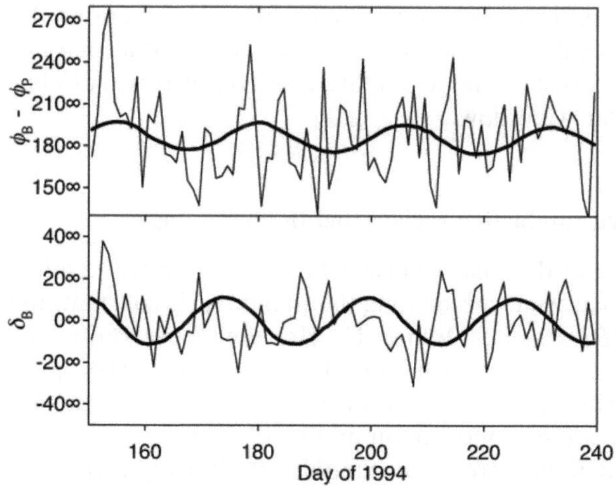

Figure 4.22. Magnetic field directions compared with predictions of the Fisk model. Daily averages of the azimuth angle minus the Parker spiral angle are shown in the upper panel as points connected by straight lines. The solid curve shows the result of calculations based on the Fisk model carried out by Zurbuchen, Schwadron, and Fisk (1997). The lower panel contains averages of the elevation angle and calculated variations predicted by the Fisk model assuming the same parameters as in the upper panel. This data interval was chosen because it was the only one while Ulysses was in the south polar cap to contain apparent quasi-periodic variations. The observed and calculated variations do not appear to be in phase although a shift of the calculations by ~7 days would improve the agreement. This interval was also of interest because it occurred when low-energy electrons accelerated by CIR shocks had access to high latitudes. (Forsyth *et al.*, 2002)

field line changes co-latitude, a B_θ component is generated. Since ϕ varies periodically, both B_θ and B_ϕ will exhibit periodic variations that are out-of-phase.

There have been several attempts to test the Fisk model using Ulysses magnetic field observations. Zurbuchen, Schwadron, and Fisk (1997) and Forsyth, Balogh and Smith (2002) have produced plots that compare observations with theory. Figure 4.22 shows $\phi_B - \phi_P$ and δ_B over a 90-day interval in 1994 as Ulysses was traveling to the south polar cap and the energetic particles were being observed at high latitudes. The figure also contains solid curves derived from the model for a specific set of model parameters. The variability in the observed angles and the relatively small variations in the model make a direct comparison difficult but a correspondence does appear to be present. This time interval has produced the best possible correlation between the data and the model yet found, although more comparisons are likely in the future.

It is difficult to quarrel with the basic assumptions of the Fisk model that incorporates observed features that were not known at the time the Parker model was developed (i.e., the dipole tilt, super-radial expansion of the field, and rigidly rotating polar coronal holes). A problem in comparing the predictions of the model with data is that the deviations in the B_θ and B_ϕ components tend to be small, as can be seen by substituting representative values into the equations. That contributes to the problem especially since the comparisons at high latitude have to be carried out in the presence of persistent, large-amplitude Alfvén waves that characterize the fast, high-latitude solar wind flow.

4.4.5 Corotating rarefaction regions and the spiral angle

Simple expansion of the solar wind would not be expected to affect the spiral angle. As long as the solar wind speed along the streamline is constant, the Parker equation holds with $\tan(B_T/B_R) = -\Omega r \sin\theta / V_r$. However, observations of the spiral angle inside CRRs typically reveal large departures from the Parker spiral even though the observed solar wind speed is used in the above equation (Figure 4.23). Deviations of 30 to 45 degrees are not uncommon and involve changes in the field direction that make it more radial than predicted (Smith *et al.*, 2000b; Murphy, Smith, and Schwadron, 2002).

It was customary originally to assume a value of Ω equal to the equatorial rotation rate because the PCH were rotating at that rate and the wind was assumed to come either from the edge of a PCH or, alternatively, from the equatorial streamer belt. Allowance for differential rotation of high-latitude field lines could produce more radial directions but an unrealistically small value was needed to accommodate the observed departures.

The solution to this problem was to incorporate a change in solar wind speed along the field line. A model was developed which included differential rotation of the field lines through the PCH and a change in outflow speed as they crossed the trailing boundary (Schwadron, 2002). This assumption was consistent with the well-known result that when CRRs are extrapolated back to the Sun they originate at the same, or nearly the same, solar longitude (Figure 4.24). The model involves several parameters, three being Ω and the fast and slow solar wind speeds. In addition, a fourth parameter turned out to be important, a finite width in longitude over which the speed changed (or, alternatively, a gradient in speed). The equation for the spiral angle again differs from the Parker equation:

$$\tan\phi_B = -(\Omega - \omega)\frac{r\sin\theta}{\left[V_R + \left(\frac{r\omega}{V}\right)\left(\frac{\delta V}{\delta\phi_o}\right)\right]}$$

where $\Omega - \omega$ is the rotation rate of the field lines through the coronal hole, ϕ_o is the Carrington longitude of the trailing edge of the coronal hole, and V_R is the solar wind speed. The boundary of the coronal hole is characterized by width, $\delta\phi_o$, and by a change in solar wind speed from $V + \delta V/2$ to $V - \delta V/2$. The boundary widens with distance to become a CRR.

The distinctive feature is the velocity shear, $\delta V/\delta\phi_o$, causing the speed to vary along the field line as it moves through the boundary of the coronal hole. That results in a turning of the field toward the radial direction across the CRR as observed. The Parker spiral is recovered by letting ω and $\delta V/\delta\phi_o$ vanish showing the difference

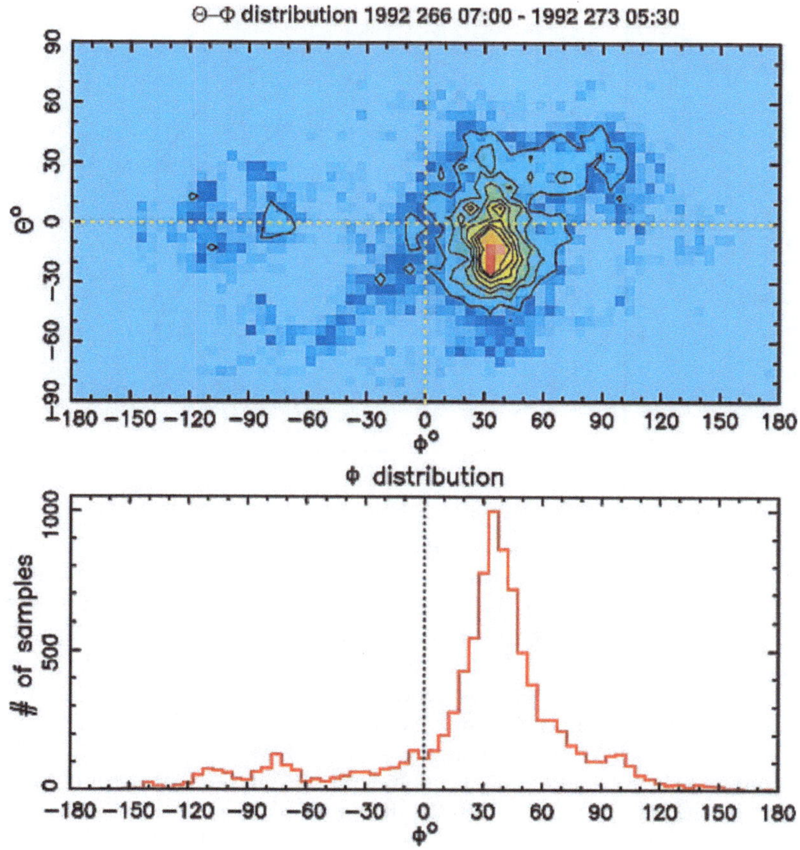

Figure 4.23. Departure of the magnetic field direction from the Parker spiral. Ulysses data acquired within a Corotating Rarefaction Region (CRR) during the interval at the top of the figure were averaged over 1 minute and rotated from RTN coordinates into a coordinate system with the x-axis along the Parker spiral based on hourly measurements of the solar wind speed. The components were converted to the latitude angle, called θ here, and azimuthal angle, ϕ, with the Parker spiral coinciding with $\phi = 0$. The top panel contains contours of constant probability with the maximum probability shown in red. The bottom panel is the corresponding histogram of the ϕ angles. Both displays show a significant departure from the Parker spiral of about 30° and a slight southward displacement. The average of ϕ is deviated toward the radial direction. (Murphy *et al.*, 2002)

between the observed spiral and the Parker spiral even when the local solar wind speed as measured is used to find ϕ_P.

The field lines derived from this model no longer follow the Parker spiral but deviate toward the radial direction within the CRR as observed (a "sub-Parker spiral"). Deviations of up to 45 degrees are achieved with quite reasonable parameters ($\omega = 0.15\Omega$, $V = 600\,\text{km/s}$, $\delta V = 275\,\text{km/s}$, and $\delta\phi_o = 5^o$ with $r = 5.2\,\text{AU}$). Although the context in which this model was developed is different than the origin

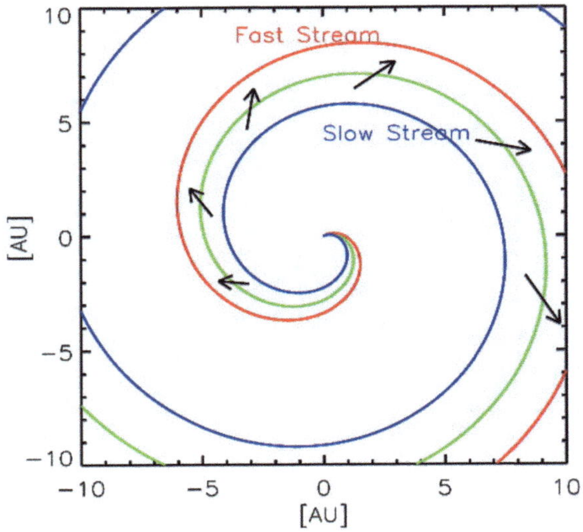

Figure 4.24. The field directions inside a CRR based on a model in which the solar wind speed varies along field lines. This figure illustrates a model developed by Schwadron (2002). The model is based on a changing solar wind velocity as a field line moves through a polar coronal hole because of differential rotation (polar holes rotate at the equatorial rate). The red spiral corresponds to the trailing edge of the fast stream from the coronal hole. The green spiral corresponds to the solar wind speed at the center of the coronal hole boundary in which the speed decreases gradually from fast to slow wind (the spiral shown in blue). The black arrows are the direction of the field at points along the mean solar wind speed inside the CRR from within the coronal hole boundary. The direction deviates from the Parker spiral because the speed changes as the field line moves through the boundary of the coronal hole. The angle between the black arrows and the green spiral shows the field is deviated toward the radial direction. (Murphy *et al.*, 2002)

of the Fisk model, it obviously incorporates some of the same features (and extends them somewhat) and so can be considered further evidence in support of that model.

4.4.6 Magnetic field strength and flux deficit

Since B_R, B_T, and B_N basically agree with the Parker model, the field magnitude, B, would be expected to agree also and provide no new information on the validity of the model. However, as spacecraft have traveled farther away from the Sun into the outer heliosphere, B has become a parameter of interest. At increasing radial distances, B_R decreases more rapidly than B_T—as it becomes the dominant component, is the most easily measured, and approaches B fairly rapidly. Thus, the field magnitude becomes the parameter of choice in testing the Parker model at large distances.

The field strength can be expressed as a function of the spiral angle, $B = B_R(1 + \tan^2 \phi_P)^{1/2}$. The spiral angle depends on $\sin\theta$ and is largest near the equator and approaches zero at high latitudes. Therefore, B should be stronger at the equator than at the pole. An early prediction was that the excess magnetic pressure at low latitudes would create a "flux deficit" by pushing some of the magnetic flux to higher latitudes (Suess and Nerney, 1975). There were no observations to test this proposal as long as measurements were restricted to 1 AU but the first spacecraft to the outer heliosphere, Pioneer 10 followed by Pioneer 11, produced supporting evidence (Slavin, Smith, and Thomas, 1984; Smith, 1989). Extrapolations of the field at 1 AU outward to 5 AU (the orbit of Jupiter) were found to be too large compared with the observations by about 5% implying a gradient of about 1%/AU. As Pioneer 11 continued outward to 10 AU (the orbit of Saturn) and beyond, the flux deficit continued to grow reaching ~30% at 30 AU (Figure 4.25, Winterhalter *et al.*, 1990; Smith, 1993). As evidence accumulated over a solar cycle, it became apparent that the flux deficit was time-dependent and was largest near solar minimum. This departure from the Parker model is relatively unimportant at 1 AU but becomes significant for observations made at large heliocentric distances. Furthermore, observation of the deficit caused the explanation to be reexamined.

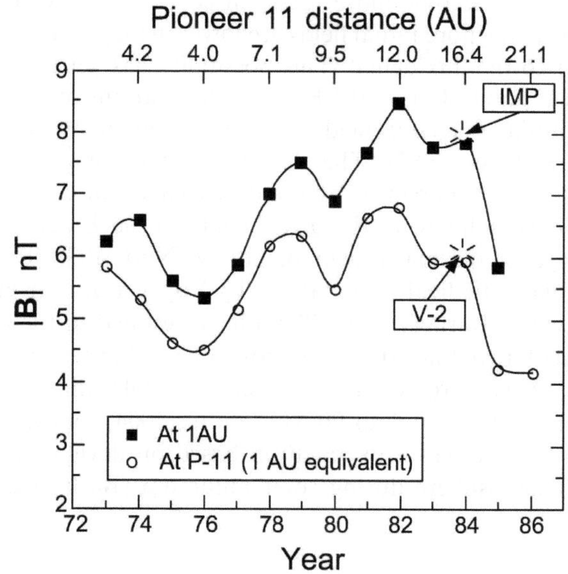

Figure 4.25. Variation in the HMF magnitude with distance and evidence of a deficit in flux relative to 1 AU. The circles and filled squares are yearly averages of B at Pioneer 11 and 1 AU, respectively. The averages are plotted as a function of time with the Pioneer 11 distances from 1 to 20 AU shown above the figure. The Pioneer averages have been converted to corresponding values at 1 AU based on the Parker model. Compared with average B measured by spacecraft at 1 AU, a systematic deviation toward lower values with time/distance is evident. A specific comparison of Pioneer 11 with IMP at 1 AU and Voyager at 16 AU is indicated by the star-like symbols. The differences are significant amounting to 25% at 16 AU or ~1%/AU overall. (Winterhalter *et al.*, 1990)

The plasma measurements at large distances made it evident that significant heating by the forward–reverse shocks was occurring at low latitudes in the outer heliosphere. An alternative explanation for the flux deficit was that excess plasma pressure as well as magnetic pressure was responsible (Suess, Thomas, and Nerney, 1985; Pizzo and Goldstein, 1987). A detailed mathematical model showed that the observed deficit could be explained in this way using quite reasonable parameters and that the plasma pressure was dominant.

As with the earlier explanation, flux was being pushed to higher latitudes and away from the equatorial region not only by excess magnetic pressure but also by increased plasma pressure associated with increases in density and temperature. Ulysses again had a unique opportunity to investigate this hypothesis by measuring the magnetic flux, $r^2 B_R$, as a function of latitude over multiple passages between the equator and the poles. In a study of four transits in latitude, a modest decrease in flux at low latitudes compensated by a small increase in flux at mid-latitudes was indeed obvious on three occasions (Smith *et al.*, 2000b). A numerical estimation of the flux lost at low latitudes equaled that gained at high latitudes consistent with the model.

Subsequent developments led to the recognition and explanation of the tendency for the magnetic field to become more radial in corotating rarefaction regions, as described above. Since the more radial fields are lower in magnitude, they are another possible contributor to the flux deficit. This possibility has yet to be studied.

From a historical point of view, the Pioneer observations of the flux deficit were not confirmed by initial magnetic field observations made by Voyagers 1 and 2 (Klein, Burlaga, and Ness, 1987). The displacement of magnetic flux to higher latitudes should also be accompanied by the development of corresponding north–south deflections of the solar wind (by a few degrees). Such deflections have been identified in the Voyager plasma measurements (McNutt, 1988; Richardson *et al.*, 1996). Subsequent magnetic field observations by Voyager at ever-greater distances and over the solar cycle (Burlaga *et al.*, 2002) have resulted in the recognition of a decrease in B that is larger than the predictions of the Parker model extrapolated outward from 1 AU. This decrease is most evident near solar minimum and implies a gradient of about 1%/AU from 1 to 90 AU (Smith, 2004). However, the Voyager investigators prefer an alternative explanation based on development of a "vortex street" in the outer heliosphere during solar minimum (Burlaga and Richardson, 2000).

4.5 NORTH–SOUTH ASYMMETRY OF THE SOLAR DIPOLE AND ITS SOLAR CYCLE VARIATION

Ulysses observations at solar minimum during the first Fast Latitude Scan provided convincing evidence of a significant north–south asymmetry in heliospheric structure. The asymmetry was first seen in measurements of the latitude gradients in galactic and anomalous cosmic rays (Simpson, Zhang, and Bame, 1996; McKibben *et al.*, 1996; Heber *et al.*, 1996). Both data sets revealed flux minima, not on the solar equator, but displaced southward by about 10°. The observations implied a corresponding southward displacement of the heliospheric magnetic equator (i.e., the heliospheric current sheet). However, a search for supporting evidence in the corresponding magnetic field measurements proved confusing because the values of open flux, $r^2 B_R$, were very nearly the same in both hemispheres contrary to expectation for such an offset (Erdös and Balogh, 1998).

This apparent contradiction was resolved by the realization that the spatial variations were being influenced by simultaneous temporal variations. When Ulysses was at high latitudes in the southern hemisphere, the HCS was offset southward as indicated by the energetic particle measurements. However, by the time Ulysses reached high latitudes in the north hemisphere, the offset had disappeared and the magnetic field observations gave no indication of an offset. Furthermore, at high latitudes, Ulysses was located below then above the HCS and unable to observe both sectors during a solar rotation.

In-ecliptic magnetic field observations by the WIND spacecraft were then analyzed during the Ulysses south–north transit and B_R was found to be significantly different in the two sectors as expected when an offset is present (Smith *et al.*, 2000a). A simple model (Figure 4.26) shows that the different values of B_R in the WIND measurements were consistent with a current sheet displacement of $-10°$. The model

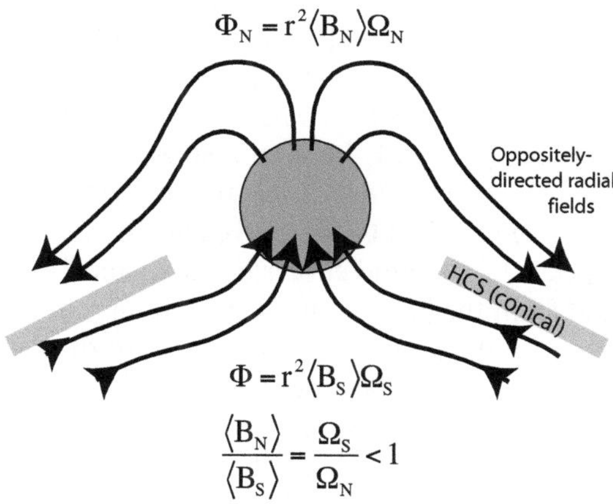

Figure 4.26. Diagram of an asymmetric current sheet and its effect on the heliomagnetic field. In the diagram the asymmetry is caused by a southward displacement of the solar magnetic dipole (magnetic equator) as observed by Ulysses as a difference in the north–south fluxes of galactic and anomalous cosmic rays. As a result, the current sheet (shown shaded) is deflected southward to form a cone. The effect on the radial magnetic field in the two hemispheres and the two magnetic sectors (B_N and B_S) is shown by the equations. The open magnetic flux is assumed to be equal in the north and south magnetic hemispheres and is expressed in terms of the radial field component, the square of the radial distance, r, and the solid angle (Ω) above and below the current sheet. The greater spreading of the field lines (larger Ω) in the northern hemisphere causes the radial component above the HCS to be less than below it. This implication was used to confirm the southward displacement of the HCS by comparing in-ecliptic measurements of B_R in the two sectors. (Smith *et al.*, 2000a)

simply relates the solid angles above and below the offset HCS to the open flux, $r^2 B_R$, that expands to fill the two asymmetric hemispheres.

Further supporting evidence was provided by Earth-based magnetograph observations of the north and south polar caps that showed a significant difference in field strength with stronger fields in the north (Figure 2.6). It had been presumed that the north–south asymmetry was associated with such a difference or, alternatively, an equivalent offset of the solar magnetic dipole from the solar equator.

The Ulysses observations made a significant contribution by supporting earlier studies that provided less direct evidence of the asymmetry. About 10 years before the Ulysses observations, a study of the relative widths of Away (positive) and Toward (negative) sectors (Tritakis, 1984) revealed an annual variation in addition to the semi-annual variation caused by the $\pm 7.25^\circ$ excursion in heliographic latitude of Earth and in-ecliptic spacecraft—the Rosenberg–Coleman (1969) effect. The annual variation was attributed to a displacement of the HCS from the solar equator. The author was prescient in predicting that Ulysses could provide a definitive test of this hypothesis.

Luhmann, Russell, and Smith (1988) compared B_R at the orbits of Venus (Pioneer Venus Orbiter/PVO) and Earth (International Sun Earth Explorer-3/ISEE3) and found a departure from the expected r^{-2} dependence that was correlated with the separation of the two spacecraft in heliolatitude. The results were interpreted as implying a north–south asymmetry in the HMF without a specific explanation.

The Ulysses observations of the HCS displacement also stimulated further studies of the N–S asymmetry that revealed a previously unsuspected solar cycle dependence. Zieger and Mursula (1998) compared annual variations in solar wind speed, geomagnetic activity (K_p) and the sector structure observed by near-Earth spacecraft over 3.5 sunspot cycles. All three parameters showed an asymmetry between the spring and fall equivalent to the N–S excursion in latitude. The minimum in annual solar wind speed, V_{min} (found earlier by Zhao and Hundhausen, 1981) was systematically displaced from the helio-equator in a manner that was correlated with the solar cycle as well as with the solar–HMF polarity. The displacement was greatest near solar minimum and appeared to be absent near solar maximum. When the north polar cap was positive, V_{min} was displaced northward and was offset to the south when the polarity reversed—that is, it was characteristically located in the north *magnetic* hemisphere.

The correlation between solar wind speed and geomagnetic activity was used in a subsequent study to investigate the annual variation and its phase over a much longer interval than just the last 3.5 solar cycles (Mursala and Zieger, 2001). The correlation of K_p with solar polarity persisted unchanged back into the mid-1800s when a reversal in phase appeared to occur. It was speculated that the offset was associated with a relic (primordial) solar magnetic field that oscillated in its location relative to the solar dipole with a long ($\sim$100-year) period.

Mursala and Hitula (2003) reexamined the relation between the displacements in V_{min} and the HCS using the ratio of sector widths—$T =$ Toward and $A =$ Away and $T/(T + A)$—in the spring and in the fall. A pattern was found in $T/(T + A)$ representing the displacement of the HCS that turned out to be southward for both solar polarities. However, V_{min} was only displaced southward for P(−) cycles and was found to be northward for P(+) cycles. Thus, in successive cycles, the HCS and V_{min} first coincide and then in intervening cycles are displaced into opposite hemispheres! This situation is curious especially since the HCS has usually been associated with the location of V_{min}.

The persistent southward displacement of the HCS is consistent with the solar magnetic dipole being offset southward irrespective of magnetic polarity. Since an offset dipole is equivalent to a centered dipole plus a quadrupole, observations of the HMF may be revealing information about the quadrupole component of the solar magnetic dynamo.

Finally, Zhao, Hoeksema, and Scherrer (2005) used a PFSS model to study the displacement of the neutral line/HCS from the solar equator. The southward displacements were replicated in 1985 and 1995, the latter agreeing with the Ulysses/WIND observations. Their investigation of the solar magnetic field/solar wind expansion factor showed no change between hemispheres when the HCS was offset implying the offset was likely associated with either the area of the polar coronal holes or with the polar cap field strengths (equivalent to a quadrupole component).

This succession of investigations confirms that studies of the HMF by Ulysses and in-ecliptic spacecraft are revealing new and important information about the global structure of the heliosphere as well as providing basic information about the Sun's global magnetic field. Ulysses is presently executing another complete latitude scan and should contribute once again to this developing area of research.

4.6 TEMPORAL VARIATIONS—CORONAL MASS EJECTIONS

The magnetic fields inside coronal mass ejections (CMEs) near the Sun are very different than the Parker fields associated with normal solar wind. The same is true of their counterparts at larger distances in the heliosphere, called interplanetary coronal mass ejections (ICMEs) although heliospheric might be a better designation (or HCMEs) since they are a three-dimensional phenomenon not restricted to the space between the planets. The difference is that solar wind field lines are open (very few, if any, counterexamples having been observed) and the fields inside ICMEs are closed. The latter can either have both ends connected to the Sun or closed to form a disconnected loop. Which possibility occurs in a specific event is a long-standing problem that has defied simple solution. Various magnetic field configurations have been proposed or inferred from limited observations. Examples include a series of parallel loops, a single field wrapped several times around a central axis, and a magnetic flux rope perhaps with twisted fields throughout it.

ICME fields are closed because they originate in closed-field regions on the Sun. At the source of normal solar wind, the magnetic field is unable to restrain the outflow of the plasma and, in fact, may aid it. For example, in the Sun's polar caps, field lines tend to be radial and parallel to the outflow. In closed-field regions, on the other hand, the fields are transverse to radial and are able to overcome the outward plasma pressure and restrain the outflow. The breakdown of restraint and the eruption of a CME appear to depend on a major reconfiguration of the field or plasma pressure. Suggested causes are the emergence of new magnetic flux from below the solar surface, reconnection of oppositely directed fields, and sudden increased heating of the plasma.

Images of the solar disk and corona have identified the sources of CMEs. Approximately three-quarters have solar prominences as their source. These are loop-like features extending above the edge of the solar disk or are observed as dark "filaments" that disappear abruptly when a CME occurs. Polar crown prominences are a source of high-latitude CMEs. Many CMEs originate at low latitudes in the streamer belt or coronal disk, particularly near solar minimum. These observations make it possible to relate CMEs near the Sun to ICMEs near and beyond 1 AU in spite of a large gap in spatial coverage. The speed and direction of the CME obtained from images is very helpful in this identification.

When observed near the Sun, the volume occupied by a CME is a fraction of that of the Sun. However, by the time it reaches 1 AU, the CME volume has grown enormously often by a factor of $\sim 10^6$. This large expansion has two causes. The leading edge of the CME typically travels much faster than the trailing edge and simply outruns the latter. Commonly, the speed decays monotonically from front to back. A second reason is that the pressure of the plasma and magnetic field greatly exceeds that of the surrounding solar wind causing the ICME to expand. For example, the field magnitude of a prominence is about 100 gauss, whereas the solar wind typically arises in a field of ~ 5 gauss. In approximately 10–30% of ICMEs, the observed internal magnetic pressure exceeds the plasma pressure. This occurrence has given rise to a distinctive class of ICMEs called magnetic clouds (MCs). When the major contributor to the expansion is internal pressure rather than a speed gradient, the ICME is said to be "over-expanding" (Gosling *et al.*, 1994).

Since the speed of the ICME usually exceeds the speed of the surrounding slow solar wind, low-latitude ICMEs have to push their way through the wind. The speed differential is typically large enough for a shock wave to form that is detached from the ICME and precedes it. The solar wind plasma between the shock and the oncoming front of the ICME is compressed as well as diverted around the ICME. Usually, this interaction gradually slows the ICME as it propagates into the heliosphere. However, the expansion continues for a long time and the distance from the front to the back of an ICME can change from ~ 0.3 AU (~ 60 solar radii) at 1 AU to a separation of ~ 1 AU at the orbit of Jupiter.

Ulysses observations have made major contributions to the study of ICMEs (see Forsyth and Gosling, 2001). Ulysses is uniquely qualified to investigate high-latitude ICMEs, determine their properties, and compare the observations with those made at low latitudes (Reisenfeld *et al.*, 2003). In addition, high-latitude ICMEs are frequently caused by erupting high-latitude polar crown prominences and their magnetic fields can be compared with the prominence fields to see how they are related (Rees and Forsyth, 2003).

4.7 HMF AT SOLAR MAXIMUM AND ITS SOLAR CYCLE VARIATION

4.7.1 Introduction to solar maximum and the Hale cycle

The state of the heliosphere at solar maximum involves profound changes. The simple dipole-like structure is replaced by a more complex structure that is also time-dependent. The solar magnetic dipole gradually weakens and disappears then reappears with the opposite magnetic polarity. This event designates the solar cycle as lasting 22 years, the so-called Hale cycle. The solar field is dominated by smaller scale but much stronger magnetic fields distributed over much of the Sun. Curiously, spacecraft observations far from the Sun show the continuous presence of a magnetic dipole that appears to rotate from one hemisphere to the other and that dominates many aspects of the HMF. That means that the HCS is continuously present and changes inclination along with the dipole. This behavior is still closely related to the solar open flux extending into the heliosphere during solar maximum. CMEs also become much more common. The observations at solar maximum place those at minimum into the broader context of a systematic variation over the solar activity cycle and can be re-visited from that point of view.

4.7.2 Solar magnetic field at solar maximum

The solar activity cycle is actually a magnetic cycle driven by cyclic changes in the Sun's magnetic dynamo. The solar dynamo, like that of Earth and other planets, operates in a convecting electrically conducting fluid. At Earth, the dynamo is located in a central fluid core, whereas on the Sun the dynamo occupies a spherical shell about three-quarters of the distance from the center to the surface where convection replaces radiation as the dominant mechanism for transferring heat from the thermonuclear interior to the surface.

Hydromagnetic theory indicates that the solar and planetary dynamos operate by alternating between a dipole-like, axial "poloidal" field and an equatorial "toroidal" field (Parker, 1979: Priest, 1982). At the Sun, the transformation from one into the other takes 11 years and produces the sunspot cycle. At solar minimum, sunspots are absent and the magnetic field is poloidal or dipole-like. Differential solar rotation is distorting the poloidal field lines by stretching them out parallel to the equator and eventually the toroidal fields so generated break through the surface near 30 degrees latitude to form a sunspot. This event signals the beginning of a new solar cycle. As time progresses, the toroidal component grows at the expense of the poloidal field as it declines in strength. The solar field becomes complex with strong sunspot fields appearing at latitudes (below 30°) and longitudes simultaneously and the previously dominant dipole field weakening and eventually disappearing at solar maximum. The many strong field sites are centers of activity associated with CMEs, flares, radio-emissions, etc. that are characteristic of maximum solar activity.

The dynamo then causes the dominant toroidal field to transform into a new poloidal field so the Sun's dipole field reappears after a delay of months. When the polar cap fields representing the axial dipole reappear, their polarity has reversed.

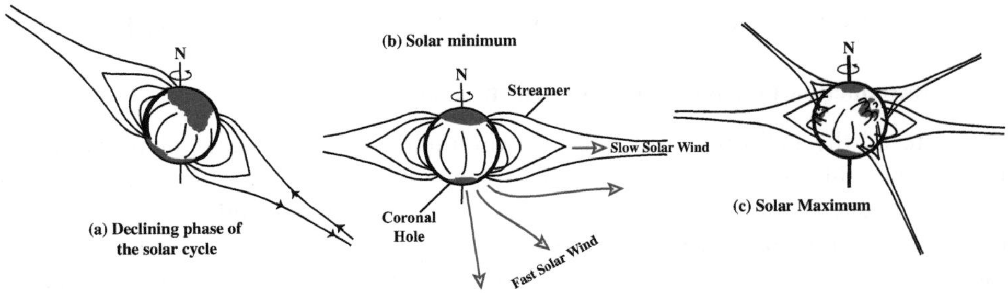

Figure 4.27. Changes in the solar magnetic field during the solar cycle. Panel (a) is a diagram of the tilted dipole magnetic field with fast wind from the polar coronal holes and slow wind from the streamer belt. This configuration is typical of the descending or ascending phases of the solar cycle. Panel (b) is the configuration at solar minimum when the tilt angle becomes very small (essentially zero degrees). In this phase, the fast wind is restricted to high latitudes only and fast–slow stream interactions are weak or absent. Panel (c) refers to solar maximum when the polar coronal holes are small and streamers originate all over the Sun including at high latitudes. Actually, the current sheet is still present but highly inclined or nearly perpendicular to the solar equator (see following figures). (Suess *et al.*, 1998)

If the field in the north polar cap had been outward, it is redirected to be inward and *vice versa* in the south polar cap. This change in polarity means that the total solar cycle takes 22 years—it is called the Hale cycle.

As the strength of the dipole slowly grows, the toroidal field component undergoes a decline as does the level of solar activity. Several years afterward, sunspots are disappearing and the dipole field is once more becoming dominant. This trend continues until a minimum in activity is reached and a new cycle is ready to begin.

Naturally, these profound changes affect the solar wind and HMF (Figure 4.27). The disappearance of the polar cap fields is accompanied by the disappearance of the polar coronal holes and with them the fast solar wind so that the wind at maximum is slow on average (McComas, Elliott, and von Steiger, 2002a). Lower latitude coronal holes occur intermittently and serve as sources of solar wind. Furthermore, the increased structure in the solar field leads to multiple streamers at various latitudes that can also act as sources of both slow solar wind and CMEs. Active regions consisting of large magnetic loops associated with sunspot magnetic fields have also been found to be a source of solar wind as well as CMEs (Levine, 1977; Neugebauer *et al.*, 2002).

It might be expected that the simple HMF structure characteristic of solar minimum would be replaced by a much more fragmented field without a recognizable sector structure. However, observations in and out of the ecliptic during solar maximum showed instead that the sector structure persists (Figure 4.28). The sector structure is not totally disrupted by CMEs or other aspects of solar activity (Smith, Slavin, and Thomas, 1986). The field magnitude increases but only by a factor of about 2 (King, 1979; Slavin, Jungman, and Smith, 1986). All in all, the field looks surprisingly similar during both maximum and minimum.

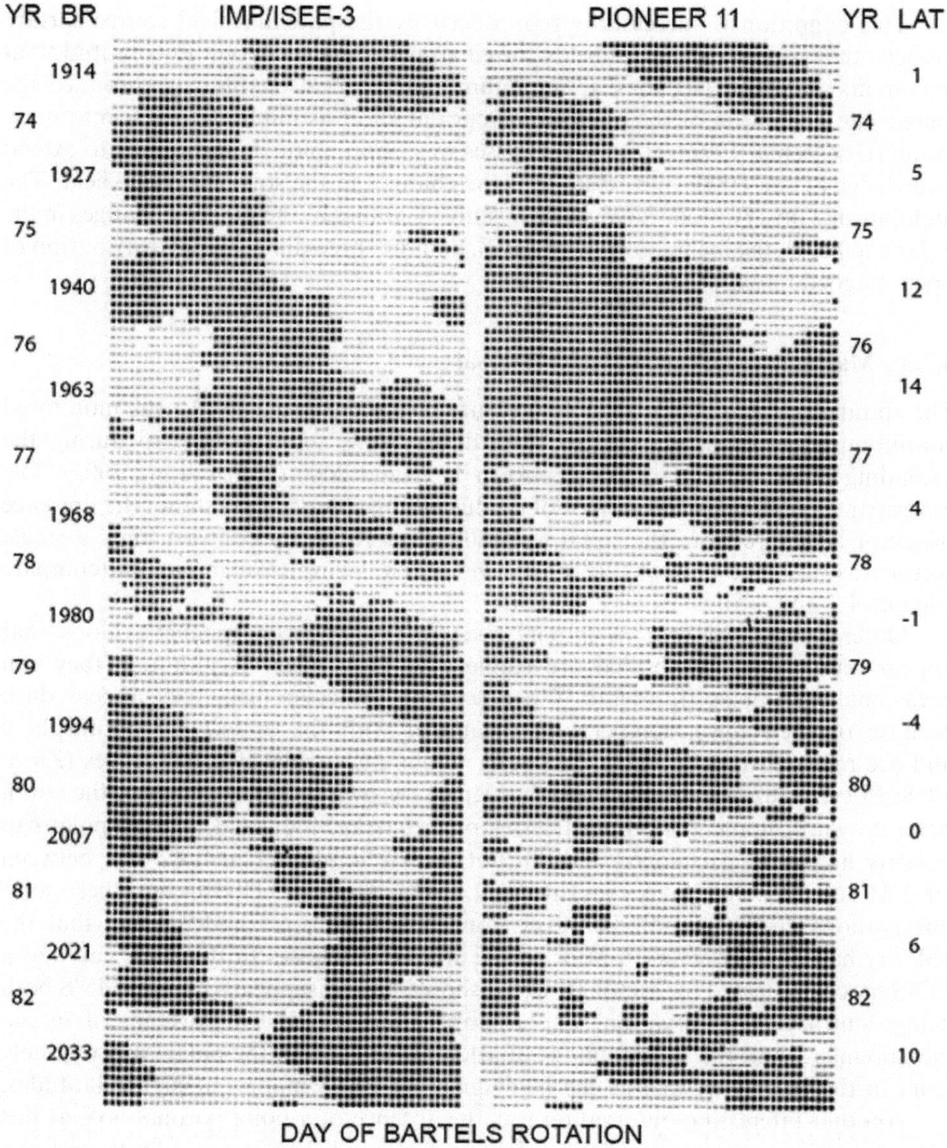

Figure 4.28. The magnetic sector structure before, during, and after solar maximum. This figure uses IMP/ISEE-3 and Pioneer 11 data. The rows show the sector structure over successive solar rotations identified on the left by year (74) and Bartels Rotation Number (1914). The right side shows the year and the latitude of Pioneer 11 which traveled from 5 to beyond 10 AU in this interval. Negative polarities are identified by minus signs and positive polarities by filled symbols. The relatively few blanks denote mixed or indeterminate polarities. In spite of the large separation in distance and varying latitudes, there is a close correspondence in the two sector structures. The lower half figure (1978–1982) covers the period of solar maximum. There is little if any disruption of sector structure when solar activity is at a maximum and CMEs are commonly occurring. (Smith, Slavin, and Thomas, 1986)

This condition is successfully reproduced by the potential field source surface models that show the open fields originating primarily from low latitudes at maximum and high latitudes at minimum. They are also able to reproduce the increasing inclination of the neutral line/current sheet as minimum turns into maximum (Hoeksema, 1992). However, Ulysses was again uniquely positioned to extend knowledge of the HMF and relate it to the changes in the solar magnetic field. The high-latitude observations have, in fact, enabled studies of the polarity changes in the polar cap fields, the inclination of the HCS, and the three-dimensional distribution of open magnetic flux.

4.7.3 Magnetic dipole and polarity reversal

The spiral angle provides a reliable measure of the field polarity. It was monitored continuously as Ulysses ascended toward the Sun's south polar cap during the ascending phase of solar activity between 1997 and 2000 (Smith *et al.*, 2001). The lower panel of Figure 4.29 is a plot of ϕ_B during this interval that shows the presence of sector structure from the equator to 80°S and 70°N. By contrast, only a single sector was observed above ~30° north and south during solar minimum (compare Figures 4.3 and 4.5).

Other solar structures of importance are prominences, magnetic loops that appear on the solar limb that are stable over long intervals although they can occasionally erupt as transients. When seen on the solar disk, they appear dark because of their lower temperature compared with the underlying photosphere and are referred to as filaments. At solar minimum, a ring of prominences (Zirin, 1988; Priest, 1982) surrounds the polar cap or "crown". Observations of the south polar crown prominences during Ulysses polar passage indicated that the polar cap polarity had not yet changed and did not reverse until later in the cycle, between 2002.31 and 2002.46 (Harvey and Receley, 2002). Thus, the Ulysses and these solar observations are consistent. However, some solar observers had inferred that the polarity had already reversed prior to the arrival of Ulysses in the polar cap and a PFSS model indicated the sector structure should have disappeared around 45°S both being contrary to observation. This example illustrates the importance of *in situ* measurements and the shortcomings of attempts to observe the polar caps remotely from in the ecliptic as well as the inadequacy of PFSS models at higher latitudes.

Another interesting observation was the absence of a polar coronal hole at this time and the disappearance of the polar cap fast wind (McComas, Elliott, and von Steiger, 2002a). The study of the polar cap field reversal that agreed with the Ulysses observations also shows that the polar coronal holes typically disappear 1.1 to 1.8 years before the reversal occurs (Harvey and Receley, 2002).

Ulysses returned equator-ward and proceeded to the north polar cap between November 2000 and October 2001 (the "Fast Latitude Scan"). The measured ϕ angle is shown in the upper panel of Figure 4.29 (Smith *et al.*, 2003). Above 70°N, a positive unipolar sector was observed that persisted up to and over the polar cap. According to Harvey and Receley (2002), the reversal was observed between 2001.19 and 2001.34 in agreement with Ulysses observations. The north polar coronal hole

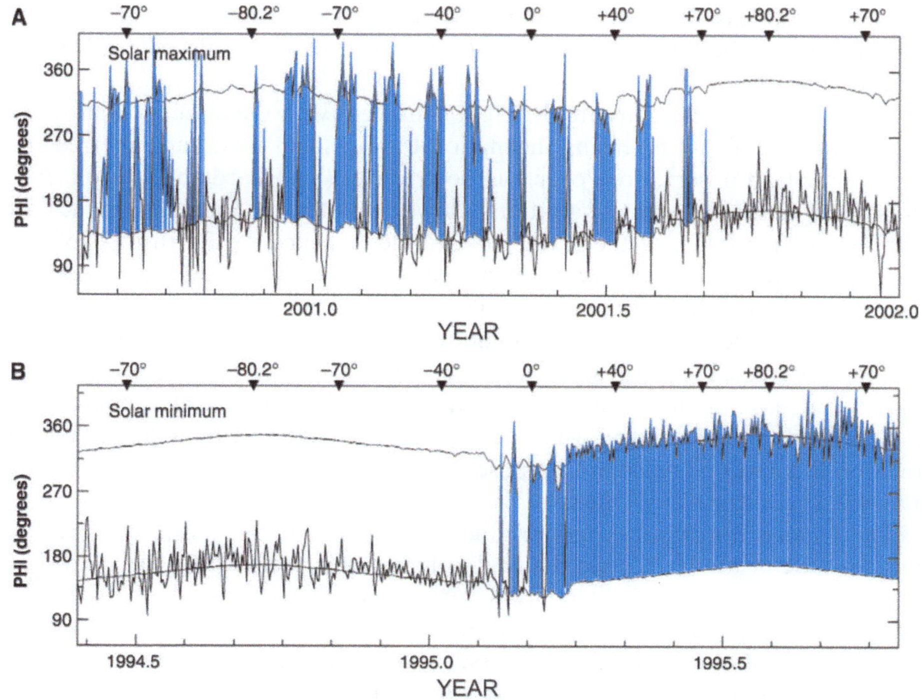

Figure 4.29. The spiral angle as a function of solar latitude at solar minimum and maximum. The angle, $\phi_B = \text{PHI}$, is plotted versus time (shown along the lower scales) and Ulysses latitude (along the upper scales). The lower panel contains data acquired during the minimum phase superposed on Parker spirals corresponding to inward (white) and outward (blue) polarities. A general agreement with the Parker spiral is obvious throughout both phases of the solar cycle. During minimum, only a single polarity is seen below about $-20°$ and above $+20°$ indicating the inclination of the HCS was low so that crossings only occurred at low latitudes. During solar maximum, however, polarity reversals (current sheet crossings) extend to almost $-80°$ although the negative polarity of the south polar cap was observed for a solar rotation. Current sheet crossings then occurred from $-75°$ to $+70°$ with the polarity in the north polar cap having reversed to inward rather than outward (compare the white region at maximum and the blue region at minimum). These results confirm the presence of the current sheet during solar maximum with a very large inclination to the solar equator (the HCS is nearly aligned with the Sun's rotation axis). (Smith *et al.*, 2003)

had reappeared by 2001.4 and Ulysses found fast wind in the polar cap in association with the negative polarity as expected (Smith *et al.*, 2003; McComas *et al.*, 2002b, 2003).

4.7.4 Inclination of the HCS and solar dipole

Figure 4.29 contains other information than the change in polar cap polarity. The sector structure is present at essentially all latitudes in the southern hemisphere and extends up to at least 70° in the north hemisphere. Since the abrupt changes in sector polarity identify the HCS, the current sheet is seen to be highly inclined in both hemispheres.

The Ulysses observations are therefore consistent with the discussion above and earlier evidence of the increasing inclination during solar maximum inferred from geomagnetic field observations over four solar cycles (Svalgaard and Wilcox, 1975). Actually, although the annual variations showed that the HCS inclination increased following solar minimum, the restricted latitudes reached by Earth (±7.25°) precluded determining the inclination at solar maximum. The method depended on a difference in the two polarities at higher latitudes but that difference became increasingly smaller as the current sheet inclination increased. However, as source surface modeling progressed through solar maximum, the inclination of the neutral line was found to increase monotonically toward a limit of 70° above which photospheric magnetic field observations on which the models were based were no longer reliable. However, there was no reason to suppose the inclination did not continue to increase up to 90° as confirmed by Ulysses.

The Ulysses north polar pass produced two more determinations of the maximum latitude of the HCS since the field polarity had reversed and the north polar coronal hole reappeared. The two Ulysses orbits provide a "calibration" of the PFSS predictions throughout most of the solar cycle (Smith, 2006). Figure 4.30 is a plot of the PFSS inclinations on which the Ulysses observations of maximum current sheet latitude are superposed. In general, the agreement is reasonable except near solar maximum when the predicted values differ from the observations by many tens of degrees. An additional conclusion that followed from the comparison was that the "classic" Stanford model agreed better than the "radial" model (both available at *http://quake.stanford.edu/~wso*).

Although spacecraft in the solar wind observe only open fields from the Sun, it is natural to try to relate the observations to the underlying photospheric fields. The high inclination of the neutral line/current sheet can be explained by referring to a resultant solar dipole consisting of an axial and an equatorial component. As described above, at solar maximum, the axial (polar) fields are decreasing while the equatorial fields are simultaneously increasing. The equatorial component can be viewed as the resultant of the several magnetic dipoles associated with sunspots that are distributed over the solar surface (Wang, Lean, and Sheeley, 2000). In general, these vector dipoles add to yield a resultant component. During solar maximum, the increasing equatorial dipole and decreasing axial dipole constitute a dipole that rotates equator-ward, and the HCS, as the magnetic equator, rotates pole-ward.

This simple picture has won wide acceptance and is often extended to explain the change in polarity of the polar caps by invoking a rotation of the resultant dipole through 180°. The Ulysses data have, in fact, been interpreted in this way by using the width in longitude of the two sectors and the latitude of the spacecraft to infer the dipole inclination (Figure 4.31, Jones, Balogh, and Smith, 2003). Additional support for this phenomenological model is the increase in the HMF magnitude by a factor of about 2, as though measuring a dipole field first at the equator and then at the pole. In

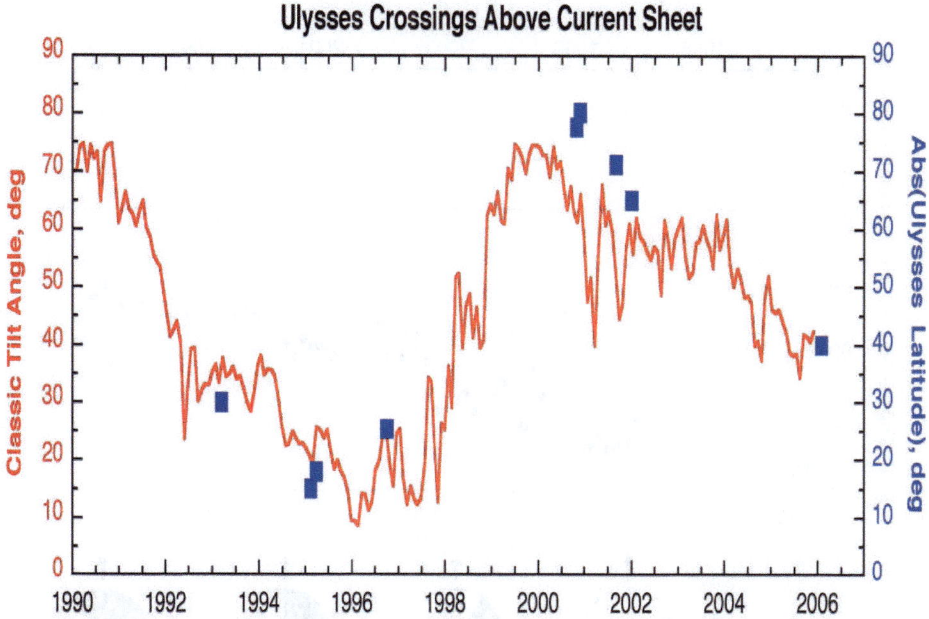

Figure 4.30. Ulysses crossings of the HCS at the highest latitudes compared with the inclinations predicted by a potential field source surface model. The tilt angle of the current sheet obtained from the Wilcox Solar Observatory "classic" model is plotted in red. The blue squares show the latitudes of the last (first) crossings of the HCS as Ulysses ascended (descended) in latitude. The interval, 1990 to 2006, begins at solar maximum, continues into solar minimum, the succeeding minimum and maximum—that is, a Hale cycle during which the polarity of the magnetic dipole reversed. In general, the agreement between the observed "tilts" and the predicted tilts (essentially, the maximum latitude attained by the potential field neutral sheet each solar rotation) is satisfactory but better at minimum than maximum (in fact, the model has difficulty correctly predicting sector structure at latitudes above 45° and when the polarity reversal occurs). The alternative "radial" model agrees less well with the Ulysses observations. (Smith, 2007)

addition, there is evidence that the source surface dipole tends to occur at a "preferred" solar longitude (Neugebauer *et al.*, 2000).

It is difficult, however, to reconcile this simple model with the observed behavior of photospheric magnetic fields most of which are closed. The changes in surface and polar cap fields during solar maximum were formulated long ago by Babcock (1959), developed soon after magnetograph observations first became available. It has withstood the test of time extremely well and is still generally accepted (e.g., Foukal, 1990).

The model exploits magnetic "annihilation", "reconnection", or "merging" of adjacent fields with opposite polarities. Figure 4.32 shows the Sun and various magnetic regions during the erosion of the polar cap magnetic fields (and, in principle, shows how the polar caps reverse polarity). The polar caps are positive in the north and negative in the south. The smaller areas containing + and −

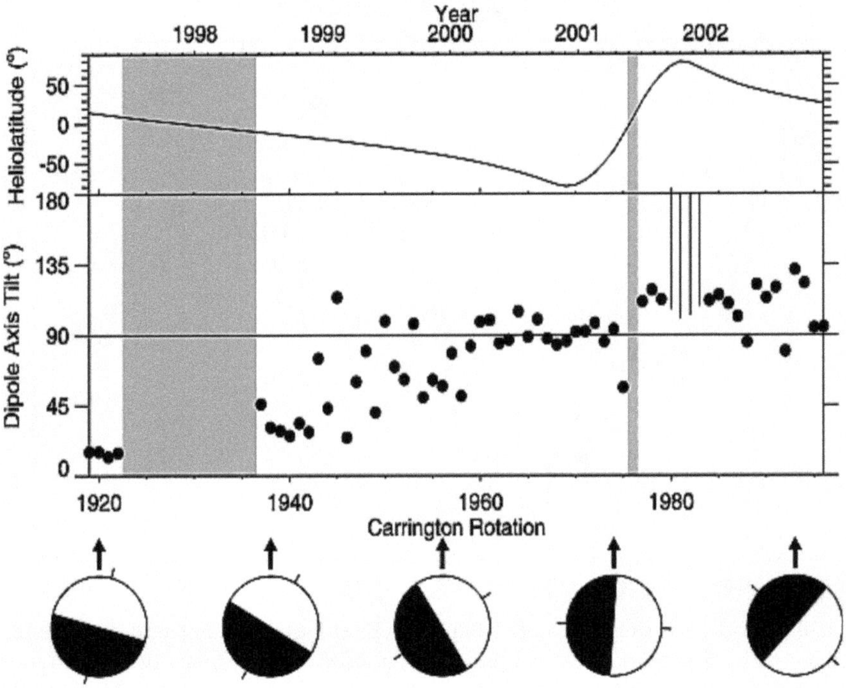

Figure 4.31. Evidence of solar dipole rotation during the reversal in polar cap field polarities. The upper panel shows the latitude of Ulysses during the five years surrounding the recent solar maximum (the year is shown along the top scale). The middle panel contains estimates of the dipole tilt angle obtained from magnetic field measurements by comparing the relative duration of positive and negative sectors during successive Carrington rotations (numbered along the lower scale). The angle begins near 15° (the dipole almost aligned with the rotation axis) gradually increases to 90° (the dipole perpendicular to the solar axis or the HCS almost "vertical") and then proceeds to values >90° corresponding to a reversed polarity. The interpretation is given in the schematic at the bottom. The rotation axis is diagrammed as an arrow and the polarity (positive is white and negative is black) gradually reverses. The boundary between white and black is the HCS that rotates from an equatorial to an axial orientation. This figure shows how the magnetic dipole, in effect, rotates as observed in the heliosphere far from the Sun. In fact, the changes in solar magnetic fields are complex and a magnetic dipole does not exist that simply rotates through 180°. (Jones, Balogh, and Smith, 2003)

polarities develop from sunspot magnetic fields that appear at mid-latitudes with positive polarities in "leading" spots and negative polarities in the "trailing" spots. The pair of "unipolar" magnetic regions gradually grow and drift in opposite directions. The trailing unipolar regions move pole-ward while the leading regions drift equator-ward. When the sunspots appear at the beginning of a new solar cycle, the polarities of the two regions are as shown—in particular, the pole-ward traveling regions have the opposite polarity to the polar caps. The equator-ward regions have

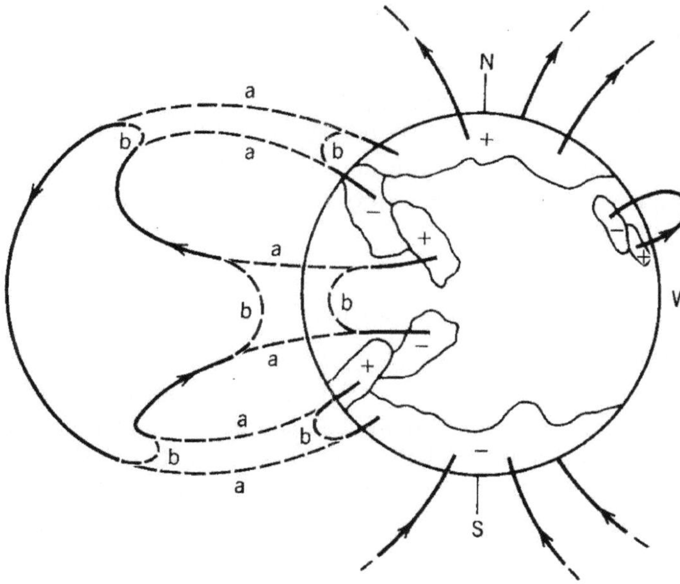

Figure 4.32. The Babcock model and the reversal of the polar cap magnetic polarities at solar maximum. According to this model, the erosion of the polar cap field occurs because magnetic fields originating in the trailing regions of sunspots gradually drift pole-ward. These fields have the opposite polarity of the polar cap fields and neutralize the polar cap fields by magnetic merging or reconnection of the oppositely directed fields ultimately eliminating the polar cap. The trailing fields continue to arrive and regenerate the polar cap fields with the reversed polarity. The leading regions of the sunspots drift toward the equator and neutralize each other. (Foukal, 1990)

opposite polarities in the north and south hemispheres. When the unipolar regions reach the polar caps, they gradually neutralize them, decreasing the area until the polar cap vanishes. As more unipolar regions arrive from lower latitudes, the polar caps reform but with reversed polarities. The process has been reproduced successfully using a computer model (Wang, Nash, and Sheeley, 1989).

The unipolar regions on opposite sides of the equator gradually approach each other and reconnection forms magnetic loops and then eventually erodes the two unipolar regions. The left side of the figure shows magnetic fields that originally extended between polar caps and between the unipolar regions. The letters show where reconnection occurs so that the connecting dashed field lines (a) are replaced by a pair of loops (b). The outermost loop no longer connects to the Sun and is lost into space (presumably as a CME).

In this model, there is no magnetic dipole that simply rotates along the solar surface from one hemisphere to the other. The polar cap fields and polar crown prominences are actually present throughout most of solar maximum. Another problem for the rotating dipole interpretation is that the magnetic poles do not disappear and reappear at the same time. Characteristically, one polar cap field disappears first and can then reappear with the opposite polarity before the other polar cap changes. For example, the recent sequence described above shows that the north polar cap changed polarity between 2001.19 and 2001.34 while the south polar cap changed sign between 2002.31 and 2002.46. There is no doubt, however, that from a heliospheric point of view, the open field lines appear to behave as though they were derived from a rotating dipole without any additional complications being needed.

4.7.5 The radial component at solar maximum

Ulysses observations near the recent solar maximum provided the opportunity to investigate the latitude dependence of the radial component and compare its behavior with that found at solar minimum. Figure 4.33 is $r^2 B_R$ as a function of latitude at minimum and maximum (Smith and Balogh, 2003). In spite of an increase in the variability of the individual measurement averages, $r^2 B_R$ is still essentially independent of latitude at maximum as well as at minimum. Furthermore, the average value of $r^2 B_R$ averaged over the two intervals is virtually the same at both phases. The total open flux appears to be unchanged.

The absence of a latitude dependence of $r^2 B_R$ at maximum is even more surprising than at minimum. The increased complexity of the solar field at all latitudes and the weakness of the polar cap fields might have been expected to eliminate the simple structure resulting from the dynamics and over-expansion associated with the strong dipole field prevailing at minimum. For example, the magnetic fields in active regions are typically much larger than the polar cap fields and it might have been supposed that the open flux would be enhanced at low latitudes.

Such is not the case and the interpretation is the same as that used to explain the field configuration at minimum. The strong localized fields evidently dominate the pressure of the plasma and cause an expansion that leads to a uniform distribution of flux and eliminates the strong gradients in magnetic pressure (Smith and Balogh, 1995; Suess *et al.*, 1996).

Important implications follow from these observations. The magnetic field and solar wind must be expanding non-radially near the Sun. Within a few solar radii, the flow outward into the heliosphere will become radial. However, the non-radial part of the expansion is important—for example, in attempts to find the source regions of solar wind observed by spacecraft (e.g., Neugebauer *et al.*, 2002).

Potential field source surface models are one of the means of describing the behavior of solar wind magnetic fields near the Sun. Such models ignore currents and stresses below and in the corona—a serious restriction that is now being corrected by MHD models. Nevertheless, how well do they describe the over-expansion? The models have been used to derive the total open flux for comparison with measurements of B_R, a reasonable test of their validity (Wang, Lean, and Sheeley, 2000). A reasonably good correlation has been found. The reason is that the models characteristically lead to an over-expansion of fields that reach the source surface. The source surface boundary condition actually causes some photospheric fields to

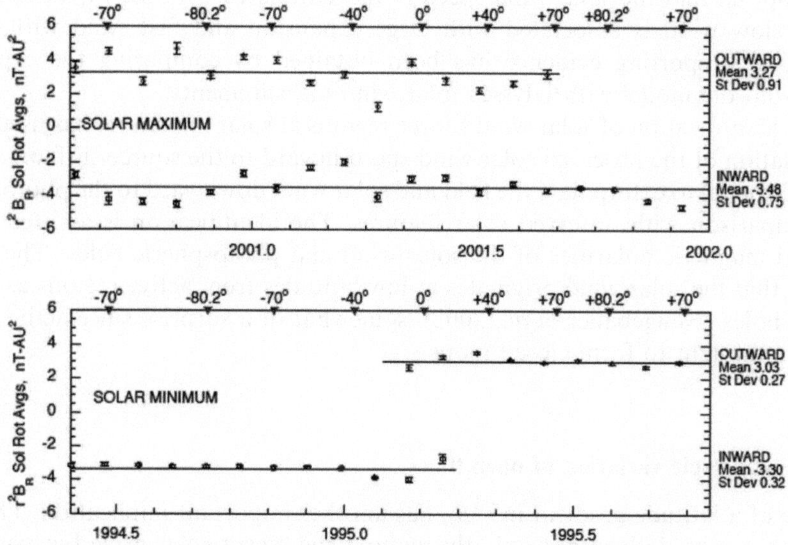

Figure 4.33. The open magnetic flux as a function of latitude and time (solar minimum and maximum). The figure contains solar rotation averages of open flux ($r^2 B_R$) measured by Ulysses as it ascended to the south polar cap (latitude is shown along the upper scales), descended to cross the equator, and then ascended to the north polar cap. In the bottom panel, the data were obtained at solar minimum and data obtained at maximum appear in the top panel. The data points are shown with bars representing standard deviations. Averaging was carried out for the two polarities when both were present. Positive $r^2 B_R$ (outward) is plotted above negative values (inward). At minimum, as also seen in Figure 4.28, unipolar fields are present at mid- to high latitudes. At maximum, both polarities are often present simultaneously because the HCS is highly inclined. Mean values north and south and at minimum and maximum are shown by the solid lines parallel to the horizontal axes and are displayed in the right-hand column. Two important conclusions to be drawn are that $r^2 B_R$ is independent of latitude at both minimum and maximum and has very nearly the same values at minimum and maximum. (Smith and Balogh, 2003)

become radial or nearly radial well below the source surface (at altitudes of $\sim 1\ R_\odot$). The field caused by the source surface currents not only cancels the non-radial components there but reduces them significantly and causes the fields to become more radial between the source surface and the photosphere as required by the Ulysses observations.

Only a relatively small fraction of photospheric fields reach the source surface and they have generally expanded by more than the increase in surface area between the photosphere and the source surface. The "expansion factor", introduced by Wang and Sheeley (1990), is the ratio of the radial field strengths at the source surface and the photosphere compared with a geometrical decrease of r^{-2}. For many fields, the expansion factor exceeds 1 but factors less than 1 also occur (i.e., the field is compressed rather than over-expanded). One reason for interest in the expansion is their proposal that the solar wind speed is anti-correlated with the expansion factor so that slow wind is associated with large expansion and fast wind with under-expansion. Supporting evidence has been obtained by comparing the expansion factor from the model with Ulysses solar wind measurements.

The identification of solar wind source regions at solar maximum has used radial extrapolation of the observed solar wind speed inward to the source surface, a PFSS or MHD model to extrapolate the field and solar wind downward to the photosphere, and comparison with observed solar features. The identification is assisted by the observed magnetic polarities of the solar wind and photospheric fields. The results indicate that the solar wind originates at low latitudes from active regions as well as coronal holes (Neugebauer *et al.*, 2002), somewhat of a surprise since active region fields are thought to form closed loops.

4.7.6 Solar cycle variation of open flux

Absence of a latitude gradient in $r^2 B_R$ has another important implication. The total open flux can be derived not only throughout the recent solar cycle but past solar cycles from measurements of B_R at any latitude including in-ecliptic measurements obtained over a much longer time interval. That makes it possible to examine a proposed invariance of the total open flux with time (Fisk and Schwadron, 2001).

The Ulysses measurements of $r^2 B_R$ in Figure 4.33 indicate that the average value was essentially the same at maximum and minimum. The timing of these observations coincided with a prediction that the open flux was likely to be invariant. This proposal was based on a model of the solar wind and solar magnetic field developed by Fisk and Schwadron (2001) that emphasized the "diffusion" of magnetic field lines in the photosphere by reconnection of already open fields with adjacent closed-field lines. Since the reconnection produced another open-field line and another closed-field line, the total of each was conserved or invariant.

The Ulysses observations appeared to support this view causing more interest in this possibility (as well as motivating opposition to the idea from other investigators). A preliminary study of open-flux invariance using both Ulysses and in-ecliptic measurements of B_R through 2000 showed that B_R generally varied but by much less than a factor of 2 (Smith and Balogh, 2003). Furthermore, the Ulysses observations during solar maximum may have occurred fortuitously when the open flux happened to be unchanging. Certainly, the open flux varies by much less than the total closed flux that changes by an order of magnitude from maximum to minimum. The open flux is only about 10% of the total flux at maximum but becomes about one-half of the total flux at minimum.

The invariance of the open flux and its relation to B were re-investigated recently as several more years of Ulysses observations became available. The additional Ulysses data are consistent with the simultaneous in-ecliptic measurements of B_R at 1 AU to within statistical uncertainty. Therefore, the in-ecliptic observations were extended into the more recent interval to investigate how the open flux has changed (Figure 4.34; Zhou and Smith, in preparation).

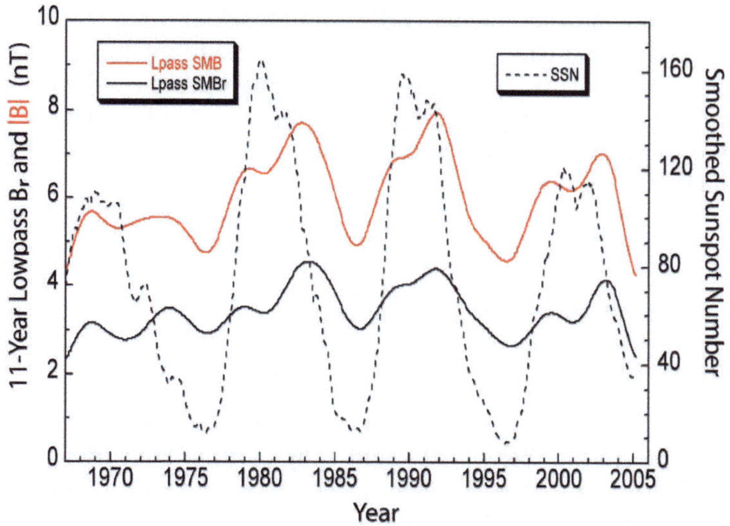

Figure 4.34. The absence of a dependence of $r^2 B_R$ on latitude means that the long series of in-ecliptic measurements represent how the open flux has varied over the last 3.5 cycles. The figure contains $|B_R|$ (black), B (red), and sunspot number (SSN, dashed) after low-pass filtering to enhance the long period variations. Both $|B_R|$ and B correlate with the smoothed sunspot numbers in a somewhat unexpected way. They have minimum values near solar minimum, and two distinct increases on either side of sunspot maximum. The interval between the two increases coincides closely with sunspot maximum (and the disappearance of the polar cap fields) and the largest maximum occurs during the descending phase of solar activity when the dipole reappears and reaches maximum strength. B and $|B_R|$ are highly correlated which would be expected on the basis of Parker's model but does not show an obvious contribution from CMEs. However, the latter also vary with the solar cycle and their effect may be suppressed by filtering. (Zhou and Smith, 2007)

During the recent solar maximum, B_R was found to vary systematically over the past three cycles: B_R is low at minimum, increases toward maximum, decreases again near maximum, and then increases to its highest value before declining toward the next minimum. In view of this cyclic variation, the Ulysses measurements at high latitude were obtained accidentally during two intervals of nearly equal B_R. Thus, although the open flux is not strictly an invariant, it is relatively constant and changes much less than the total flux. The physical argument advanced by Fisk and Schwadron (2001) appears to have merit in explaining why the open flux tends to be so constant.

This cyclic pattern appears to follow the variations in the solar magnetic field including the changes in the polar cap field or axial magnetic dipole. The initial increase mimics the increase in the equatorial fields and their moment as the sunspots reappear and increase in number. The secondary dip near maximum is attributable to the disappearance of the polar cap fields and their recovery. The following increase to a maximum of B_R occurs when the axial dipole is gaining strength and the "equatorial" fields (and the unipolar regions approaching the polar caps) are still strong although decreasing in strength. Finally, the equatorial moment decays away and only the axial dipole is left as the sole source of open flux. These changes are accompanied by corresponding variations in the inclination of the HCS as described above from low to high inclination and a return to a low inclination at the following solar minimum.

4.7.7 Solar cycle variations in field magnitude

According to Parker's model, the field magnitude is derived from B_R. To what extent is this true in view of the solar cycle variations in B_R? It has often been suggested that CMEs make a significant contribution to B especially at solar maximum. Figure 4.34 also addresses the relation between B_R and B by plotting both over the past three solar cycles. There is an obvious high degree of correlation. A large correlation coefficient of 0.96 quantifies the excellent agreement. The ratio between B_R and B, a nearly constant factor of 2, is consistent with the Parker model when allowance is made for the contribution made by the continual presence of large fluctuations in the three components. (The fluctuations appear in the field magnitude plotted in the figure because it is the average of the instantaneous field magnitudes computed from the sums of the squares of the components. Other investigators often average the components first and then compute the average magnitude—that is, "the magnitude of the averages" rather than the "average of the magnitudes". Both approaches have advantages and which is preferred is a matter of choice but it is important to know which choice has been made.)

The magnitude of the fluctuations can be derived from the observed $B/B_R = 2$ assuming the Parker relation between B_R and B_T. The Parker equation, $B_T = -(\Omega r/V)B_R = -B_R$ at 1 AU where $\Omega r/V \approx 1$. Then, $B^2 = (B_R + \delta B_R)^2 + (B_T + \delta B_T)^2 + (\delta B_N)^2 = B_R^2 + B_T^2 + \sigma^2$ where σ^2 is the sum of the squares of the variations in the three components. Hence, $\sigma^2/B^2 = 1/2$ so that the fluctuations in the three components are a large fraction of the field magnitude and are comparable with the power in the two components when averaged over a solar rotation.

An interesting aspect of the close correlation is the apparent absence of a significant contribution to B from CMEs. This issue has been of interest for some time with some investigators anticipating that the increased rate of occurrence of CMEs at maximum would present a problem by increasing the magnetic flux in the heliosphere by a large amount. The figure shows no such large increase and, if the increase near maximum is attributed solely to CMEs, it is still modest. Although the fields within CMEs tend to be stronger than those of the surrounding solar wind, they could be closed internally and disconnected from the Sun. As such, they would make no lasting contribution to the open flux and a "flux catastrophe" need not be of concern. Alternatively, the solar cycle variation in B has been attributed solely to changes in the rate of occurrence of CMEs since it follows the solar cycle (Owens and Crooker, 2006). A characteristic time constant limiting the connection of the CMEs

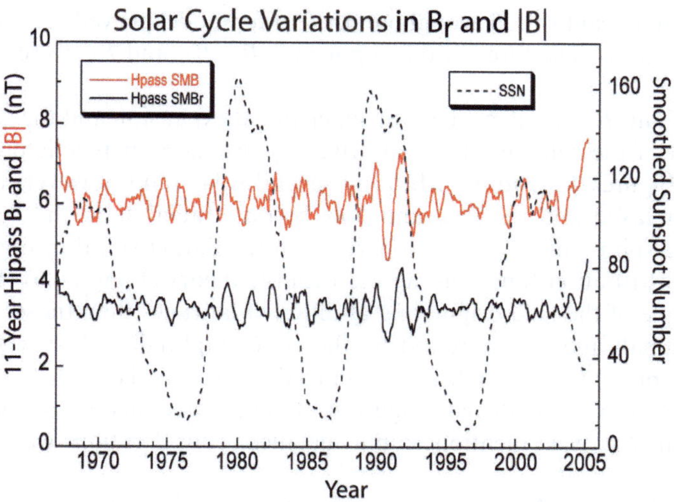

Figure 4.35. Quasi-periodic variations in B_R and B over 3.5 sunspot cycles. The radial component and field magnitude averaged over 27 days (a solar rotation) have been high-pass filtered to eliminate the larger and longer period solar cycle variations seen in Figure 4.33. For comparison, the dotted curves show the sunspot numbers. Both B_R and B exhibit the quasi-periodicities and the variations are highly correlated. Close inspection of the signal reveals that the variations are not constant in period or amplitude. When they are subjected to a new type of analysis, empirical mode decomposition, they are found to be a superposition of six distinct modes including those with quasi-periods of 155 days, 1.7 years, and two harmonics of 11 years. These quasi-periods have been identified before in various solar–heliospheric parameters but not simultaneously or in a single parameter. (Zhou and Smith, 2007)

to the Sun of $\sim$50 days has been introduced in order to limit the buildup of magnetic flux.

An incidental feature of interest appears in Figure 4.35—namely, an apparent periodicity in B_R and B. A periodicity has been observed intermittently in the solar wind speed and HMF with a period of $\sim$150 days. However, the apparent quasi-period in B_R and B in Figure 4.34 is much longer—approximately 1.7 years (Smith, Zhou, and Ruzmaikin, in preparation). The variations are not very regular but vary in amplitude and period. A preliminary analysis using empirical mode decomposition indicates that the "signal" is a superposition of several quasi-periodic modes including 150 days and 1.7 years.

4.8 SUMMARY—SOLAR CYCLE VARIATIONS

Section 4.1 provides a global description of the heliospheric magnetic field emphasizing latitudinal dependences and comparisons with the Parker model. This view corresponded to observations obtained at solar minimum and avoided consideration of solar wind structure by emphasizing higher latitudes above those in which structure plays a significant role. Throughout, the Parker model proved a useful diagnostic of the observations. The three field components, B_R, B_T, and B_N, were considered in turn.

Both B_R and B_T reveal evidence of super-radial expansion and the equator-ward displacement of the field near the Sun driven by the excess magnetic pressure in the polar caps (Figures 4.3 and 4.7). Beyond several solar radii however, the magnetic stresses are relaxed and the field and flow become radial as in the Parker model. The observed spiral angle, when compared with the Parker spiral through the use of probability distribution functions, agrees with the theory (Figure 4.5). The effect on the spiral angle of the field originating at different latitudes than those at which it is observed is small and is suppressed by the large-amplitude Alfvén waves that are continuously present at high latitude. Averaged from several days to a solar rotation period, B_N is zero as predicted (Figure 4.8). The possibility of periodic or other deviations in B_N was mentioned but detailed consideration was deferred to Section 4.4.

Sections 4.3 and 4.4 presented a more comprehensive view of HMF properties during solar minimum. The various effects of solar wind structure on the Ulysses observations were considered without which a description of the HMF at solar minimum would be incomplete. The emphasis is on the open solar fields described by the Parker model, and the important role of coronal mass ejections is only mentioned briefly for the sake of completeness. Again, the behavior of B_R, B_T or the spiral angle, and B_N was considered. The tilt angle between the solar magnetic dipole and the rotation axis causes the sector structure to appear in both B_R and B_T and in the spiral angle (Figure 4.9). At the solar/heliospheric magnetic equator, the oppositely directed field lines from the north and south hemispheres are separated by the heliospheric current sheet that is embedded in the heliospheric plasma sheet. The existence of the sector structure has resulted in attempts to relate the HMF to measured magnetic fields in the solar photosphere through the use of potential field source surface (PFSS) models (Figure 4.11). Such models lead to a neutral line on the solar wind "source surface" that is identified with the HCS in addition to providing estimates of the field strength at higher latitudes. The excursions of Ulysses in latitude tested the model and various comparisons are presented. In addition, Ulysses observations provide details of the HCS/HPS such as their respective thicknesses, which are not included in PFSS models.

A major topic was the interaction of fast solar wind from high latitudes with slower wind from low latitudes resulting from the tilted dipole (Figure 4.15). The properties and structure of corotating interaction regions or CIRs were discussed in considerable detail because of their effect on the HMF, the development of shocks, and evidence that energetic particles accelerated at the CIRs were able to access higher latitudes because of departures from the Parker model.

Large departures of the observed angle from the Parker spiral of tens of degrees occur inside corotating rarefaction regions (Figure 4.23). These departures are explainable by a model that allows the solar wind speed to vary along the field line as it moves through a polar coronal hole, an effect not contemplated in the original Parker model but one that can be incorporated (Figure 4.24).

Another departure from the Parker model is considered: the dependence of B on radial distance. Ulysses and the Pioneer and Voyager spacecraft that travel farther distances into the heliosphere provide evidence of a more rapid decrease in B with increasing distance than predicted by the model (Figure 4.11). Theoretical arguments support such a "flux deficit" although the observations have proven controversial when sought in Voyager data.

The presence of a north–south asymmetry in the heliosphere at solar minimum was discussed in Section 4.5. The asymmetry involves a displacement of the magnetic equator/the cosmic ray equator/the HCS southward by about 10° (Figure 4.26). The Ulysses and in-ecliptic observations by the WIND spacecraft provide convincing evidence of such an "offset", which was inferred prior to Ulysses from studies of how the sector structure varied annually as Earth traveled between ±7.25° in latitude over the solar cycle.

Finally, some properties of CMEs and their internal magnetic fields including observations by Ulysses at solar minimum were discussed briefly in Section 4.6.

Section 4.7 was devoted to the unique observations carried out by Ulysses during solar maximum. The sector structure persists throughout the change to maximum solar activity and the decay during the descending phase (Figure 4.28). However, the sector structure is changed significantly in that it extends all the way to the polar cap (Figure 4.29). In other words, the HCS rotates from being nearly equatorial to being nearly aligned with the Sun's rotation axis (Figure 4.31). The maximum latitude reached by the HCS provided another opportunity to test or calibrate PFSS modeling, this time during solar maximum and the agreement was found to be much less satisfactory (Figure 4.30). In fact, the models showed that the polarity of the south polar cap field had reversed long before it actually occurred.

A special effort was made to follow the apparent rotation of the solar dipole and the reversal in the dipole polarity or the sign of the polar cap fields. The Ulysses orbit did not prove optimum for such observations and the south polar cap still had not reversed when Ulysses reached its highest latitude and the polarity had already reversed by the time Ulysses reached the north polar cap (Figure 4.29). Nevertheless, the Ulysses observations were consistent with the timing of the polar cap reversals based on the disappearance and reappearance of polar crown prominences. Although the reversals seem simply to result from a rotating dipole as seen in the heliosphere, it was shown that the Sun's surface magnetic fields behave very differently than such a model would imply (Figure 4.32). There is no persistent dipole that simply rotates from one pole through the equator to the other pole.

The open flux, given by $r^2 B_R$, was found to be independent of latitude at solar maximum (Figure 4.33). Although the configuration of the solar fields was significantly changed with strong sunspot magnetic fields dominating low latitudes and the polar cap fields decreasing in strength, the observations show that magnetic pressure gradients were still driving non-radial solar wind flow and the non-radial expansion of the open magnetic fields until a uniform field distribution was produced near the Sun.

The average value of $r^2 B_R$ was very near the same as at minimum. That supported the theoretical suggestion that total open flux was an invariant. Comparison with B_R measured in the ecliptic over 3.5 sunspot cycles showed that the agreement with Ulysses measurements was accidental and that, in general, B_R did vary over the solar cycle but by less than a factor of 2 (Figure 4.34).

In addition to B_R, the solar cycle variation of B was studied (Figure 4.34). Both B and B_R are highly correlated as implied by the Parker model. Averages over a solar rotation show little, if any, evidence that CMEs affect B_R or B, the latter a puzzling result. In addition to systematic slow variations over the solar cycle, B_R and B exhibit variations of $\sim$150 days, 1.7 years, and harmonics of 11 years that emerge from a new type of analysis that separates a data stream into a number of quasi-periodic modes of oscillation (Figure 4.35).

With these considerations as background, the solar cycle variation of the different components and aspects of the HMF can now be summarized as follows:

1 Beyond several solar radii, the radial component, B_R, is independent of latitude in both minimum and maximum phases. However, the magnitude of B_R varies systematically with the solar cycle. At 1 AU, it appears to return to a value of 3 nT at successive minima and then increases as the source shifts from polar and low-latitude coronal holes to sunspot fields and the active regions and unipolar magnetic regions that evolve from them. The continued contribution of polar cap fields is manifested by secondary minima in both B_R and B when the polar cap fields disappear and then reappear with the opposite polarity. The increasing strength of the polar cap fields over-compensates for the declining influence of fields associated with sunspots as their number declines and this results in maximum values for B_R and B during the declining phase of solar activity. Because the field magnitude derived from Ulysses measurements is the average of instantaneous magnitudes (rather than the magnitude of averages taken over the same time interval), it appears indirectly that the contribution to B from ever-present magnetic fluctuations also waxes and wanes with the solar cycle.

2 Studies of the spiral angle at solar minimum, supplemented by similar investigation of $\phi_B - \phi_P$ at solar maximum, show a close agreement in both phases. The spiral angle is only affected slightly by either non-radial flow near the Sun or solar cycle variations in solar wind speed. Therefore, the solar cycle has little effect on the spiral angle. (At solar minimum, large departures from the Parker spiral are observed on the scale of a solar rotation in corotating rarefaction regions.) Curiously, at one time, a solar cycle variation was believed to be present as a departure of the angle between the opposite polarity fields from 180°. However, when the effect of CMEs was removed from the magnetic field measurements, the effect disappeared.

3 The same conclusion—that there is little if any effect of the solar cycle—also applies to the north–south component, B_N, or the equivalent angle, δ_B. Averages of B_N over fairly long intervals such as days invariably yield a null result whether at solar minimum or maximum. Consistent with the statement made above regarding the solar cycle variation of the magnetic field fluctuations, an early study showed that fluctuations in B_N are largest at solar maximum.

4 The largest variation of all the HMF properties is the change in current sheet inclination. From one minimum to the next, the HCS effectively rotates through 180°. At minimum, the HCS has its lowest inclination and then gradually rotates to higher latitudes during the ascending phase until it is essentially aligned with the Sun's rotation axis at maximum. Since the sector structure endures throughout the solar cycle, the HCS is also continuously present.

5 An alternate interpretation is based on the solar magnetic dipole and the magnetic poles rather than the magnetic equator/current sheet. The magnetic dipole can be considered as the resultant of an axial and an equatorial dipole with the equatorial dipole vanishing at minimum and the axial dipole vanishing at maximum. The axial dipole is associated with polar cap fields while the equatorial dipole is the resultant of the magnetic poles associated with sunspot fields. The changes in the two dipoles are out-of-phase with one growing while the other is decreasing leading to the apparent rotation of the resultant.

6 The Ulysses results reveal the variation in inclination over the solar cycle in terms of the highest and last crossing of the HCS independent of PFSS models. Studies of annual variations of the in-ecliptic sector structure are unable to determine the inclination when the HCS becomes highly inclined. The duration of the positive and negative sectors become equal within statistical error and information on changes at high inclination becomes unavailable. Basically, both the Ulysses observations and PFSS modeling lead to a good correlation between inclination and sunspot number.

7 The large-scale structure of the HMF varies significantly with the solar cycle because it is correlated with solar cycle changes in the fast–slow solar wind. In the absence of a dipole tilt, the fast wind would be confined to high latitudes and the slow wind to low latitudes without an interaction. That proposition ignores irregularities in the shape of polar coronal holes that can depart from a well-defined polar cap to gross changes in shape such as long channels/lanes that lead from high latitudes to the equator. Such configurations are often seen near the descending phase and allow fast wind to interact with slow wind in and near the ecliptic. They are partly responsible for a maximum in geomagnetic activity (magnetic storms) during the descending phase rather than at maximum. Another contributor to the increase in geomagnetic storms at this time is the concurrence of two effects, the increasing area of the polar coronal holes and the decreasing tilt of the current sheet. This conjunction occurs over an interval of a year or two to produce enhanced structure and activity at low latitudes. The formation of CRRs and CIRs is also favored by this configuration including the development of higher pressures at the stream interfaces within CIRs.

8 As solar minimum approaches, the low inclination of the HCS (small tilt of the dipole) tends to keep fast and slow wind separated and stream–stream interactions and CIRs become weak. The gradual disappearance of fast wind as solar maximum approaches and the area of the polar coronal holes decreases continues to produce weak interaction regions. Rapid temporal changes in the structure of the solar magnetic field interfere with the periodicity of CIRs so that they appear and disappear from one solar rotation to the next. The solar wind structure is disrupted by frequently occurring CMEs that are the dominant source of geomagnetic activity.

9 Overall, the variations in solar wind and HMF structure at times other than solar maximum are spatial differences that change slowly with the solar cycle.

4.9 ACKNOWLEDGMENTS

The results reported here represent one aspect of research carried out by the California Institute of Technology Jet Propulsion Laboratory for the National Aeronautics and Space Administration.

4.10 REFERENCES

Altschuler, M. D., and G. Newkirk, Jr. (1969), Magnetic fields and the structure of the solar corona I: Methods of calculating coronal fields, *Solar Phys.*, **9**, 131.

Babcock, H. D. (1959), The Sun's polar magnetic field, *Astrophys. J.*, **130**, 364.

Banaszkiewicz, M., W. I. Axford, and J. F. McKenzie (1998), An Analytic Solar Magnetic Field Model, *Astron. Astrophys.*, **337**, 940.

Borrini, G., J. T. Gosling, S. J. Bame, W. C. Feldman, and J. M. Wilcox (1981), Solar wind helium and hydrogen structure near the heliospheric current sheet: A signal of coronal streamers at 1 AU, *J. Geophys. Res.*, **86**, 4565.

Burlaga, L. F., and J. D. Richardson (2000) North–south flows at 47 AU: A heliospheric vortex street?, *J. Geophys. Res.*, **105**(A5), 10501.

Burlaga, L. F., N. F. Ness, Y.-M. Wang, and N. R. Sheeley, Jr. (2002), Heliospheric magnetic field strength and polarity from 1 to 81 AU during the ascending phase of solar cycle 23, *J. Geophys. Res.*, **107**(A11), SSH-20-1.

Burton, M. E., E. J. Smith, A. Balogh, R. J. Forsyth, S. J. Bame, J. L. Phillips, and B. E. Goldstein (1996), Ulysses out-of-ecliptic observations of interplanetary shocks, *Astron. Astrophys.*, **316**, 313–322.

Clack, D., R. J. Forsyth, and M. W. Dunlop (2000), Ulysses observations of the magnetic field structure within CIRs, *Geophys. Res. Lett.*, **27**, 625.

Crooker, N., J. A. Joselyn, and J. Feyman (eds.) (1997), *Coronal Mass Ejections*, American Geophysical Union, Washington.

Crooker, N. G., G. L. Siscoe, S. Shodhan, D. F. Webb, J. T. Gosling, and E. J. Smith (1993), Multiple heliospheric current sheets and coronal streamer belt dynamics, *J. Geophys. Res.*, **98**, 9371.

Davis, L., Jr. (1966), Models of the Interplanetary Fields and Plasma Flow, in *The Solar Wind* (R. J. Mackin, Jr. and M. Neugebauer, eds.), Jet Propulsion Laboratory, Pasadena.

Dungey, J. W. (1961), Interplanetary magnetic field and auroral zones, *Phys. Rev. Lett.*, **6**, 47.

Erdös, G., and A. Balogh (1998), The symmetry of the heliospheric current sheet as observed by Ulysses during the fast latitude scan, *Geophys. Res. Lett.*, **25**, 245.

Ferraro, V. C. A., and C. Plumpton (1961), *An Introduction to Magneto-Fluid Mechanics*, Oxford University Press, London.

Fisk, L. A. (1996), Motion of the footpoints of heliospheric magnetic field lines at the Sun: Implications for recurrent energetic particle events at high heliographic latitudes, *J. Geophys. Res.*, **101**, 15547.

Fisk, L. A. (2003), Acceleration of the solar wind as a result of reconnection of open flux with coronal loops, *J. Geophys. Res.*, **108**(A4), SSH 7-1.

Fisk, L. A., and N. A. Schwadron (2001), The behavior of the open magnetic flux of the Sun, *Astrophys. J.*, **560**, 425.

Forsyth, R. J., and J. T. Gosling (2001), Corotating and transient structures in the heliosphere, in *The Heliosphere near Solar Minimum* (A. Balogh, R. G. Marsden, and E. J. Smith, eds.), Springer/Praxis, Chichester, UK.

Forsyth, R. J., A. Balogh, and E. J. Smith (2002), The underlying direction of the heliospheric magnetic field through the Ulysses first orbit, *J. Geophys. Res.*, **107**, 10102.

Foukal, P. (1990), *Solar Astrophysics*, p. 387, Wiley-Interscience, New York.

Gloeckler, G., T. H. Zurbuchen, and J. Geiss (2003), Implications of the observed anticorrelation between solar wind speed and coronal electron temperature, *J. Geophys. Res.*, **108**(A4), SSH 8-1.

Gosling, J. T., A. J. Hundhausen, and S. J. Bame (1976), Solar wind stream evolution at large heliocentric distances: Experimental demonstration and test of a model, *J. Geophys. Res.*, **81**, 2111.

Gosling, J. T., and V. J. Pizzo (1999), Formation and evolution of corotating interaction regions and their three-dimensional structure, *Space Sci. Rev.*, **89**, 21.

Gosling, J. T., G. Borrini, J. R. Asbridge, S. J. Bame, W. C. Feldman, and R. T. Hansen (1981), Coronal streamers in the solar wind at 1 AU, *J. Geophys. Res.*, **86**, 5438.

Gosling, J. T., S. J. Bame, D. J. McComas, J. L. Phillips, E. E. Scime, V. J. Pizzo, B. E. Goldstein, and A. Balogh (1994), A forward–reverse shock pair in the solar wind driven by over-expansion of a coronal mass ejection: Ulysses observations, *Geophys. Res. Lett.*, **21**, 237.

Gosling, J. T., W. C. Feldman, D. J. McComas, J. L. Phillips, V. J. Pizzo, and R. J. Forsyth (1995), Ulysses observations of opposed tilts of solar wind corotating interaction regions in the northern and southern hemispheres, *Geophys. Res. Lett.*, **22**, 3333.

Harvey, K. L., and F. Receley (2002), Polar coronal holes during cycles 22 and 23, *Solar Physics*, **211**, 31.

Heber, B., W. Droge, P. Ferrando, L. J. Haasbroek, H. Kunow, R. Muller-Mellin, C. Paizis, M. S. Potgeiter, A. Raviart, and G. Wibberenz (1996), Spatial variation of >40 MeV/n nuclei fluxes observed during the Ulysses rapid latitude scan, *Astron. Astrophys.*, **316**, 538.

Hoeksema, J. T. (1992), Large-scale structure of the heliospheric magnetic field: 1796–1991, in *Solar Wind Seven* (E. Marsch and R. Schwenn, eds.), p. 191, Pergamon Press, Oxford, UK.

Howard, R., and J. Harvey (1970), Spectroscopic Determinations of Solar Rotation, *Solar Phys.*, **12**, 23.

Howard, R. A., and M. J. Koomen (1974), Observations of sector structure in the outer corona: Correlation with interplanetary field, *Solar Phys.*, **37**, 469.

Hundhausen, A. J. (1972), *Coronal Expansion and Solar Wind*, Springer Verlag, New York.

Hundhausen, A. J. (1973), Nonlinear model of high speed solar wind streams, *J. Geophys. Res.*, **78**, 1528.

Hundhausen, A. J. (1977), An interplanetary view of coronal holes, in *Coronal Holes and High-Speed Wind Streams* (J. B. Zirker, ed.), p. 225, Colorado Associated University Press, Boulder.

Hundhausen, A. J. (1985), Some macroscopic properties of shock waves in the heliosphere, in *Collisionless Shocks in the Heliosphere: A Tutorial Review* (R. J. Stone and B. T. Tsurutani, eds.), Geophysical Monograph 34, American Geophysical Union, Washington, DC.

Jokipii, J. R., and E. N. Parker (1969), Stochastic aspects of magnetic lines of force with applicaton to cosmic-ray propagation, *Astrophys. J.*, **155**, 777.

Jokipii, J. R., and B. T. Thomas (1981), Effect of drifts on the transport of cosmic rays, IV: Modulation by a wavy interplanetary current sheet, *Astrophys. J.*, **243**, 1115.

Jokipii, J. R., E. H. Levy, and W. B. Hubbard (1977), Effects of drift on the transport of cosmic rays, I, General properties, application to solar modulation, *Astrophys. J.*, **213**, 861.

Jones, G. H., A. Balogh, and E. J. Smith (2003), Solar magnetic field reversal as seen at Ulysses, *Geophys. Res. Lett.*, **30**(19), ULY 2-1.

King, J. H. (1979), Solar cycle variations in IMF intensity, *J. Geophys. Res.*, **84**, 5938.

Klein, L. W., L. F. Burlaga, and N. F. Ness (1987), Radial and latitudinal variations of the interplanetary magnetic field, *J. Geophys. Res.*, **92**, 9885.

Kota, J., and J. R. Jokipii (1995), Corotating Variations of Cosmic Rays near the South Heliospheric Pole, *SCIENCE*, **268**, 1024.

Landau, L. D., and E. M. Lifschitz (1960), *Electrodynamics of Continuous Media*, Addison-Wesley, New York.

Levine, R. H. (1977), Large scale solar magnetic fields and coronal holes, in *Coronal Holes and High Speed Streams* (J. B. Zirker, ed.), p. 103, Colorado Associated University Press, Boulder.

Luhmann, J., C. T. Russell, and E. J. Smith (1988), Asymmetries of the interplanetary field inferred from observations at two heliocentric distances, in *Proceedings of 6th International Solar Wind Conference* (V. J. Pizzo, T. E. Holzer, and D. G. Sime, eds.), p. 323, National Center for Atmospheric Research, Boulder.

McComas, D. J., H. A. Elliott, and R. von Steiger (2002a), Solar wind from high-latitude coronal holes at solar maximum, *Geophys. Res. Lett.*, **27**, 28-1.

McComas, D. J., H. A. Elliott, J. T. Gosling, D. B. Reisenfeld, R. M. Skoug, B. E. Goldstein, M. Neugebauer, and A. Balogh (2002b), Ulysses second fast-latitude scan: Complexity near solar maximum and the reformation of polar coronal holes, *Geophys. Res. Lett.*, **29**(9), 4-1.

McComas, D. J., H. A. Elliott, N. A. Schwadron, J. T. Gosling, R. M. Skoug, and B. E. Goldstein (2003), The three-dimensional solar wind around solar maximum, *Geophys. Res. Lett.*, **30**(10), 1517.

McKibben, R. B., J. J. Connell, C. Lopate, A. J. Simpson, and M. Zhang (1996), Observations of galactic cosmic rays and the anomalous helium during Ulysses passage from the south to the north solar pole, *Astron. Astrophys.*, **316**, 547.

McNutt, R. L., Jr. (1988), Possible explanation of north–south plasma flow in the outer heliosphere and meridional transport of magnetic flux, *Geophys. Res. Lett.*, **15**, 1523.

Mikić, Z., and J. A. Linker (1996), The large scale structure of the solar corona and inner heliosphere, in *Solar Wind Eight* (D. Winterhalter, J. T. Gosling, S. R. Habbal, W. S. Kurth, and M. Neugebauer, eds.), p. 104, American Institute of Physics, Woodbury, NY.

Murphy, N., E. J. Smith, and N. A. Schwadron (2002), Strongly underwound magnetic fields in co-rotating rarefaction regions: Observatons and Implications, *Geophys. Res. Lett.*, **29**, 23-1.

Mursala, K., and T. Hitula (2003), Bashful ballerina: Southward shifted heliospheric current sheet, *Geophys. Res. Lett.*, **22**, 2-1.

Mursala, K., and B. Zieger (2001), Long-term north–south asymmetry in solar wind speed inferred from geomagnetic activity: A new type of century-scale solar oscillation?, *Geophys. Res. Lett.*, **28**, 95.

Neugebauer, M., E. J. Smith, A. Ruzmaikin, J. Feynman, and A. H. Vaughan (2000), The solar magnetic field and the solar wind: Existence of preferred longitudes, *J. Geophys. Res.*, **105**(A2), 2315.

Neugebauer, N., P. C. Liewer, E. J. Smith, R. M. Skoug, and T. H. Zurbuchen (2002), Sources of the solar wind at solar activity maximum, *J. Geophys. Res.*, **107**(A12), 1035.

Owens, M. J., and N. U. Crooker (2006), Coronal mass ejections and magnetic flux buildup in the heliosphere, *J. Geophys. Res.*, **111**, A10104.

Parker, E. N. (1963), *Interplanetary Dynamical Processes*, Wiley-Interscience, New York.

Parker, E. N. (1979), *Cosmical Magnetic Fields*, p. 532, Clarendon Press, Oxford, UK.

Parker, E. N. (1996), The alternative paradigm for magnetospheric physics, *J. Geophys. Res.*, **101**, 10578.

Pizzo, V. J. (1991), The evolution of corotating stream fronts near the ecliptic plane in the inner solar system 2. Three-dimensional tilted dipole fronts, *J. Geophys. Res.*, **96**, 5405.

Pizzo, V. J. (1994), Global quasi-steady dynamics of the distant solar wind. 1. Origin of north–south flows in the outer heliosphere, *J. Geophys. Res.*, **99**, 4173.

Pizzo, V. J., and B. E. Goldstein (1987), Meridional transport of magnetic flux in the solar wind between 1 and 10 AU: A theoretical analysis, *J. Geophys. Res.*, **92**, 7241.

Pneuman, G. W., and R. A. Kopp (1971), Gas–magnetic field interaction in the solar corona, *Solar Phys.*, **18**, 258.

Priest, E. R. (1982), *Solar Magneto-Hydrodynamics*, p. 325, D. Reidel, Dordrecht, The Netherlands.

Rees, A., and R. J. Forsyth (2003), Magnetic clouds with East/West orientated axes observed by Ulysses during solar cycle 23, *Geophys. Res. Lett.*, **30**(19), 8030, doi:10.1029/2003GL017296.

Reisenfeld, D. B., J. T. Gosling, R. J. Forsyth, P. Riley, and O. St. Cyr (2003), Properties of high-latitude CME-driven disturbances during Ulysses second northern polar passage, *Geophys. Res. Lett.*, **30**(19), 8031, doi:10.1029/2003GL017155.

Richardson, J. D., J. W. Belcher, A. J. Lazarus, K. I. Paularena, J. T. Steinberg, and P. R. Gazis (1996), Non-radial flows in the solar wind, in *Solar Wind Eight* (D. Winterhalter, J. T. Gosling, S. R. Habbal, W. S. Kurthl, and M. Neugebauer, eds.), p. 479, American Institute of Physics, Woodbury, NY.

Rosenberg, R. L., and P. J. Coleman, Jr. (1969), Heliographic latitude dependence of the dominant polarity of the interplanetary magnetic field, *J. Geophys. Res.*, **74**, 5611.

Schatten, K. H., J. M. Wilcox, and N. F. Ness (1969), A model of interplanetary and coronal magnetic fields, *Solar Phys.*, **70**, 5793.

Schatten, K. H., J. M. Wilcox, and N. F. Ness (1996), A model of interplanetary and coronal fields, *Solar Phys.*, **6**, 442.

Schwadron, N. A. (2002), An explanation for strongly underwound magnetic fields in co-rotating rarefaction regions and its relation to footpoint motion on the Sun, *Geophys. Res. Lett.*, **29**(14), doi:10.1029/2002GL015028.

Schwenn, R., and E. Marsch (eds.) (1990), *Physics of the Inner Heliosphere*, Vols. 1 and 2, Springer-Verlag, Berlin.

Simpson, A. J., M. Zhang, and S. Bame (1996), A solar polar north–south asymmetry for cosmic ray propagation in the heliosphere: The Ulysses pole-to-pole rapid transit, *Astrophys. J. Lett.*, **465**, L69.

Siscoe, G. L. (1972), Structure and orientation of solar wind interaction fronts: Pioneer 6, *J. Geophys. Res.*, **77**, 27.

Slavin, J. A., E. J. Smith, and B. T. Thomas (1984), Large scale temporal and radial gradients in the IMF: HELIOS 1, 2, ISEE-3 and Pioneer 10, 11, *Geophys. Res. Lett.*, **11**, 279.

Slavin, J. A., G. Jungman, and E. J. Smith (1986), The interplanetary magnetic field during solar cycle 21: ISEE-3/ICE observations, *Geophys. Res. Lett.*, **13**, 513.

Smith, E. J. (1989), Interplanetary magnetic field over two solar cycles and out to 20 AU, *Adv. Space. Res.*, **4**, 159.

Smith, E. J. (1993), Magnetic fields throughout the heliosphere, *Adv. Space Res.*, **13**(6), 5.

Smith, E. J. (2001), The heliospheric current sheet, *J. Geophys. Res.*, **106**, 15819.

Smith, E. J. (2004), Magnetic Field in the Outer Heliosphere, in *Physics of the Outer Heliosphere* (V. Florinski, N. V. Pogorelov, and G. P. Zank, eds.), p. 213, American Institute of Physics, Melville, New York.

Smith, E. J. (2006), The Heliospheric Current Sheet and Galactic Cosmic Rays, in *Physics of the Inner Heliosheath* (J. Heerikhuisen, V. Florinski, G. Zank, and N. V. Pogorelov, eds.), p. 104, AIP Conference Proceedings, Vol. 858, Melville, New York.

Smith, E. J., and A. Balogh (1995), Ulysses observations of the radial magnetic field, *Geophys. Res. Lett.*, **22**, 3317.

Smith, E. J., and A. Balogh (2003), Open Magnetic Flux: Variation with Latitude and Solar Cycle, in *Solar Wind Ten: Proceedings of the 10th International Solar Wind Conference* (M. Velli, R. Bruno, and F. Malara, eds.), p. 67, American Institute of Physics, Melville, New York.

Smith, E. J., and J. H. Wolfe (1976), Observations of interaction regions and shocks between one and five AU: Pioneers 10 and 11, *Geophys. Res. Lett.*, **3**, 137.

Smith, E. J., and J. H. Wolfe (1979) Fields and plasma in the outer solar system, *Space Sci. Rev.*, **23**, 217.

Smith, E. J., B. T. Tsurutani, and R. L. Rosenberg (1978), Observations of the interplanetary sector structure up to heliographic latitudes of 16 degrees: Pioneer 11, *J. Geophys. Res.*, **83**, 717.

Smith, E. J., J. A. Slavin, and B. T. Thomas (1986), The heliospheric current sheet: 3-dimensional structure and solar cycle changes, in *The Sun and the Heliosphere in Three Dimensions* (R. G. Marsden, ed.), p. 267, D. Reidel, Dordrecht, The Netherlands.

Smith, E. J., X.-Y. Zhou, and A. Ruzmaikin (in preparation), Quasiperiodic modes in the heliospheric magnetic field.

Smith, E. J., M. Neugebauer, A. Balogh, S. J. Bame, G. Erdös, R. J. Forsyth, B. E. Goldstein, J. L. Phillips, and B. T. Tsurutani (1993), Disappearance of the Heliospheric Sector Structure at Ulysses, *Geophys. Res. Lett.*, **20**, 2327.

Smith, E. J., A. Balogh, M. E. Burton, G. Erdös, and R. J. Forsyth (1995), Results of the Ulysses fast latitude scan: Magnetic field observations, *Geophys. Res. Lett.*, **22**, 2327.

Smith, E. J., A. Balogh, M. E. Burton, R. J. Forsyth, and R. P. Lepping (1997a), Radial and azimuthal components of the heliospheric magnetic field: Ulysses observations, *Adv. Space Res.*, **20**, 47.

Smith, E. J., M. Neugebauer, B. T. Tsurutani, A. Balogh, R. Forsyth, and D. McComas (1997b), Properties of Hydromagnetic Waves in the Polar Caps: Ulysses, *Adv. Space Res.*, **20**, 55.

Smith, E. J., J. R. Jokipii, J. Kota, R. P. Lepping, and A. Szabo (2000a), Evidence of a north–south asymmetry in the heliosphere associated with a southward displacement of the heliospheric current sheet, *Astrophys. J.*, **533**, 1084.

Smith, E. J., A. Balogh, R. J. Forsyth, B. T. Tsurutani, and R. P. Lepping (2000b), Recent observations of the heliomagnetic field at Ulysses: Return to low latitude, *Adv. Space Res.*, **26**, 823.

Smith, E. J., A. Balogh, R. J. Forsyth, and D. J. McComas (2001), Ulysses in the south polar cap at solar maximum: Heliospheric magnetic field, *Geophys. Res. Lett.*, **28**(22), 4159.

Smith, E. J., R. G. Marsden, A. Balogh, G. Gloeckler, J. Geiss, D. J. McComas, R. B. McKibben, R. J. MacDowell, L. J. Lanzerotti, N. Krupp *et al.* (2003), The Sun and Heliosphere at Solar Maximum, *SCIENCE*, **302**, 1165.

Suess, S. T., and S. F. Nerney (1975), The global solar wind and predictions for Pioneers 10 and 11, *Geophys. Res. Lett.*, **2**, 75.

Suess, S. T., B. T. Thomas, and S. F. Nerney (1985), Theoretical interpretation of the observed interplanetary magnetic field radial variation in the outer solar system, *J. Geophys. Res.*, **90**, 4378.

Suess, S. T., E. J. Smith, J. Phillips, B. E. Goldstein, and S. Nerney (1996), Latitudinal dependence of the radial IMF component—interplanetary imprint, *Astron. Astrophys.*, **316**, 304.

Suess, S. T., J. L. Phillips, D. J. McComas, B. E. Goldstein, M. Neugebauer, and S. Nerney (1998), Solar wind—inner heliosphere, *Space Sci. Rev.*, **83**, 75.

Svalgaard, L. (1975), On the use of Godhavn H component as an indicator of the interplanetary sector polarity, *J. Geophys. Res.*, **80**, 2717.

Svalgaard, L., and J. M. Wilcox (1975), Long term evolution of solar sector structure, *Solar Phys.*, **41**, 461.

Thomas, B. T., and E. J. Smith (1980), The Parker spiral configuration of the interplanetary magnetic field between 1 and 8.5 AU, *J. Geophys. Res.*, **85**, 6861.

Tritakis, V. P. (1984), Heliospheric current sheet displacements during the solar cycle evolution, *J. Geophys. Res.*, **89**, 6588.

Tu, C. Y., and E. Marsch (1995), *MHD Structures, Waves and Turbulence in the Solar Wind*, Kluwer Academic Publishers, Dordrecht, The Netherlands.

Wang, Y. M., and N. R. Sheeley, Jr. (1990), Solar wind speed and coronal flux tube expansion, *Astrophys. J.*, **355**, 726.

Wang, Y. M., J. Lean, and N. R. Sheeley, Jr. (2000), The long-term variation of the Sun's open magnetic flux, *Geophys. Res. Lett.*, **27**, 505.

Wang, Y.-M., A. G. Nash, and N. R. Sheeley, Jr. (1989), Magnetic Flux Transport on the Sun, *SCIENCE*, **245**, 712.

Wilcox, J. M., and N. F. Ness (1965), Quasi-stationary corotating structure in the interplanetary medium, *J. Geophys. Res.*, **70**, 5793.

Winterhalter, D. E., E. J. Smith, J. H. Wolfe, and J. A. Slavin (1990), Spatial gradients in the heliospheric magnetic field: Pioneer 11 observations between 1 and 24 AU and over solar cycle 21, *J. Geophys. Res.*, **95**, 1.

Winterhalter, D. E., E. J. Smith, M. E. Burton, N. Murphy, and D. J. McComas (1994), The heliospheric plasma sheet, *J. Geophys. Res.*, **99**, 6667.

Wolfson, R. (1985), A coronal magnetic field model with volume and sheet currents, *Astrophys. J.*, **288**, 769.

Zhao, X.-P., and A. J. Hundhausen (1981), Organization of solar wind plasma properties in a tilted, heliomagnetic coordinate system, *J. Geophys. Res.*, **86**, 5423.

Zhao, X. P., J. T. Hoeksema, and P. H. Scherrer (2005), Prediction and understanding of the north–south displacement of the heliospheric current sheet, *J. Geophys. Res.*, **110**, A10101.
Zhou, X.-Y., E. J. Smith, D. Winterhalter, D. J. McComas, R. M. Skoug, B. E. Goldstein, and C. W. Smith (2005), Morphology and evolution of the Heliospheric Current Sheet and Plasma Sheet from 1 to 5 AU, in *Proceedings of Solar Wind 11/ SOHO 16*, p. 659, ESA Publication SP-592, Noordwijk, The Netherlands.
Zhou, X.-Y., and E. J. Smith (in preparation), Solar cycle variations in open magnetic flux and field magnitude.
Zieger, B., and K. Mursala (1998), Annual variation in near-Earth solar wind speed: Evidence for persistent north–south asymmetry related to solar magnetic polarity, *Geophys. Res. Lett.*, **25**, 841.
Zirin, H. (1988), *Astrophysics of the Sun*, p. 303, Cambridge University Press, Cambridge, UK.
Zurbuchen, T. H., N. A. Schwadron, and L. A. Fisk (1997), Direct observational evidence for a heliospheric magnetic field with large excursions in latitude, *J. Geophys. Res.*, **102**, 24175.

5

Heliospheric energetic particle variations

D. Lario and M. Pick

5.1 ENERGETIC PARTICLE POPULATIONS IN THE INNER HELIOSPHERE

As observed from the ecliptic plane and at a distance of 1 AU from the Sun, the energetic particle population of the heliosphere drastically changes from solar maximum to solar minimum. The energetic particle populations in the inner heliosphere include:

(1) Galactic cosmic rays (GCRs) originated in the interstellar medium and able to penetrate into the heliosphere.
(2) Anomalous cosmic rays (ACRs) that originate as interstellar neutral atoms traveling into the heliosphere, ionized by solar UV and carried out as pickup ions in the solar wind to be finally accelerated to energies as high as ~100 MeV/nucleon presumably close to the solar wind termination shock or in the heliosheath.
(3) Solar energetic particles (SEPs) that originate near the Sun in association with solar flares and/or large coronal mass ejections (CMEs). As CMEs expand outward from the Sun, they may be able to drive interplanetary shock waves that can reaccelerate SEPs to form large gradual SEP events. Occasionally, SEP events are observed at very high energies reaching ~GeV for protons and ~100 MeV for electrons.
(4) Energetic particles accelerated by other shocks and disturbances in the solar wind such as shocks formed in the solar wind stream interaction regions (SIs) or corotating interaction regions (CIRs).
(5) Energetic particles accelerated in planetary magnetospheres, such as Jovian electrons observed in the inner heliosphere at energies from a few hundred keV to less than about 30 MeV.

The Ulysses spacecraft, with its eccentric orbit over the solar poles, and its more than 15 years in space (Figure 7.1), allows us to study the characteristics of these particle populations at low and high latitudes and their variations over the solar cycle. The intensities of all these populations are affected by variations in the level of solar activity, the characteristics of the solar wind, and the properties of the interplanetary magnetic field that enables energetic particle propagation through the heliosphere. These changes result in short-term and long-term modulations of GCRs and ACRs, variations in latitudinal and radial gradients of particle intensities, and changes in the energy spectra and composition of the heliospheric energetic particle population. The study of these particle populations at different latitudes and under different heliospheric conditions provides information about the global structure of the heliosphere during solar-minimum and solar-maximum conditions and the mechanisms of particle propagation in the heliosphere. In this chapter we deal with Ulysses observations of populations (2), (3), and (4), whereas Chapter 6 deals with populations (1), (2), and (5).

Three instruments onboard Ulysses have continuously scanned these particle populations: the Energetic Particle Composition Experiment (EPAC) (Keppler *et al.*, 1992), the Heliosphere Instrument for Spectra, Composition, and Anisotropy at Low-Energies (HI-SCALE) (Lanzerotti *et al.*, 1992), and the telescopes of the Cosmic Ray and Solar Particle Investigation (COSPIN) program (Simpson *et al.*, 1992). These three sets of instruments cover a wide range of energies and species allowing us to distinguish the above five particle populations and their variations over the solar cycle.

5.2 SOLAR MINIMUM ORBIT (1992–1998)

An overview of the solar-minimum measurements by the low-energy particle instrumentation on Ulysses is shown in Figure 5.1. The intensities of 40–65 keV electrons and 1.8–4.7 MeV ions from HI-SCALE (Lanzerotti *et al.*, 1992), and 71–94 MeV protons from the High Energy Telescope (HET) of COSPIN (Simpson *et al.*, 1992) are plotted in the upper three panels of Figure 5.1, respectively, as a function of time throughout the solar minimum orbit. The fourth panel of Figure 5.1 shows the solar wind speed, whereas the bottom panel shows the heliographic latitude (blue line) and the heliocentric radial distance (red dashed line) of Ulysses together with the monthly sunspot number (green hatched area). Figure 5.1 spans from 22 August 1992 when Ulysses was at the heliocentric radial distance $R = 5.28$ AU and heliographic latitude $\Lambda = 15.8°$S to 30 October 1998 when Ulysses was at $R = 5.29$ AU and again at $\Lambda = 15.8°$S, therefore this period includes the first perihelion at 1.34 AU on 12 March 1995 and the second aphelion at 5.41 AU on 17 April 1998. The yellow vertical shading areas mark the polar passes of Ulysses in 1994 (southern polar pass) and 1995 (northern polar pass) defined as those periods when Ulysses was at heliographic latitudes above 70°. The maximum southern heliographic latitude at $\Lambda = 80.2°$S was reached on 13 September 1994 and the maximum northern heliographic latitude at

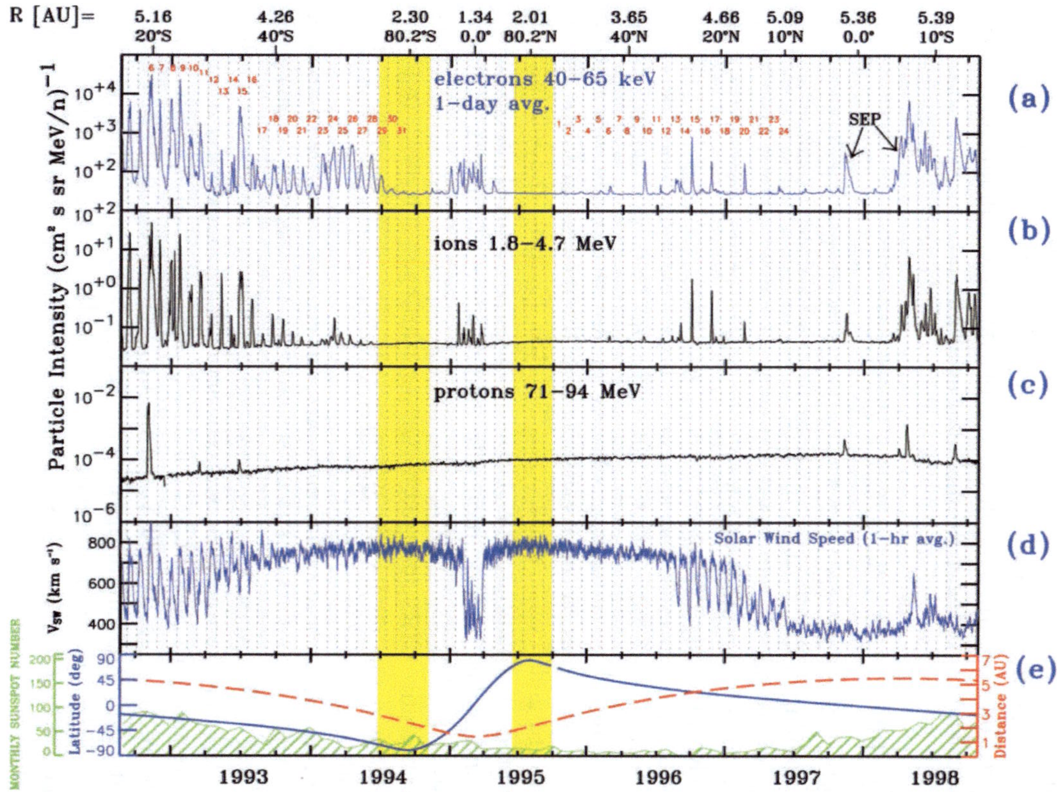

Figure 5.1. Daily averages of (a) 40–65 keV electron intensities; (b) 1.8–4.7 MeV ion intensities; (c) 71–94 MeV proton intensities; (d) solar wind speed; and (e) monthly sunspot number (green hatched area) with Ulysses heliographic latitude (blue line) and Ulysses heliocentric radial distance (red dashed line). The yellow vertical shading areas mark the polar passes (heliographic latitudes above 70°). Thin dotted lines are 26 days apart and mark the solar rotation period. The rotation numbering scheme in panel (a) has been adopted from Bame *et al.* (1993) and Roelof *et al.* (1997). Time interval extends from day 235 of 1992 to day 303 of 1998.

$\Lambda = 80.2°$N on 30 July 1995. Thin dotted lines in Figure 5.1 are 26 days apart and mark the solar rotation period.

Energetic particle observations over the first polar orbit of Ulysses have been thoroughly analyzed by several authors (e.g., Simnett *et al.*, 1994; Roelof, Simnett, and Armstrong, 1995; Rooelof *et al.* 1997; Sanderson *et al.*, 1995, 1999; Keppler, 1998a) and summarized in Lanzerotti and Sanderson (2001); we refer the reader to these works for a detailed description. The following is an outline of both the solar-minimum observations and their implications for the understanding of particle transport and acceleration in the solar-minimum heliosphere.

5.2.1 Summary of the Ulysses solar-minimum observations

After departing Jupiter in February 1992, Ulysses began its journey out of the ecliptic plane. At the beginning of this journey, the level of solar activity was relatively high and Ulysses was still completely immersed in a slow solar wind regime. Moderately high fluxes of electrons and protons with no regular patterns were observed throughout this period (Roelof *et al.*, 1992; Sanderson *et al.*, 1995).

When Ulysses reached 13°S, the spacecraft began entering, once per solar rotation (~26 days), a fast solar wind flow emanating from the southern polar coronal hole (Figure 5.1d). A regular sequence of low-energy ion and electron intensity increases was observed in association with the passage of CIRs, mostly bounded by forward and reverse shock pairs (FS–RS). Electron and ion intensity enhancements at these heliolatitudes (i.e., below about 30°S) occurred approximately simultaneously and peaked in association with the passage of the shocks, mainly at the reverse shocks in the case of near-relativistic electrons. Bame *et al.* (1993) numbered the consecutive CIRs observed by Ulysses starting in July 1992. In Figure 5.1a we have followed this numbering system and labeled each recurrent particle enhancement with a consecutive number.

When Ulysses reached about ~36°S, the spacecraft became completely immersed in the high-speed solar wind flow. CIRs continued to be observed, propagating poleward, but only with reverse shocks associated with them. Poleward of ~42°S (rotation 19), reverse shocks were observed only sporadically (Gosling *et al.*, 1995). However, quasi-regular particle increases continued to be observed. These increases were seen in the protons up to ~70°S and in the electrons up to 80°S. Particle enhancements observed poleward of the streamer belt ($\Lambda > 30°$S) had the peculiarity that maximum ~50 keV electron intensities were considerably delayed (up to ~4 days) with respect to the ~1 MeV proton maximum intensities (Roelof, Simnett, and Tappin, 1996).

The regular pattern of events observed in the southern hemisphere was disrupted by the occurrence of transient events of solar origin as in rotations 6, 15, 23, and 24, or by the arrival of the interplanetary counterparts of CMEs (i.e., ICMEs) as in rotations 6, 14, 23, 24, and 26 (Sanderson *et al.*, 1995). Interspersed with the regular CIR-associated particle intensity increases, Roelof, Simnett, and Armstrong (1995) identified also the occurrence of small inter-events between rotations 7–8, 10–11, 11–12, 15–16, 16–17, and 18–19, as events of solar origin able to fill the rarefaction regions formed in the high-speed solar wind streams. The most intense transient SEP events, such as the events in November 1992 or June 1993 (Reuss *et al.*, 1995; Pick *et al.*, 1995a), were also observed at the high-energy channels as shown in the 71 MeV–94 MeV proton channel (Figure 5.1c).

A few sporadic transient SEP events were also observed at high latitudes, overlaid on the recurrent CIR events (Bothmer *et al.*, 1995). Curiously, a very weak electron event was observed at $\Lambda = 74°$S associated with a type III radio burst and a radioheliograph source at 6°S (Pick *et al.*, 1995b). These authors suggested coronal propagation of the electrons to distribute from their low-latitude solar source to the higher latitudes where Ulysses was presumably connected. This weak electron event was the highest latitude SEP event observed throughout the solar-minimum orbit.

The passage from the south to the north pole, covering the full range of latitudes (from $\Lambda = 80.2°$S to $\Lambda = 80.2°$N) in only 10 months, and comprised between the two yellow bars in Figure 5.1, is known as the fast latitude scan (FLS). Ulysses kept observing recurrent electron events at very high latitudes, whereas the first ion event of the FLS after the south polar pass was not observed until 46°S at the end of 1994 (Roelof *et al.*, 1997). Ulysses observed slow solar wind from the streamer belt again at 22°S (February 1995); and for three solar rotations, two low-energy ion and electron intensity peaks were observed in each rotation, due to the spacecraft encounter with CIRs from both the northern and southern coronal hole solar wind flows once per solar rotation (Sanderson *et al.*, 1995). Two transient SEP events were observed during the solar minimum FLS on day 82 of 1995 and in late April 1995 (Buttighoffer *et al.*, 1996). Ulysses emerged from the streamer belt into the northern hemisphere in late March 1995. No recurrent electron or ion events were observed throughout the north polar pass.

As Ulysses descended from northern latitudes, recurrent electron events reappeared in October 1995 at $\Lambda = 64°$N, in the rotation labeled 1 in the top panel of Figure 5.1 following the numbering system introduced by Roelof *et al.* (1997). The electron intensity increases were not as recurrent as observed in the southern hemisphere. Ion recurrences appeared at lower latitudes and only for a few rotations. Roelof *et al.* (1997) attributed the variability of the northern recurrences during 1996 to temporal changes of the near-Sun polar magnetic field configuration, whereas in 1993–1994 the recurrent southern hemisphere observations resulted from a nearly constant corotating magnetic field configuration for both CIRs and the high-latitude heliospheric magnetic field. Sanderson *et al.* (1999) studied the effects that the heliospheric current sheet (HCS) produce on the recurrent particle intensity enhancements, and concluded that in 1995–1996 the HCS was much flatter than during the Ulysses southern hemisphere excursion, producing less intense CIR events.

The rest of the descent from northern polar regions to the equator in 1996 and 1997 was characterized by nearly recurrent CIR events with the sporadic occurrence of SEP events in December 1996, and February, April, and May 1997 that contributed to increase the intensity of the concurrent CIR events (Lario *et al.*, 2000a). Ulysses entered full immersion in slow-speed flow in July 1997 and started to observe intense SEP events, such as the events in November 1997 and the series of events in April–May 1998 coinciding with the rising phase of the solar cycle 23 (Lario *et al.*, 2000b). These events were energetic enough to produce enhancements in the 71–94 MeV proton intensities (Figure 5.1c). Note that the background level intensity of this high-energy channel gradually increased throughout the time interval shown in Figure 5.1 owing to the increasing number of GCRs penetrating into the inner heliosphere during solar minimum, and only started to decrease after the occurrence of the SEP events in November 1997 and April 1998 (Chapter 6).

5.2.2 Energetic particle origin, transport, and acceleration processes in the solar-minimum inner heliosphere

The well-organized structure of the solar-minimum heliosphere with fast solar wind at high latitudes, slow solar wind at low latitudes, CIRs formed only in the vicinity ($<30°$) of the Sun's equatorial plane, and relatively flat HCS, together with the scarcity of solar events, led to energetic particle intensities dominated by CIR events and their processes of particle acceleration. The analysis of the CIR-associated energetic particle events observed by Ulysses led to the following developments in the study of particle acceleration and transport in the solar-minimum heliosphere.

(1) CIR effects were observed at latitudes well beyond the disappearance of the associated forward and reverse shocks, especially in the southern heliolatitudes (Simnett *et al.*, 1994). This observation indicates that energetic particles must be transported from the remote CIRs at low latitudes to Ulysses at high latitudes. Simnett and Roelof (1995) argued for a direct magnetic connection between Ulysses and the low-latitude distant CIRs. This led Fisk (1996) to develop a model of the heliospheric magnetic field that allows magnetic connection between Ulysses at high latitudes and CIRs at lower latitudes and larger distances (>10 AU). On the other hand, Kóta and Jokipii (1995) argued for a latitudinal propagation of energetic particles in a diffusive transport across the average magnetic field due to a random walk or braiding of the field lines in the latitudinal direction.
(2) The highest intensities of $\sim$50 keV electrons and $\sim$1 MeV protons were associated with reverse shocks (when these shocks were observed, i.e., $\Lambda < 45°$S) rather than the forward shocks. This observation is a simple consequence of the fact that seed particles for the mechanisms of shock acceleration have higher upstream energies because of the higher solar wind velocity at the reverse shocks (Giacalone and Jokipii, 1997). The association between the recurrent electron intensity enhancements and electron acceleration at the CIR reverse shocks led to the development of several theories of electron acceleration at CIR shocks (Scholer *et al.*, 1999; Treumann and Terasawa, 2001; Mann *et al.*, 2002, *op. cit.*).
(3) Anisotropy ion flows measured during CIRs and in the solar wind frame (i.e., corrected for the Compton–Getting effect) have components perpendicular to the local magnetic field that are effectively zero (indicating that there is no net flow of particles across the magnetic field), whereas the parallel components exhibit significant contributions either aligned or anti-aligned with the magnetic field. Figure 5.2 shows the evolution of the 1.12–1.87 MeV ion anisotropy

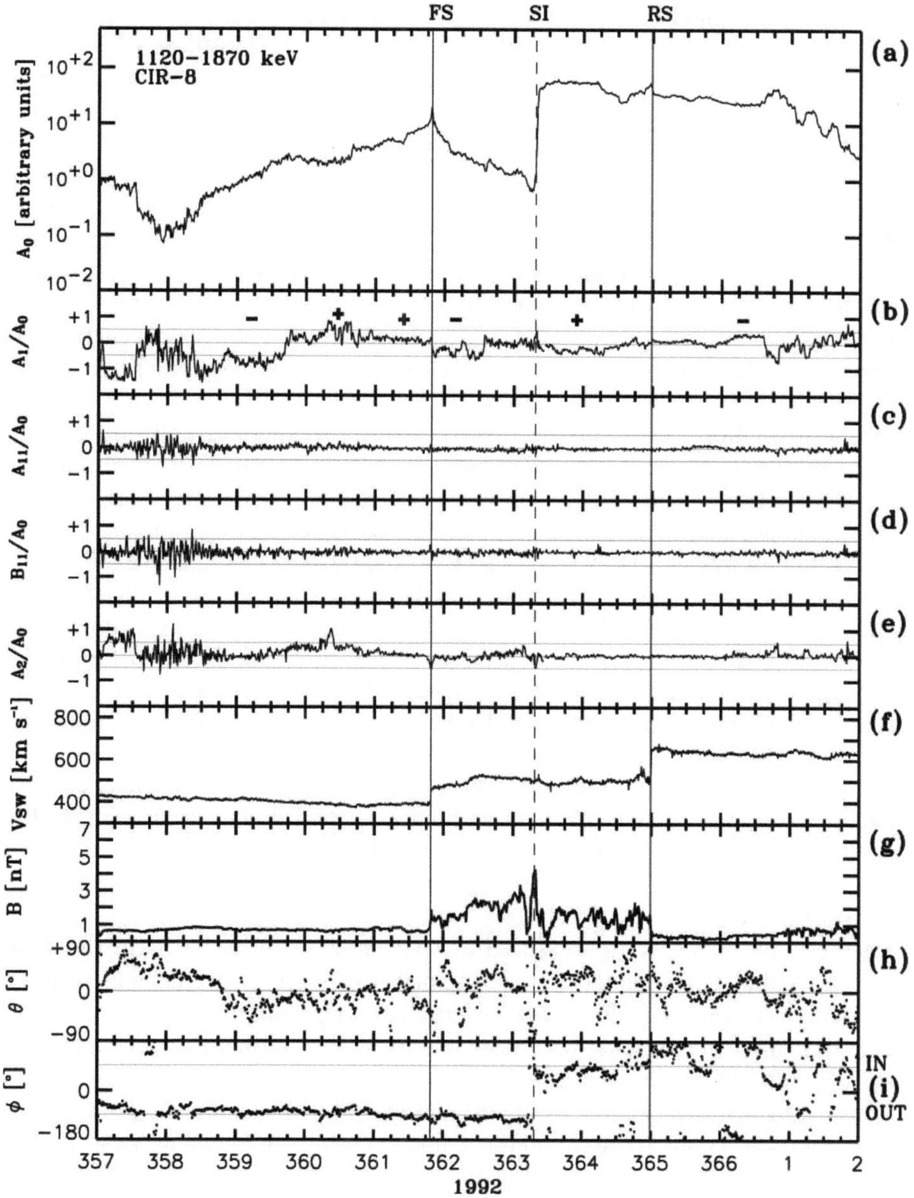

Figure 5.2. Anisotropy flow coefficients in the solar wind frame for the CIR #8. (a) Zero-order isotropic coefficient A_0. (b) First-order parallel anisotropy coefficient. (c–d) First-order perpendicular anisotropy coefficients. (e) Second-order anisotropy coefficient. (f) Solar wind speed. (g) Magnetic field magnitude. (h) Magnetic field altitude angle in the Ulysses RTN coordinate system. (i) Magnetic field azimuth angle in the Ulysses RTN coordinate system (Forsyth *et al.*, 1995). Solid vertical lines and dashed vertical line mark the arrival of interplanetary shocks (FS–RS) and of a stream interface (SI), respectively, as identified by Wimmer-Schweingruber, von Steiger, and Paerli (1997).

coefficients measured in the solar wind frame for CIR #8 as observed by the LEMS30 and LEMS120 telescopes of HI-SCALE (Lanzerotti *et al.*, 1992) and computed following the method described by Lario *et al.* (2004a). A_0 is the isotropic component, A_1/A_0 represents the first-order anisotropy resolved along the magnetic field direction, A_{11}/A_0 and B_{11}/A_0 represent the flow transverse to the magnetic field and are practically zero throughout the CIR event, and A_2/A_0 represents bidirectional flows when first-order coefficients are close to zero and $A_2 > 0$. In the second panel of Figure 5.2, we indicate whether particle flows are directed outward (i.e., anti-sunward) along the field (indicated by a plus symbol) or inward (i.e., sunward) along the field (indicated by the minus symbol). Particles stream away from the CIR-related shocks, consistent with shock particle acceleration. No evidence of particle diffusion across the field lines is observed within the CIR. The anisotropies shown in Figure 5.2 are also consistent with those measured by the Anisotropy Telescopes (AT) of COSPIN (Laxton, 1997). Anisotropies measured by EPAC during CIR events at high heliolatitudes ($\Lambda > 30°$S) show inward field-aligned flows suggesting that the particle sources are located beyond the Ulysses spacecraft (Franz *et al.*, 1997).

(4) Energetic ion intensities (50 keV–5 MeV) at the CIRs below <32°S exhibit peaks close to the forward and reverse shocks (with the peak at the FS being weaker and briefer) and minima between the forward and reverse shock (sometimes in the vicinity of stream interfaces as shown in the case of CIR #8 in Figure 5.2). Intriligator *et al.* (2001) analyzed magnetic field variations within CIRs and concluded that fluctuations normal to the average field increased near the stream interfaces (SI) but in planar structures that, together with the effects of shear and compression, reduce the random walk of field lines near SI. Whereas energetic particles propagate along field lines within the CIRs, field line mixing (that allows energetic particles to fill the region between the SI and the RS) is reduced near SIs, and thus leads to a decrease in particle intensities in the vicinity of stream interfaces.

(5) Energetic particle spectra measured at CIR shocks are harder than those predicted by the theory of the diffusive shock acceleration mechanism, assuming particle acceleration from a solar wind thermal population. This observation suggests either the presence of a more energetic seed population than those used in the models of particle acceleration or enhanced levels of magnetic field fluctuations near the edges of the CIRs (Desai *et al.*, 1999). The sporadic occurrence of SEP events also contributes to fill the heliosphere with energetic particles that can be reaccelerated by CIRs (Sanderson *et al.*, 1995; Lario *et al.*, 2000a). In the absence of SEP contamination, Keppler (1998b) argued that the variation of the energetic ion peak intensity of CIR events is basically a function of the radial distance with a latitudinal-dependent feature superimposed on it with decreasing intensities with increasing latitude.

(6) Energetic particle abundances at CIR shocks are in agreement with solar wind thermal abundances with the exception of He that is overabundant at CIR reverse shocks (Franz *et al.*, 1999). Proton-to-helium ratios at ~1 MeV/amu are <10 near CIR reverse shocks (versus values >20 observed in SEP events),

consistent with a large contribution of accelerated pickup He^+ from high-speed solar wind streams (Gloeckler *et al.*, 1994; Simnett, Sayle, and Roelof, 1995).

(7) ACR fluxes are modulated by CIRs (Reuss, Fränz, and Keppler, 1996). Magnetic structures associated with the CIRs may act as moving diffusion barriers that hinder the inward propagation of ACRs (McKibben *et al.*, 1999; Kissmann, Fichtner, and Ferreira, 2004; *op. cit.*). Low-energy ($\lesssim$10 MeV/nucleon) ACR fluxes were higher in the northern hemisphere than in the southern hemisphere, suggesting that ACRs can penetrate into the heliosphere easily during periods of low-level solar activity and flat HCS (Maclennan and Lanzerotti, 1998).

(8) The sporadic observation of SEP events by Ulysses during the solar-minimum south polar pass (at $> 48°$S) was associated with the passage of ICMEs (Bothmer *et al.*, 1995). The existence of propagation channels embedded within CIRs or within transient ICMEs was suggested as an appropriate conduit for particle propagation toward large heliocentric distances and high heliolatitudes (Pick *et al.*, 1995a; Maia *et al.*, 1998).

5.3 SOLAR MAXIMUM ORBIT (1998–2004)

Figure 5.3 shows, with the same format as Figure 5.1, an overview of the energetic particle measurements by Ulysses during the solar-maximum orbit. Figure 5.3 spans from 30 October 1998 when Ulysses was at the heliocentric radial distance $R = 5.29$ AU and heliographic latitude $\Lambda = 15.8°$S to 4 January 2005 when Ulysses was again at $R = 5.29$ AU and $\Lambda = 15.8°$S; therefore, this period includes the second perihelion at 1.34 AU on 23 May 2001 and the third aphelion at 5.41 AU on 30 June 2004. The yellow vertical shading areas mark the polar passes of Ulysses at the end of 2000 (southern polar pass) and 2001 (northern polar pass) defined as those periods when Ulysses was at heliographic latitudes above 70 degrees. The maximum southern heliographic latitude at $\Lambda = 80.2°$S was reached on 27 November 2000 and the maximum northern heliographic latitude at $\Lambda = 80.2°$N on 13 October 2001.

Energetic particle observations over the solar-maximum polar orbit of Ulysses have been thoroughly analyzed and compared with the solar-minimum observations by several authors (Simnett, 2001; Lario *et al.*, 2001a; McKibben *et al.*, 2003; Maclennan, Lanzerotti, and Gold, 2003; Marsden, 2004; Sanderson, 2004); we refer the reader to these works for a detailed description.These observations have clear implications for both the identification of the sources of energetic particles and the mechanisms of particle propagation in the complex solar-maximum heliosphere.

The most notable signature of the Ulysses solar-maximum orbit (in contrast to the solar-minimum orbit) is the lack of any regular pattern in energetic particle intensities and in solar wind data (Figure 5.3). With the exception of the northern polar pass in September–December 2001, Ulysses observed an irregularly structured mixture of slow ($\sim$350 km s^{-1}) and intermediate-speed ($\sim$600 km s^{-1}) solar wind flows (McComas, Gosling, and Skoug, 2000). The periods with fast ($>$700 km s^{-1}) solar wind flow were scarce and mainly concentrated at northern polar longitudes owing to the reconstruction of the northern polar coronal hole (McComas, Elliott,

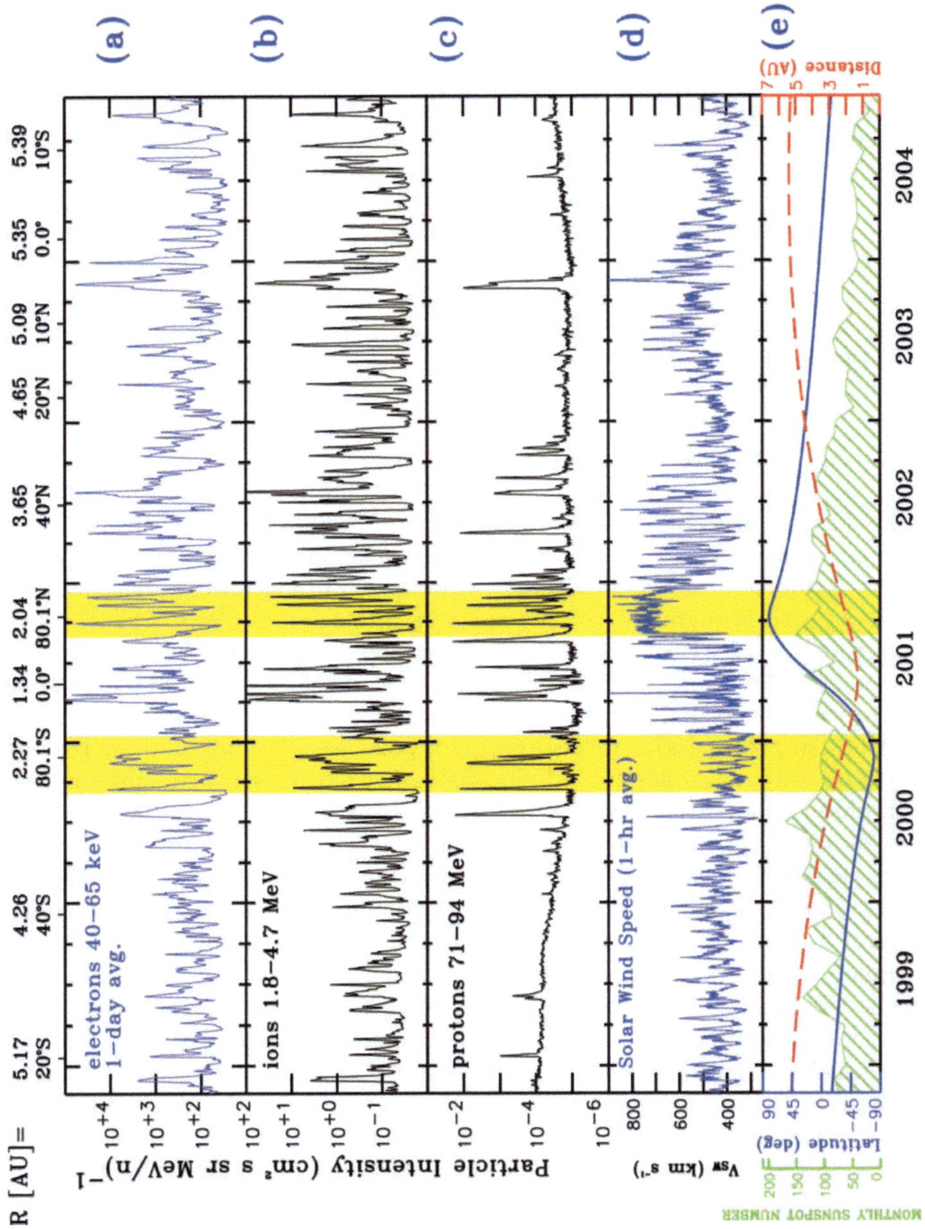

Figure 5.3. The same as Figure 5.1 but from day 303 of 1998 to day 4 of 2005.

and von Steiger, 2002; McComas *et al.*, 2002). The interaction between slow solar wind streams and either intermediate or fast-flow streams resulted in SIs, many of which were bounded by FS–RS pairs (McComas *et al.*, 2000), and in a few cases appeared recurrently at roughly the solar rotation period over a few consecutive rotations (see examples in McComas, Elliott, and von Steiger, 2002 and Lario *et al.*, 2001a, b, 2003a, b). The highest solar wind speeds ($>900\,\mathrm{km\,s^{-1}}$) throughout the time interval shown in Figure 5.3d were observed on days 110–111 of 2001 and days 319–321 of 2003 in association with the passage of fast ICMEs, comparable also with the fast ICME observed in November 1992 (Figure 5.1d).

In contrast to the solar-minimum orbit, particle intensities fluctuated without any consistent pattern. Low-energy ($<20\,\mathrm{MeV}$) ion and near-relativistic electron intensities were elevated throughout the solar-maximum orbit and only occasionally declined to background levels (Figures 5.3a, b). In contrast to the solar-minimum orbit when the lowest electron intensities were observed at high latitudes ($\Lambda > 70°$S or $\Lambda > 40°$N), several intense events were observed, even at heliolatitudes as high as 80°S or 80°N. The high intensities observed throughout the solar-maximum orbit, independently of heliolatitude and heliocentric radial distance, indicate that the entire heliosphere was essentially populated by energetic particles at all heliolatitudes and heliolongitudes.

Figure 5.4 shows the 40–65 keV electron intensity during those time intervals that Ulysses spent at latitudes above 65°S (Figure 5.4a) and 65°N (Figure 5.4b). Particle intensities were higher (up to four orders of magnitude) during the solar-maximum passes (black traces) than during solar minimum (gray traces). Whereas a solar-minimum south polar pass presented the recurrent pattern associated with CIRs, solar-maximum intensities did not show any regular pattern. The recurrent electron intensity increases in the first orbit were observed only during the south polar pass and up to 80°S (Figure 5.4a) but not during the north polar pass, because of the north–south asymmetry in the CIR pattern, the flattening of the HCS, and the global decay of CIR intensities in the heliosphere. The occurrence of intense solar events during both solar-maximum polar passes led to very high SEP intensities even at the highest latitudes.

The conditions under which Ulysses observed the events in the south or north solar-maximum polar passes differ significantly (Lario *et al.*, 2003c). Figure 5.5 shows energetic ion intensities, solar wind speed, magnetic field magnitude and directions (in the spacecraft RTN centered coordinate system; see Forsyth *et al.* (1995) for a definition) measured by Ulysses during the solar-maximum orbit at heliographic latitudes above 75°S (left panel) and75°N (right panel). During the southern polar pass, Ulysses observed low-speed solar wind ($\sim 400\,\mathrm{km\,s^{-1}}$, even below $300\,\mathrm{km\,s^{-1}}$ at the highest latitudes on days 326–330) with occasional streams of faster ($\sim 500\,\mathrm{km\,s^{-1}}$) wind (McComas, Elliott, and von Steiger, 2002). The interactions between both types of solar wind have been labeled SIRs in the left panel of Figure 5.5. The last two SIRs in this time interval were bounded by forward and reverse shocks (solid vertical lines in Figure 5.5),whereas the first SIR was preceded just by a forward shock. These SIRs were accompanied by low-energy ion intensity enhancements, peaking at the arrival of the shocks. Four SEP events (labeled 1S, 2S, 3S, and

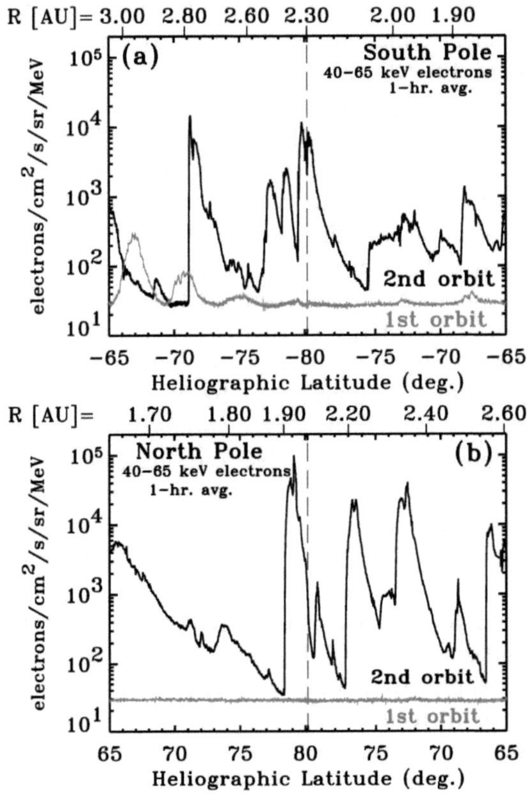

Figure 5.4. Hourly averages of 40–65 keV electron intensities measured by the LEFS150 telescope of HI-SCALE above heliographic latitudes of 65 over the south (a) and north (b) solar poles during the solar-minimum (gray trace) and solar-maximum (black trace) Ulysses orbits. Measurements taken after the passage above the highest heliographic latitude attained by Ulysses have been mirrored with respect to 80.2° (dashed line). Panel (a) covers the period from day 144 to 324 of 1994 for the solar-minimum orbit and from day 214 of 2000 to day 29 of 2001 for the solar-maximum orbit. Panel (b) covers the period from day 159 to 296 of 1995 for the solar-minimum orbit and from day 230 of 2001 to day 4 of 2002 for the solar-maximum orbit.

4S in Figure 5.5) were observed (Lario *et al.*, 2003c). The onsets of events 1S and 3S were affected by the presence of SIRs; event 4S developed over a high pre-existing background level; and event 2S occurred in relatively steady conditions and showed clear signatures of velocity dispersion.

By contrast, at high northern latitudes (right panel of Figure 5.5), Ulysses was immersed in the high-speed (>700 $km\,s^{-1}$) polar solar wind stream and only an inward magnetic field polarity was observed (McComas *et al.*, 2002). Four ICMEs were clearly observed in the solar wind plasma even at these high latitudes (Lario *et al.*, 2004a). Three SEP events were observed during this time interval (labeled 1N, 2N, and 3N in Figure 5.5). The passage of ICMEs under these conditions

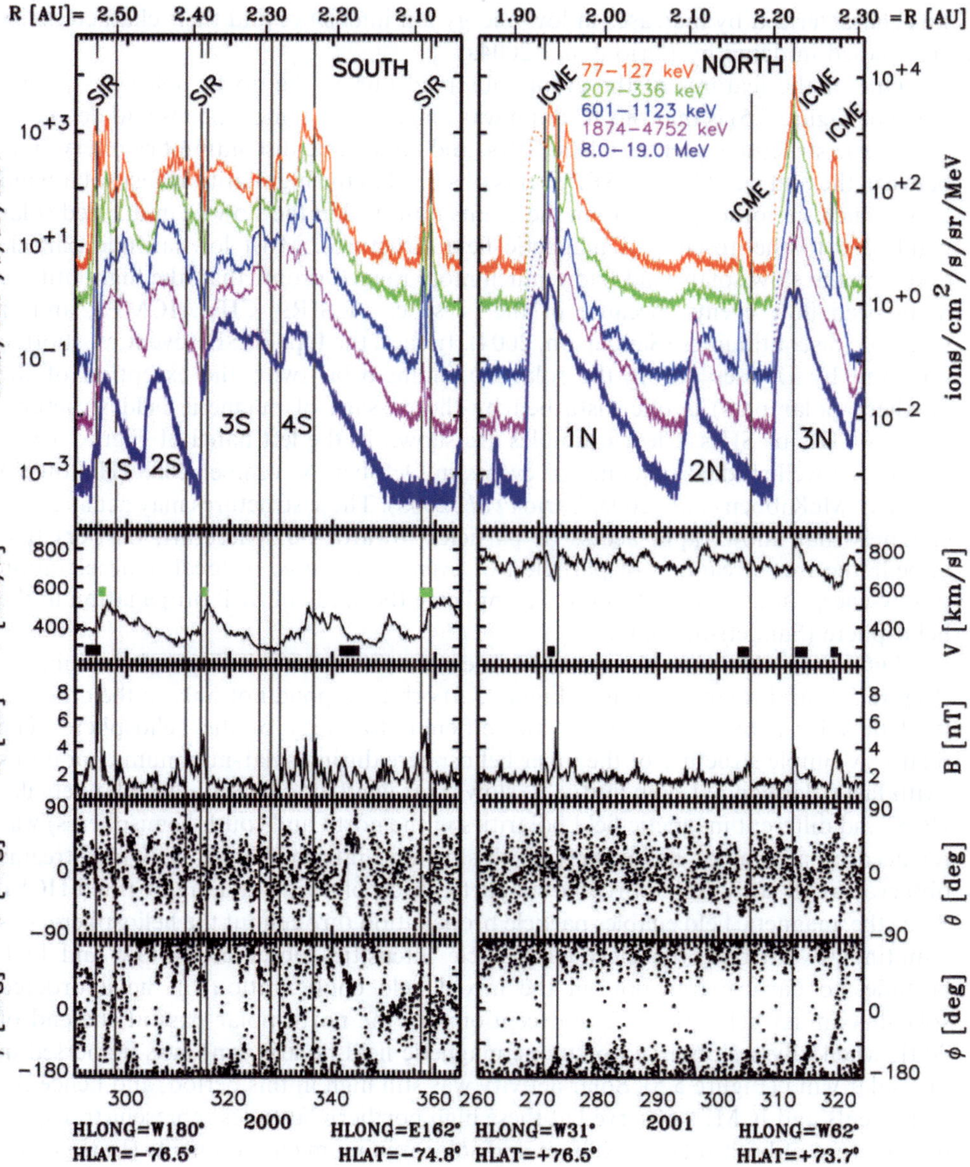

Figure 5.5. Hourly averages of (from top to bottom) ion fluxes measured by the HI-SCALE/LEMS120 (four top traces) and the Low-Energy Telescope (LET) of COSPIN (two lower traces), the solar wind speed measured by SWOOPS (Bame *et al.*, 1992), magnetic field magnitude and orientation in the RTN coordinate system measured by VHM–FGM (Balogh *et al.*, 1992), for the time intervals 290–365 of 2000 (left) and 260–324 of 2001 (right). Solid vertical lines mark the arrival of interplanetary shocks, black rectangles the passage of ICMEs, and green rectangles the passage of SIRs. The high intensity of electrons during the rising phases of the events 1N and 3N produced contamination of the ion channels of the HI-SCALE/LEMS120 detector. We have indicated those periods by dotted traces.

was characterized by increases in low-energy ion intensities and their characteristics are studied in detail by Lario *et al.* (2004a).

One of the features of the events observed during the north polar pass (right panel of Figure 5.5) is that this medium was remarkably homogeneous and devoid of large-scale structures such as CIRs, SIRs, and large-scale discontinuities intervening between the Sun and Ulysses. SEP events observed in the high-latitude fast solar wind had a much smoother profile than the events observed in the slow or mid-speed solar wind, due to the absence of magnetic field structures. Most low-latitude particle events in the slow solar wind had a much more ragged profile than the high-latitude high-speed flow events, because of the presence of SIRs, CIRs, ICMEs, and/or magnetic discontinuities (Sanderson, 2004). In fact, the type of SEP event most often observed by Ulysses during the solar-maximum orbit (with the exception of the northern polar pass) is one disturbed by the passage of magnetic field structures such as CIRs or SIRs. Clear examples are shown in the left panel of Figure 5.5 or during the well-studied sequence of events in October–November 2003 (Halloween Storms) (McKibben *et al.*, 2005; Lario *et al.*, 2005). These structures may act as either channels that allow rapid access of particles toward the spacecraft, barriers that impede the free streaming of particles, or even as a source of local re-acceleration of low-energy ions. These structures complicate the study of SEP propagation in the heliosphere (Sanderson, 2004).

Differences between the particle intensities observed during solar-minimum (Figure 5.1) and solar-maximum (Figure 5.3) orbits respond not only to the different level of solar activity but also to the different topology of the heliosphere. The relatively simple structure of the inner heliosphere during solar-minimum conditions (with fast solar wind at high latitudes, slow solar wind at low latitudes, relatively flat HCS, and different magnetic field polarities in the north and south hemispheres) was replaced by a complex heliosphere (with slow and intermediate solar wind streams observed at all latitudes, unordered magnetic field polarities, and highly tilted HCS). Since the magnetic field enables particle propagation throughout the heliosphere, it is tempting to attribute the elevated particle intensities observed at low and high latitudes to particle transport in the mixed field configuration of the unordered solar-maximum heliosphere. The exception was the north polar pass at the end of 2001, when Ulysses observed only one magnetic field polarity and was immersed in fast solar wind (Figure 5.5). Solar activity was still high in this period, and hence the SEP events and ICMEs observed at these high northern latitudes represent transport in a mixture of both solar-maximum and solar-minimum environments (Lario *et al.*, 2003c).

The transition from solar minimum to solar maximum is observed also in the variation of the elemental abundances throughout the Ulysses mission. In order to understand how SEPs are ubiquitously observed at high and low heliolatitudes, it is essential to determine both the properties of the particle sources and particle transport in the complex solar-maximum heliosphere. Solar observations, composition analyses, particle anisotropy observations, and multi-spacecraft detection of SEP events help us to understand the processes of particle acceleration and transport in the inner heliosphere and the transition from solar minimum to solar maximum.

5.4 COMPOSITION ANALYSES (1990–2005)

Low-energy ($\lesssim$2 MeV/nucleon) ion population in the inner heliosphere at solar minimum is mainly dominated by CIR processes whereas at solar maximum transient events of solar origin increase their contribution. Elemental abundances measured in the ecliptic at 1 AU during CIR events are well-differentiated from the abundances measured during SEP events, in particular the H/He decreases in CIRs compared with SEP events, and the C/O ratio which is ~1 in CIRs—roughly a factor of 2 higher than in SEP events (Mason and Sanderson, 1999).

Ulysses has also observed a solar cycle dependence of the H/He ratio. Whereas solar-minimum CIR events show low (<10) 0.5–1.0 MeV/nucleon H/He values, the solar-maximum heliosphere is characterized by high (>20) 0.5–1.0 MeV/nucleon H/He values (Lario *et al.*, 2003a). These high solar-maximum H/He ratios seem to be independent of the heliographic latitude and heliocentric distance of Ulysses (Lario *et al.*, 2003a, b).

Figure 5.6 shows 27-day averages of 0.5–1.0 MeV/nucleon fluxes of C, O, and Fe ions measured by the Wart aperture of HI-SCALE (Lanzerotti *et al.*, 1992). Time interval spans from the Ulysses launch (6 October 1990) to the end of 2005. Fluxes measured during the Jupiter fly-by in February 1992 have been removed from the figure. The gray vertical shading areas mark the polar passes of Ulysses during its

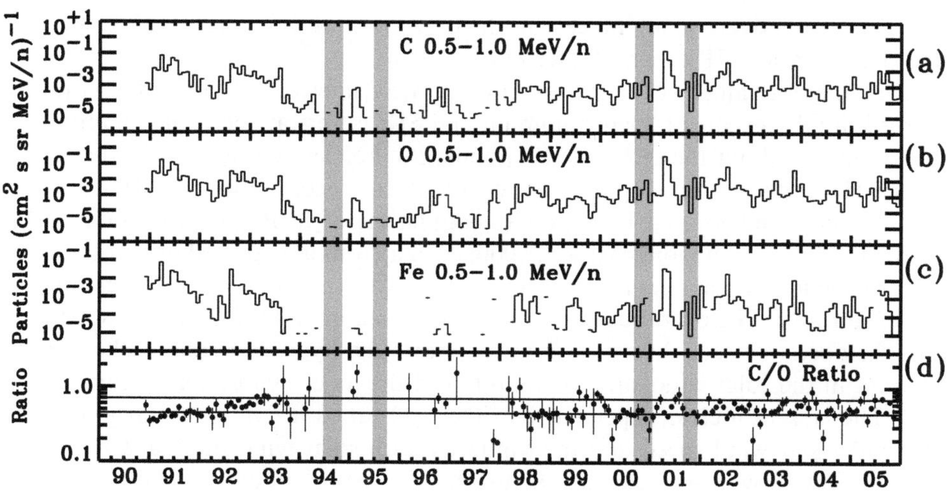

Figure 5.6. 27-day averages of 0.5–1.0 MeV/nucleon carbon, oxygen, and iron fluxes. Bottom panel: carbon-to-oxygen ratio calculated from the C and O traces shown in the two top panels whenever the C flux is above 10^{-4} (cm^2 s sr MeV/n)$^{-1}$. Horizontal lines mark the values C/O = 0.7 and 0.45, and vertical lines through each point represent the statistical error in the measurement. Gray vertical shading areas mark the polar passes of Ulysses in 1994 (S) and 1995 (N), and again in 2000 (S) and 2001 (N). The data covers from launch (6 October 1990) to the end of 2005.

first and second orbits. Ion fluxes during solar minimum (1993–1997) were very low, concentrated near the HCS, and observed mostly in association with CIRs. Note the absence of Fe counts during the solar-minimum high-latitude excursion (Maclennan and Lanzerotti, 1995). Figure 5.6d shows the 0.5–1.0 MeV/nucleon C/O ratio, where the horizontal lines mark the values 0.7 and 0.45. During the first part of the in-ecliptic Ulysses mission, the C/O ratio was low and approximately constant around ~0.4. With the beginning of the CIR recurrent events (at the end of 1992), the C/O started to increase to 0.6. In the middle of 1993, with the immersion of Ulysses in the high-speed coronal hole solar wind flow, the C/O ratios fluctuated between 0.4 and 1.1 (see also Franz *et al.*, 1995). The fluctuating high C/O values were observed also during the fast-latitude scan and the descent to in-ecliptic latitudes in 1996 and 1997. By contrast, during solar maximum, large ion fluxes were observed throughout the Ulysses orbit, regardless of its latitude. The C/O ratios remained mostly below 0.7. The few cases with large C/O ratios corresponded to periods of both low-level solar activity (e.g., at the end of 1999), and the occasional observation of SIR events (Lario *et al.*, 2003a).

The occasional CIR or SIR events observed during solar maximum present different elemental abundances from those measured in solar-minimum CIR events. Richardson *et al.* (1993) noted that the elemental abundances of corotating particle flux enhancements at 1 AU show a clear transition from solar maximum to solar minimum. While solar-minimum CIR events have large He/O and C/O ratios, at solar maximum the events associated with either SIRs or CIRs are more SEP-like (Richardson *et al.*, 1993). Ulysses observations show also a transition of the H/He ratios from <10 at solar-minimum CIRs to >10 at solar-maximum SIRs (Lario *et al.*, 2001b) and a relatively small increase in the C/O ratios at SIR events with respect to those measured at SEP events (Hofer *et al.*, 2003a).

The different values of the low-energy (<2 MeV/nucleon) H/He and C/O ratios observed in CIR events at solar minimum and SIR events at solar maximum raises some questions about the seed particle population accelerated in stream interaction regions. Heliospheric ion populations' candidate for acceleration in SIRs are the thermal solar wind ions, a background population of SEPs, and the interstellar and inner-source pickup ions (Mason, 2000). Pickup ions are favored over solar wind ions for injection into acceleration mechanisms because of their higher energies in the solar wind frame. The overabundance of He at solar-minimum CIR reverse shocks at radial distances between 3.0 and 5.0 AU supports the hypothesis that pickup He is dominant in the solar wind suprathermal tail of the inner heliosphere (Franz *et al.*, 1999). The lack of a clear dependence of the C/O ratios in Ulysses solar-minimum CIRs with solar wind speed and the similarity with solar wind thermal C/O abundances do not allow us to reach a conclusion as to the origin of these ions (Franz *et al.*, 1999). Note, however, that observations of CIR events at 1 AU show that the C/O ratios are a strong function of the solar wind speed, suggesting that C may also originate from pickup ions although from a different source than the interstellar pickup ions (Mason *et al.*, 1999). Corotating streams are generally slower in solar maximum, and since the pickup ion velocity scales directly with solar wind speed, the injection of these ions is less favored during solar maximum than in solar minimum.

Hence the higher values of the H/He ratio and the lower values of the C/O ratio during solar-maximum SIRs. Additionally, direct contamination of SEPs into solar-maximum SIR events also alters the elemental abundances that are more SEP-like than those observed in solar-minimum CIRs (Lario *et al.*, 2001b; Hofer *et al.*, 2001). Only in the absence of SEP contamination and in well-formed high-speed CIR events (such as those observed in the first part of 2002 after the solar-maximum north polar pass) are the H/He and C/O ratios similar to those observed in solar-minimum CIRs (Lario *et al.*, 2003b; Hofer *et al.*, 2003b).

Large SEP events during the solar-maximum orbit, even during the north polar pass, present enhancements of heavy-ion intensities with He/O, C/O, and Fe/O ratios similar to those observed in large gradual SEP events at 1 AU (Hofer *et al.*, 2003c). ^{3}He enrichments have been usually used as a signature of either impulsive SEP events or re-acceleration of a ^{3}He remnant population seeded in the interplanetary medium by an earlier impulsive event. Tranquille *et al.* (2003) analyzed enhancements in the ^{3}He/^{4}He ratio in the energy range 2–20 MeV/nucleon observed by the Low-Energy Telescope (LET) of COSPIN (Simpson *et al.*, 1992). Only five periods of enhanced ^{3}He/^{4}He (defined as ^{3}He/^{4}He ratio above 0.05) were observed during the solar-maximum orbit (until May 2002). These ^{3}He enhancements were associated with weak SEP events and were observed at mid-latitudes (between 24° and 48°). In large SEP events (such as the events at high northern latitudes in November 2001), the Ulysses instrumentation does not allow a clear distinction between ^{3}He and ^{4}He enhancements, probably because of the dominant ^{4}He contribution in large intense SEP events (Tranquille *et al.*, 2003). No evidence of ^{3}He enhancements was found at high heliolatitudes in association with solar-minimum CIRs (Biesecker, 1996; Tranquille *et al.*, 2003).

Solar cycle variations of the elemental abundance ratios at higher energies ($\gtrsim$2 MeV/nucleon) are influenced by modulation of ACR oxygen. ACRs are subject to solar modulation in the heliosphere, leading to larger intensities at solar minimum than at solar maximum. In addition, ACR oxygen ions are also subject to latitudinal gradients in the solar-minimum heliosphere, with peak intensities seen when Ulysses was at high solar-minimum latitudes (Marsden *et al.*, 1999). Figure 5.7 shows the abundance ratios He/O (black), C/O (red), and N/O (green) in the energy range 4–8 MeV/nucleon averaged over 40 days from Ulysses launch to the end of 2005. Vertical lines through each point indicate the statistical error associated with the measurements. The ratio N/O remains fairly constant throughout the complete mission, unlike the He/O and C/O ratios that are very sensitive to solar activity and latitudinal effects. The contribution from ACR ions to the ratio N/O is similar in both ion species and hence the constant ratio throughout the Ulysses mission. The He/O and C/O ratios clearly show the influence of the anomalous oxygen ions that are more abundant during solar minimum and at high latitudes. The He/O ratio is very sensitive to the ions of solar origin, with increases during the solar-minimum fast latitude scan in 1995, at the end of 1997, and throughout 1998 in the rising phase of solar cycle 23 (Tranquille, Marsden, and Sanderson, 2001). During solar maximum, the smaller fluxes of ACRs were swamped by the higher intensities of the SEPs (Maclennan, Lanzerotti, and Gold, 2003).

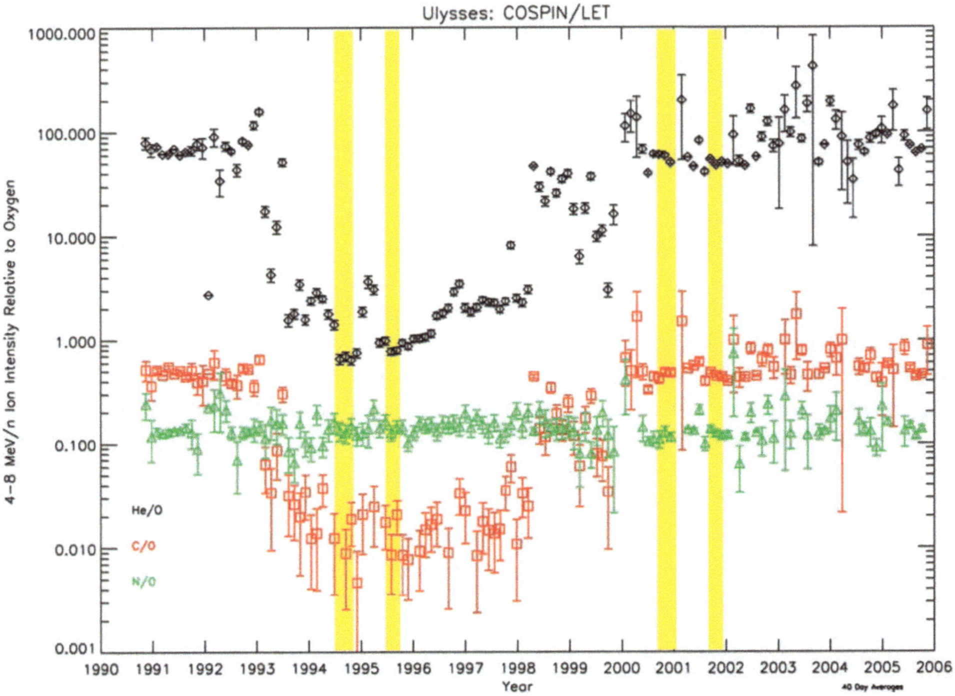

Figure 5.7. 40-day-averaged abundance ratios of He (black), C (red), and N (green) ions with respect to O in the energy range 4–8 MeV/nucleon as measured by the COSPIN/LET instrument. Yellow vertical shading areas mark the polar passes of Ulysses in 1994 (S) and 1995 (N), and again in 2000 (S) and 2001 (N). The data covers from launch (6 October 1990) to the end of 2005. Vertical lines through each point represent the statistical error in the measurement.

5.5 MULTI-SPACECRAFT OBSERVATIONS OF SEP EVENTS: ULYSSES AND NEAR-EARTH OBSERVATIONS

Simultaneous observations by near-Earth spacecraft and the Ulysses spacecraft at either low latitudes (Lario *et al.*, 2000b) or high latitudes (McKibben *et al.*, 2003) show that most events that produce large high-energy (>20 MeV) proton and near-relativistic electron flux increases near Earth also produce flux increases at Ulysses, even at the highest latitudes attained by Ulysses (Lario *et al.*, 2003c). Hypotheses to explain the concurrent observation of large SEP events regardless of the longitudinal, latitudinal, and radial separation between the spacecraft include (i) particle sources that cover a broad range of latitudes and longitudes and/or (ii) transport mechanisms that allow an efficient distribution of particles in longitude and latitude, both along and across the mean magnetic field.These transport processes include both an effective motion of energetic particles across the field lines and/or a random walk or spatial meandering of field lines allowing energetic particle transport perpendicular to the mean magnetic field. Observational evidence that discriminates between the above hypotheses is as follows:

- Particle anisotropies during SEP onsets at high latitudes are typically directed outward from the Sun and aligned with the local magnetic field (McKibben *et al.*, 2003; Lario *et al.*, 2003c; Sanderson *et al.*, 2003). The observed field-aligned anisotropies, with components perpendicular to the local magnetic field that are essentially zero, indicate that there is no net flow of particles across the local magnetic field. Therefore, particles travel mainly anti-sunward along the field lines and without crossing over them. The anti-sunward direction of the ion flows also indicates that the main source of particles is located inside the orbit of Ulysses. Particle sources may be either (i) wide enough to inject particles into a broad stretch of the heliosphere at both low and high latitudes, or (ii) located close to the flare site and able to inject energetic particles onto field lines that meander to high latitudes. The local observation of field-aligned anisotropies does not preclude the possibility that cross-field diffusion may occur close to the Sun (i.e., particles diffuse across field lines inside the orbit of Ulysses before they arrive at Ulysses). Another possibility includes the continuous injection of particles from the Sun or from CME-driven shocks. The distribution in longitude and latitude of energetic particles in this scenario may be due to magnetic field structures formed beyond Ulysses that are able to both spread the particles in longitude and latitude and scatter them back toward the Sun. The continuous injection of particles together with the focusing effect close to the Sun allows for the observed anti-sunward anisotropies. See discussion of these mechanisms in Lario *et al.* (2003c).
- Dalla *et al.* (2003) analyzed the onset time and time-to-maximum of nine high-latitude SEP events and correlated these quantities with the angular separation between the associated flare site and Ulysses, the latitudinal separation between the flare site and Ulysses, and the radial distance between Ulysses and the Sun. The best correlation was found with the latitudinal distance between the flare site and Ulysses. This implies a very effective latitudinal transport of the particles, but a very inefficient transport longitudinally. The authors concluded that cross-field diffusion was the fundamental mechanism in getting particles to high latitudes. However, these authors did not rule out the possibility that the increasing delay was due to the time taken for the CME-driven shock to reach the field lines connected to the spacecraft (Sanderson, 2004).
- Models of particle transport assuming that (i) energetic particles are injected from localized narrow sources on the Sun, (ii) energetic particles propagate only along field lines, and (iii) the footpoints of the field lines move stochastically at speeds and on timescales consistent with those of the super-granulation motion on the Sun are unsuccessful in explaining the early phase of SEP events observed concurrently by in-ecliptic near-Earth spacecraft and Ulysses at high latitudes (Giacalone, Jokipii, and Zhang, 2001; Giacalone, 2002; Zhang *et al.*, 2003).
- The concurrent observation of the SEP events during the solar-maximum northern polar pass by Ulysses (immersed in uniform solar wind coronal hole flow with only one magnetic field polarity observed) and near-Earth in-ecliptic spacecraft (immersed in slow solar wind and observing different magnetic field polarities, Lario *et al.*, 2003c), excludes the possibility that particles propagate along

field lines originating at low latitudes and that reach Ulysses at high latitudes by spatial meandering (Lario *et al.*, 2003c). Unless there was a large distortion of the field lines that organized them at northern high latitudes, a random walk of field lines from low to high latitudes is not possible. The heliospheric magnetic field proposed by Fisk (1996) during solar minimum does not account for the common SEP observations at ACE and Ulysses since magnetic connection from low to high latitudes is attained only at large heliocentric distances ($\sim$10 AU).

Analysis of the solar sources associated with the large SEP events concurrently observed by Ulysses and near-Earth spacecraft is essential to understand both the origin of the energetic particles and the processes of energetic particle transport in the solar-maximum heliosphere.

Multi-wavelength observations have been used over these last 10 years to investigate the acceleration of energetic particles at the Sun (e.g., Krucker *et al.*, 1999; Pick *et al.*, 1998; Pohjolainen *et al.*, 2001; Maia and Pick, 2004; Lehtinen *et al.*, 2005). The present understanding is that acceleration processes are associated with large-scale eruptive phenomena that include flares, filament eruptions, CMEs, and shocks which often occur simultaneously. This coincidence emphasizes the difficulty of understanding the link between solar processes and SEP events measured in the interplanetary medium. Radio observations, though restricted to investigations on the solar origin of accelerated electrons, can however bring an important contribution to the problem: they cover a broad frequency domain and observations at different frequencies sample different heights and physical conditions in the solar atmosphere, with longer wavelengths referring to higher heights above the photosphere. In the current two-class paradigm reviewed by Reames (1999), the flare processes account for acceleration in "impulsive" events, while prolonged acceleration by "CME-driven shocks" dominates in "gradual" events; some "hybrid SEP" events may however contain both particles from flares and from CME shock origin. Intense SEP events are usually associated with both major flares and large CMEs; and thus the relative roles of CME-driven shocks and flares in producing high-energy particles is not completely understood (Cliver and Cane, 2002).

There is ample evidence that coronal energy release and electron acceleration processes can last from several minutes to hours (e.g., Trottet, 1986; Akimov *et al.*, 1996; Maia *et al.*, 1999). Thus, these processes can also contribute to the production of SEPs. For example, intense, complex, and long-duration kilometric type III burst events (which are produced by beams of suprathermal electrons injected into the interplanetary medium) have a good temporal correspondence with radio emissions observed at higher frequencies. This correspondence suggests that both emissions are generated by electrons accelerated in the lower corona over extended time periods (Kundu and Stone, 1984; Reiner *et al.*, 2000). Cane, Erickson, and Prestage (2002) showed that >20 MeV proton events are associated with long-duration groups of type III bursts. These complex events are usually accompanied by the presence of several coronal non-thermal radio sources which are often located far from the flaring region and that usually spread over a large angular extent.There is a close association between these complex events and large CMEs having a width of at least 100° (Pick

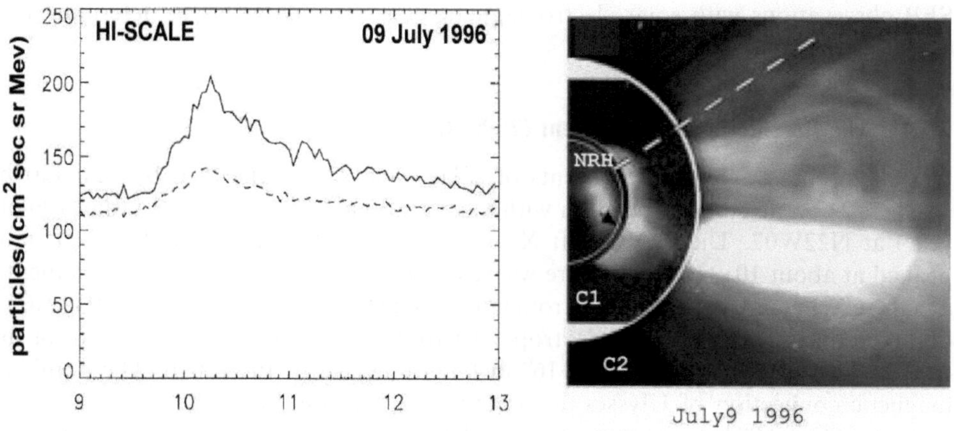

Figure 5.8. 1996 July 9. Left panel. Electron event observed by HI-SCALE. Solid line: 38–53 keV energy range. Dashed line: 53–164 keV energy range. Right panel. Composite image including: the radio sources seen by the Nançay Radioheliograph (NRH) at 164 MHz at 09:12 UT, the LASCO coronagraphs C1 at 09:23 UT and C2 at 09:28 UT. The broken line indicates the polar angle of the coronal radio source associated with the electron event (from Pick *et al.*, 1998).

and Maia, 2005). Most of the large SEP events observed by Ulysses and Earth satellites are associated with these complex events.

The first event of this class which was observed conjointly at the Sun by the LASCO coronagraphs onboard SOHO (Brueckner *et al.*, 1995), the Nançay Radioheliograph (NRH) (Kerdraon and Delouis, 1997), and in the interplanetary medium by Ulysses occurred on 1996 July 9 (day 190 of the year). This event was associated with an Hα flare at S10W30 and a radio burst starting at 09:10 UT. Ulysses was at $R = 4.06$ AU, $\Lambda = 32°$N in latitude, and 223° west of the Earth in longitude. The nominal magnetic connection of Ulysses (assuming a Parker spiral at the observed solar wind speed of 750 km s^{-1}) was 83° west. The electron event at Ulysses (shown in Figure 5.8, left panel) was relatively small. HI-SCALE measured electron intensity increases only at energies below 178 keV. The length of the path traveled by these electrons was found to be about 10 AU for a pitch angle of 60°. The magnetic connection of Ulysses was not far in latitude from the location of a radio source (N16W45) when type II (shock) and type III burst activity (electron beams) were observed. Figure 5.8 (right panel) displays a composite image including the CME seen by the LASCO coronagraphs C1 and C2 and the radio sources observed by the NRH. The dashed line indicates the polar angle of the coronal radio source associated with the electron event.The electrons detected by Ulysses had a coronal origin (Pick *et al.*, 1998).

In the following sections we present the solar observations related to two major SEP events observed during the maximum of solar cycle 23 by ACE in the ecliptic at 1 AU and Ulysses at high southern latitudes (the Bastille Day 2000 event) and high northern latitudes (the 2001 September 24 event). Comparison of multi-spacecraft SEP observations with solar electromagnetic emissions allows us to determine the extent of the sources of SEPs.

5.5.1 The Bastille flare/CME event (2000 July 14)

One of the most intense SEP events of solar cycle 23 (as observed by near-Earth spacecraft) occurred in association with a X5.7/3B flare on 2000 July 14 (day 196 of year) at N22W07. The GOES soft X-ray flux started to increase at 10:03 UT and peaked at about 10:24 UT. The flare was accompanied by the eruption of a filament, a halo CME, and by many electromagnetic signatures (Maia *et al.*, 2001). ACE observed an intense prompt anisotropic electron onset at 10:39 UT. Ulysses was at $R = 3.17$ AU and $\Lambda = 62°$S and 116° in longitude east of the Earth. The nominal magnetic connection of Ulysses (assuming a Parker spiral and the observed solar wind speed of 600 km s^{-1}) was close to the longitudinal location of the flare but separated ~80° in latitude. The left panel of Figure 5.9 shows 175–315 keV electron intensities observed by the Electron, Proton, and Alpha Monitor (EPAM) onboard ACE (i.e., the spare instrument of HI-SCALE; Gold *et al.*, 1998) (gray trace) and the HI-SCALE on Ulysses (black trace). The right panel shows 30–70 MeV proton intensities measured by the Cosmic Ray Nuclear Composition (CRNC) instrument on IMP-8 (gray trace) and by the High-Energy Telescope (HET) of COSPIN on Ulysses (black trace). The rise of particle intensities at 1 AU was rapid and anisotropic. The first indication of an increase in the 175–315 keV electron intensities above the existing background at ACE began about 10:38 UT. Particle increases at Ulysses were gradual with anti-sunward anisotropies. The first indication of a

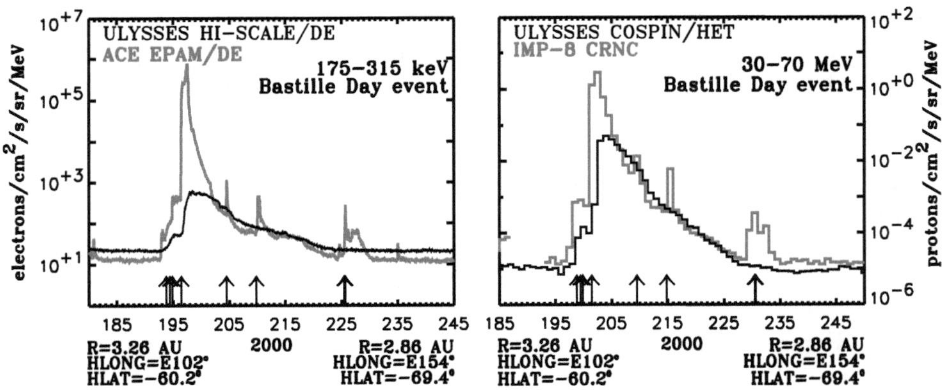

Figure 5.9. Left panel. Hourly averages of the 175–315 keV electron intensities measured on ACE by EPAM (gray trace) and on Ulysses by HI-SCALE (black trace) during the period associated with the Bastille Day 2000 event. Right panel. Daily average of the 30–70 MeV proton intensities measured on IMP-8 by CRNC (gray trace) and on Ulysses by COSPIN/HET (black trace). The arrows indicate the occurrence of the X-ray flares and CMEs associated with the major SEP events at 1 AU as identified by Smith *et al.* (2001).

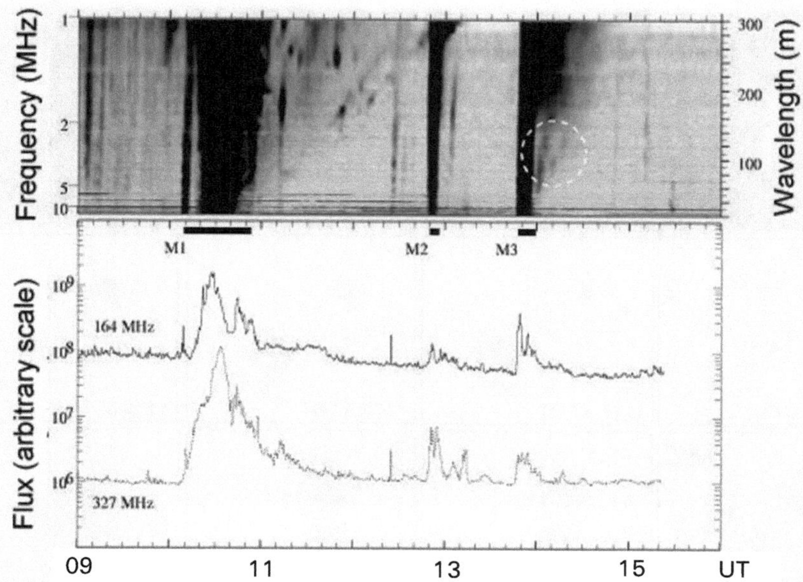

Figure 5.10. 2000 July 14. Top: Dynamic spectrum in the decametric/hectometric wavelength range as observed by WIND/WAVES. Three major outbursts M1, M2, and M3 are evident. The dashed white circle indicates a period with features of interplanetary type II radio bursts (Maia *et al.*, 2001). Bottom: Flux plots at two frequencies measured by NRH.

175–315 keV electron intensity increase at Ulysses occurred at about 16:00 UT on day 196. During the onset of this event, the Earth was not well-connected to the flare site. The estimated release time for the electrons observed at ACE is ~10:32 UT (corrected for 8 minutes for comparison with the solar events); therefore, the estimated electron injection was delayed with respect to the occurrence of the X-ray flare.

Figure 5.10 displays a WIND/WAVES dynamic spectrum (Bougeret *et al.*, 1995) and NRH flux plots at two discrete frequencies. In addition to the main event, shortly after 10:00 UT, there were two other strong occurrences at 12:50 UT and 13:48 UT. The periods of these three major events are labeled M1, M2, and M3 in Figure 5.10. These events were composed of type III bursts and evidence of type II shock-associated emission for M1 and M3. Figure 5.11 displays NRH flux images showing the evolution of the emitting sources. The initial radio emission began in the vicinity of the flare site, then in a timescale of less than 15 minutes it spanned a large extent in longitude and latitude. The anisotropic electron event seen by the EPAM/ACE detector agrees well in time with the appearance near 10:31 UT of new radio sources seen in the western quadrant at a longitude consistent with the location of the magnetic footpoint of ACE. The abrupt evolution of the western emissive region was attributed to the restructuring of the magnetic field configuration related to the passage of Hα material ejected from the flaring active region (Maia *et al.*, 2001).

The third and fourth rows of Figure 5.11 show a series of NRH images observed during events M2 and M3. The main differences between the M2 and M3 events with

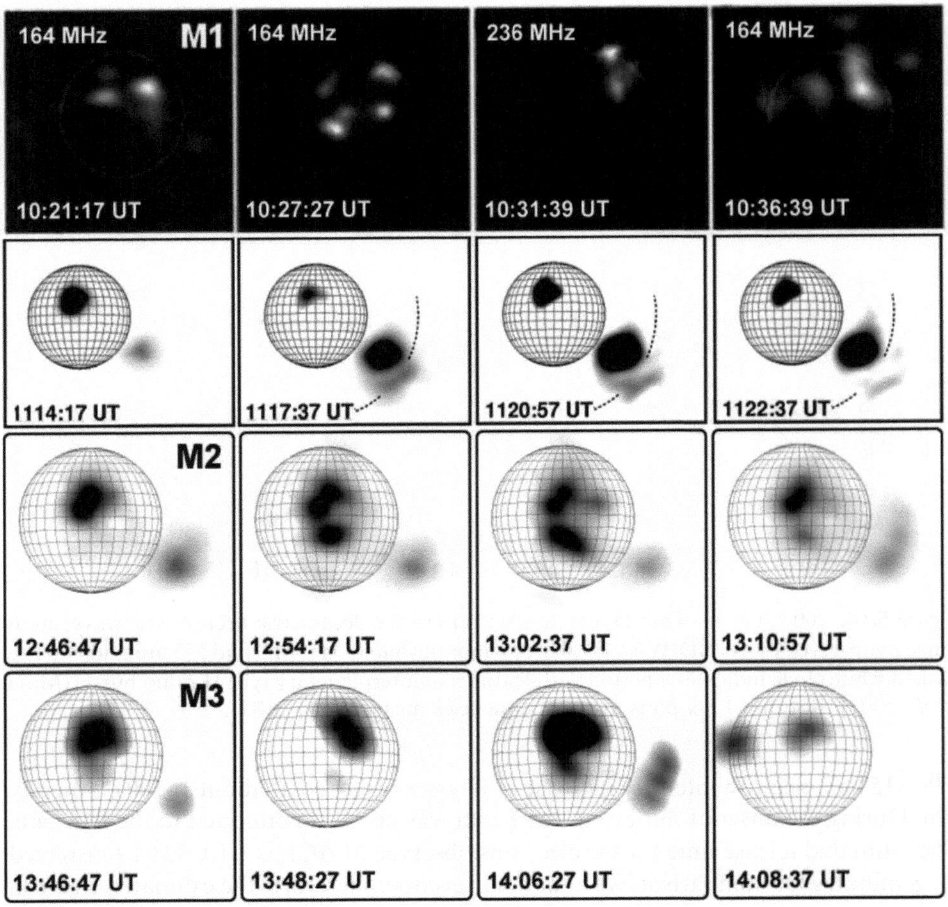

Figure 5.11. 2000 July 14. NRH images at 164 MHz illustrating the development of the period M1 (top row), the extension of an outward southward-directed moving source (second row), the period M2 with activity extending southward (third row), and the period M3 (bottom row). See Maia *et al.* (2001) for details.

respect to the M1 event are that the emissions did not extend to the east limb but they extended toward the south and west. These events developed similarly to M1 with evidence of association with CMEs. The events M2 and M3 were not detected by the LASCO coronagraphs due to energetic particles hitting the CCD. Radio-imaging observations showed that the emitting sources were moving and seen up to 2.5 solar radii from the center of the Sun. The southern extension of these emissions may be interpreted as the extension of particle sources close to the latitude of the footpoint of the nominal field line connecting Ulysses with the Sun. The delay of the particle onset at Ulysses with respect to the occurrence of the X5.7/3B flare can be naturally explained by the existence of subsequent high-latitude emissions energetic enough to produce the mechanisms for injection of SEPs and re-acceleration of SEPs from prior events. The difference in the arrival time of electrons and protons at Ulysses for this event was attributed by Zhang *et al.* (2003) to a rigidity-dependent transport between the Sun and the spacecraft that can perfectly occur along magnetic field lines without invoking cross-field diffusion.

5.5.2 The 2001 September 24 event (day 267 of year)

Figure 5.12 shows 175–315 keV electron and 30–70 MeV proton intensities measured by ACE and IMP-8 (gray traces) and by Ulysses (black traces) during part of its solar-maximum north polar pass when Ulysses remained immersed in the coronal hole solar wind flow and observed only a single magnetic field polarity (Figure 5.5). One of the most intense events during this period was the event 1N observed by Ulysses at $R = 1.90$ AU and $\Lambda = 78°$N and 34° to the west with respect to the Earth. The nominal footpoint of the Ulysses spacecraft computed assuming a Parker spiral and a solar wind speed of 800 km s^{-1} was about 90°W and at high northern latitudes. The onset of the electron event at ACE occurred at about 10:55 UT on day 267 characterized by a rapid increase and strong anti-sunward anisotropic beams. By assuming a scatter-free propagation, an estimated time (corrected by 8 minutes for comparison with the solar event) for the release of the electrons observed by ACE is about 10:49 UT. The onset of the electron event at Ulysses was observed at about 15:40 UT and characterized by a gradual enhancement and weakly anti-sunward flows (Lario *et al.*, 2003c).

The difference between the particle anisotropies at both spacecraft can be related to the medium where particles propagate. Sanderson (2004) compared energetic ion anisotropies observed during the SEP events at the north polar pass with those observed in the ecliptic plane and closer to the Sun (i.e., ~1 AU). Anisotropies associated with the events at high latitudes are small in comparison with the events

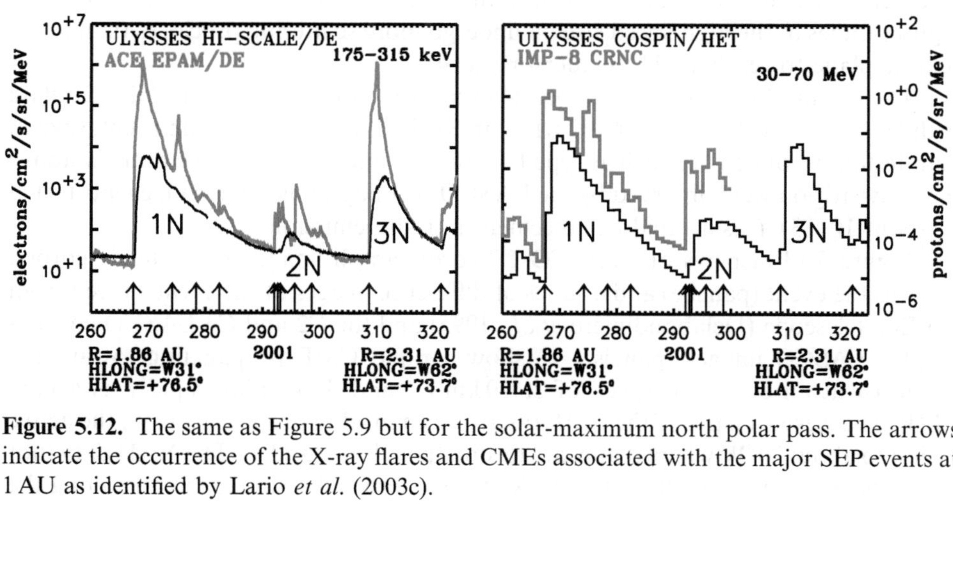

Figure 5.12. The same as Figure 5.9 but for the solar-maximum north polar pass. The arrows indicate the occurrence of the X-ray flares and CMEs associated with the major SEP events at 1 AU as identified by Lario *et al.* (2003c).

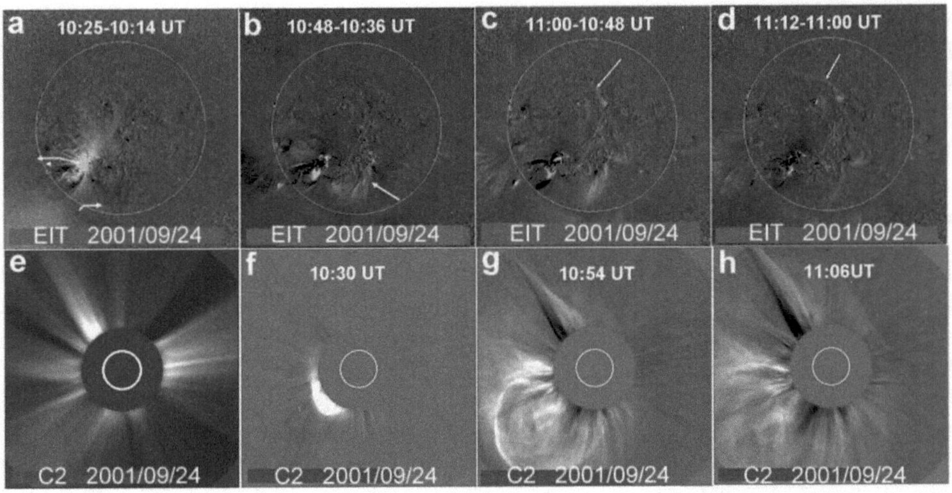

Figure 5.13. 2001 September 24. Difference images from EIT (panels a–d) and LASCO (panels f–h) showing the development of the event. Panel (e) shows the pre-event corona with a well-marked streamer in the northeast quadrant. The arrows in panel (a) indicate the direction of the type III bursts as seen by NRH. The arrows in panels (b–d) indicate the extension of the CME flanks toward the southwest (panel b) and north (panels c–d). Panels (g–h) show the extension of the CME and the distortion of the northeast streamer.

at 1 AU. Whereas the events at 1 AU show rapid onsets, the events at high northern latitudes show slow onsets. Both the gradual onsets and the small anisotropies of the SEP events at high northern latitudes suggest that particles are scattered significantly as they propagate outwards within the fast solar wind. The fast solar wind tends to be mainly homogeneous and devoid of large-scale discontinuities, but is much more turbulent than the slow solar wind (Smith, 2003), so that particles propagating to high-latitudes in the high-speed stream undergo more scattering processes than in the ecliptic plane to reach 1 AU (Sanderson, 2004).

The origin of the 2001 September 24 event was associated with an X2.6/2B flare with Hα emission starting at 09:32 UT from S16E23. Prior to the main event, several manifestations of activity such as type III emission and a considerable outflow above the active region were observed by the LASCO coronagraphs. The main event started at about 10:13 UT as a rapid enhancement in radio emission.

Figure 5.13 displays one LASCO C2 coronagraph image of the steady corona prior to the event (panel e) and a series of difference images of the event as seen from the EIT telescope (Delaboudinière *et al.*, 1995) and the C2 and C3 LASCO coronagraphs. LASCO images (panels f–h) show a rapid CME propagating toward the southeast at an estimated speed of 2,400 km s^{-1} and developing rapidly as a halo CME. The western flank of the CME (as also seen in EIT) propagates from the active region to about 20°W where it appears to stop (panels b–c and g). The NRH data (not shown here) show from 10:12 to 10:28 UT a western radio source moving along the same direction at the speed of about 470 km s^{-1} and similar to what is estimated from the EIT images (small arrow pointing westwards in panel a). The NRH emission may be interpreted as the signature of the CME expanding in the lower corona.

The east flank of the CME shows a development similar to the west flank expanding toward the north and displacing the streamer observed before the event. The NRH data (10:27–10:40 UT) show a source moving eastward and slightly toward the north at an estimated speed of about 580 km s^{-1} (eastward arrow in panel a). Coronal structures all around the CME (toward the west, north, and east) seem to be affected and distorted by the CME expansion (panels g–h).

The event at ACE

Both coronagraph and EIT data make evident the existence of a boundary region where the lateral expansion of the western flank of the CME stops (indicated by an arrow in panel b). This region is reached by the CME flank near 10:48 UT (panel g) coinciding with the estimated release time of the electrons observed by ACE. Figure 5.14 displays the magnetic field configuration derived by applying a potential field source surface (PFSS) extrapolation to magnetograph measurements of the photospheric field on that day (Schrijver and Derosa, 2003). The C2 and C3 images (Figure 5.13) and the potential magnetic field extrapolation (Figure 5.14) suggest an area of open field beyond this boundary where the CME expansion stops. These observations are consistent with the CME creating a compression region at the interface between closed and open-field line areas. Beyond this region all coronal structures appear to be distorted (panels g–h of Figure 5.13) consistent with the propagation of a compression wave or shock. Evidence of type II burst detected by the high-frequency receiver of WIND/WAVES from 7 MHz (10:40 UT) to 4 MHz (10:55 UT) (not shown here) suggests the existence of a shock. By assuming that the emission is at the second harmonic and using $f_p * \text{radius} = 20\ \text{kHz} * 1\ \text{AU}$ (where f_p is the plasma frequency and $2f_p$ the radio emission frequency), we derive a speed of 710 km s^{-1}. However, because of the absence of position information at these frequencies we cannot determine the exact location of the shock. This type II burst occurred in association with several interplanetary type III bursts and with very narrow band and slowly drifting features below 5 MHz.

Other evidence suggesting that the electrons observed by ACE originate near this boundary region are the following. The low-frequency receiver of WIND/WAVES provides the possibility to determine the location of the radio sources. From the onset of the event up to 10:40 UT, the emission comes from the ecliptic plane (within 5°) and from the east. At that time, the bursts (with evidence of type III emission) start coming from the west and close to the ecliptic (for a description of the geometry of the observations see Hoang *et al.*, 1998). We conclude that all these sets of observations agree with the origin of the electrons observed by ACE as associated with this boundary region close to the open field line, which is a region of disturbed radio emission.

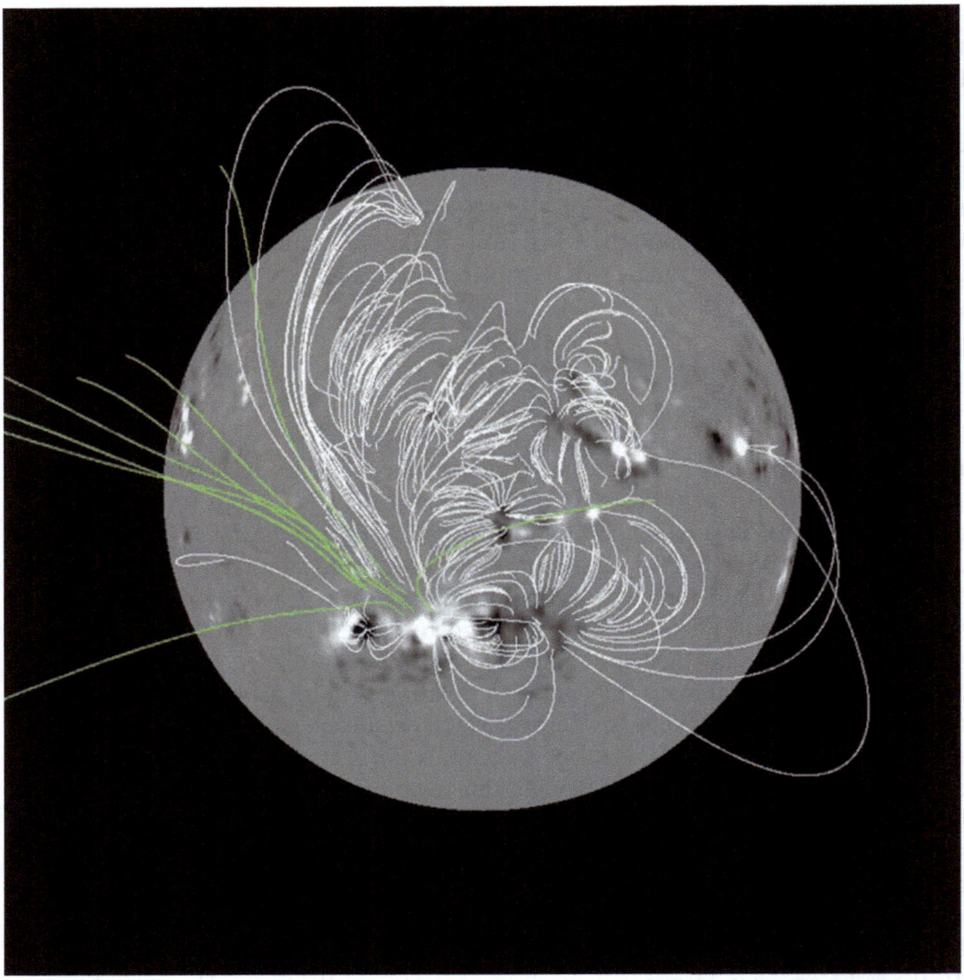

Figure 5.14. 2001 September 24. Magnetic field line configuration above the active region at S16E23 derived by applying a potential field source surface (PFSS) extrapolation to magnetograph measurements of the photospheric field. Open and closed-field lines are plotted green and white, respectively. Note the large loop connecting the flaring region with high northern latitudes. Similar examples of PFSS calculations can be found in Wang, Pick, and Mason (2006) and references therein.

The event at Ulysses

Since (1) particle anisotropies at Ulysses were field-aligned and in the anti-sunward direction with no net flow of particles across the magnetic field, (2) Ulysses observed only a single inward magnetic field polarity, and (3) Ulysses' nominal connection was at northern latitudes, the particle sources of this SEP event had to extend to high latitudes. The lateral expansion of the eastern CME flank shows an extension toward the north (panels c–d and g–h in Figure 5.13). Evidence of the effects of a shock is given by the displacement of the streamer in the northeast quadrant of the figure (i.e., panels g–h). At the same time EIT images show a bright elongated feature on the disk extending toward the north that may be located at the base of the streamer (arrows in panels c–d). The signatures of the shock in white light are progressively observed toward the northern latitudes reaching the polar regions (panel h). Similarly to the western expansion of the CME, a compression region develops in the north where the CME expansion appears to stop close to a region of open field lines associated with the reformed northern polar coronal hole. Figure 5.14 shows a large transequatorial loop system.

The energetic particles observed by Ulysses during this SEP event may result from direct particle acceleration by the CME-driven shock when it reaches the open field lines of the polar regions. Other possibilities exist for the electrons—for example, they are accelerated in the compression region where the expansion of the CME stops and then injected onto the open field lines at the polar coronal hole. We cannot exclude either that the electrons were accelerated close to the active region or propagated along the existent open field lines close to the active region and near the Sun (Figure 5.14).

The correlation found by Dalla *et al.* (2003) between the time delay of SEP event onsets and the latitudinal distance between flare site and spacecraft location can be explained by the time that the CME disturbances take to reach the field lines connected to Ulysses. However, whereas the timescale of onset delays is of the order of hours, the time that CME-associated disturbances take to spread over the corona are of the order of minutes. Therefore, transport processes need to be included to explain the delays of the SEP event onsets observed by Ulysses.

5.6 HELIOSPHERIC ENERGETIC PARTICLE RESERVOIRS

One of the discoveries made by Ulysses that affects our understanding of energetic particle propagation in the heliosphere is the observation of energetic particle reservoirs at both low and high latitudes (McKibben *et al.*, 2003; Lario *et al.*, 2003c). Particle intensities measured in the late phase of large SEP events by widely separated spacecraft often present equal intensities (to within a small ~2–3 factor) that evolve similarly in time. These periods of small radial, longitudinal, and latitudinal particle intensity gradients were first noted by McKibben (1972) and were named "reservoirs" by Roelof *et al.* (1992). Those periods of equal intensities have been observed during isolated large SEP events and also during periods of intense solar activity when a sequence of events occurs at the Sun (Roelof *et al.*, 1992; Reames, Barbier, and Ng, 1996; McKibben *et al.*, 2003; Lario *et al.*, 2003c). The formation of energetic particle reservoirs is not exclusive to protons; it has also been observed using heavy-ion and electron data (Maclennan, Lanzerotti, and Roelof, 2001; Lario *et al.*, 2003c).

Figures 5.9, 5.12, and 5.15 show 175–315 keV electron and 30–70 MeV proton intensities measured by ACE and IMP-8 (gray traces) and Ulysses (black traces)

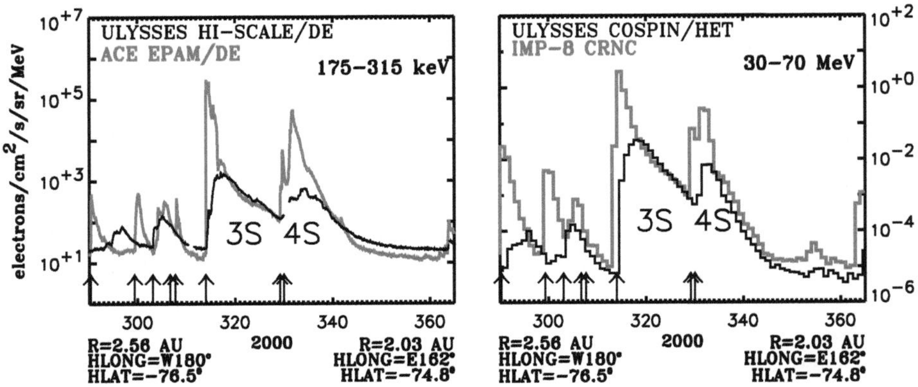

Figure 5.15. The same as Figure 5.9 but for the solar-maximum south polar pass. The arrows indicate the occurrence of the X-ray flares and CMEs associated with the major SEP events at 1 AU as identified by Lario *et al.* (2003c). SEP events labeled 3S and 4S follow the notation of Figure 5.5.

during some of the most intense SEP events of solar cycle 23. Typically, the rise to maximum intensity in SEP events is slower at Ulysses and the maximum intensity is also lower at Ulysses than at near-Earth spacecraft. Presumably both differences are a result of some combination of Ulysses' larger radial distance from the Sun and the difficulty of propagation for energetic particles that move along high-latitude field lines. However, the most striking features of Figures 5.9, 5.12, and 5.15 are (1) the relatively similar profiles of electrons and high-energy proton intensities suggesting that velocity rather than energy or rigidity is more important in determining the appearance of time–intensity profiles, and (2) the similar intensities decaying at nearly the same rate during the decay phase of the major events, independent of the longitudinal and latitudinal separation between Ulysses and near-Earth spacecraft. We emphasize that these periods of equal intensities were observed even at the highest heliographic latitudes reached by Ulysses—that is, 80°S and 80°N (Figures 5.12 and 5.15). Not all large SEP events show equal intensities at the different spacecraft. For example, during the first event shown in Figure 5.12 (event 1N), several new injections of particles were observed at ACE but were not discernible above the high intensities measured by Ulysses. The fluence of these additional events at ACE was probably too small to add significantly to the electrons injected into the reservoir by the main event.

Possible mechanisms for the formation of energetic particle reservoirs in the heliosphere have been offered in the literature: (i) McKibben (1972) and McKibben *et al.* (2001) invoked an effective cross-field diffusion to uniformly distribute particles in longitude and latitude. (ii) Roelof *et al.* (1992) considered that the outer boundaries of the reservoirs are formed by the merging of several plasma disturbances (e.g., ICMEs) launched during periods of intense solar activity. The magnetic field magnitude increases formed at these boundaries affect the transport of particles within the reservoir, delaying their escape to larger heliocentric distances and re-distributing them in latitude and longitude. This redistribution process must be efficient enough to dissipate any particle gradient within the reservoir; however, no explicit mechanism has been specified in the literature. We point out that these plasma disturbances may either exist before the occurrence of the event or be formed by the parent CME that generated the major SEP event during which we observe the reservoir. Reames, Barbier, and Ng (1996) also considered that the decay phase of the SEP events consists of particles propagating between the converging magnetic field near the Sun and a moving shell of strong scattering formed downstream of the distant traveling shocks. After formation, the reservoir slowly dissipates as a result of the nominal diffusion, convection, adiabatic cooling, and drift mechanisms that govern the propagation of SEPs.

In order to test the hypotheses for the physical processes that lead to the formation of energetic particle reservoirs it is necessary to study (i) the plasma structures that move past the spacecraft throughout the decay phase of the major SEP events, (ii) the occurrence of solar events prior to and during the occurrence of large SEP events, and (iii) the anisotropy flows throughout the duration of the SEP events that serve as a signature of energetic particle sources and cross-field diffusion processes. The field-aligned anisotropies (indicating no net flow of particles across the field lines) and the observation of this reservoir in events such as event 3N (with Ulysses immersed in uniform solar wind and unipolar field lines) advocate against a dominant role of either cross-field diffusion or excursion of field lines from low to high latitudes. However, a complete understanding of this phenomenon and the mechanism responsible for the formation of the reservoir is still under intense investigation.

One possibility of establishing the mechanism that forms reservoirs involves a third observer favorably aligned with any of the two other spacecraft allowing for the observation of traveling interplanetary structures. Figure 5.16 shows ~40 keV electron intensities measured by ACE (blue), Ulysses (black), and Cassini (red), together with the magnetic field magnitude measured by the three spacecraft, during the intense events of November 2001. Ulysses was above 70°N in high-speed solar wind, and Cassini at 6.6 AU and close to the Sun–Earth line. Vertical lines show the passage of interplanetary shocks. Equal decaying intensities at ACE and Ulysses were observed only after the interplanetary shocks moved past Ulysses. The Cassini spacecraft did not observe the prompt component of the SEP events. Lario *et al.* (2004b) interpreted this as the result of the effects of a merged interaction region (MIR) formed from multiple ICMEs prior to the events in November 2001 intervening between the Sun and Cassini. The fortuitous Sun–ACE–Cassini alignment permits the association between the shocks observed first at ACE and later at Cassini. The shock observed by ACE early on day 310 was not observed by Cassini because of a data gap. The shock at ACE on day 328 was most likely observed by Cassini on day 341. The shock at ACE was followed by an ICME on day 329. A shock followed by an ICME was also observed by Ulysses on day 330. Reisenfeld *et al.* (2003a, b) associated the origin of the ICMEs at ACE and Ulysses with the same CME on the Sun. Only after the ICME crossed over Ulysses were electron intensities similar at ACE and Ulysses, indicating that an energetic particle reservoir was formed

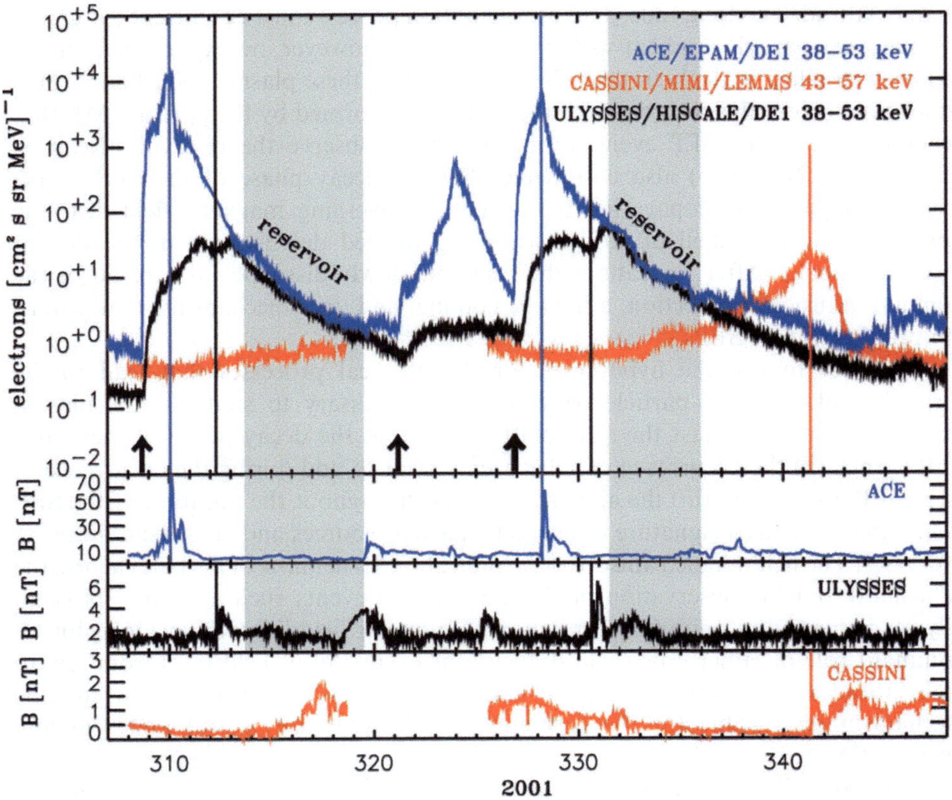

Figure 5.16. From top to bottom. Electron intensities measured by ACE, Ulysses, and Cassini during November–December 2001. The arrows indicate the onset of the parent solar events as identified by Lario *et al.* (2004a). Magnetic field magnitude observed by the three spacecraft. Vertical lines indicate the arrival of interplanetary shocks at each spacecraft. Gray-shaded areas indicate the periods with equal electron intensities at ACE and Ulysses.

behind the ICME and was observable only after the ICME moved past each spacecraft.

Energetic particle increases at Cassini were observed in association with the arrival of the interplanetary shock on day 341, presumably driven by the same ICME previously observed by ACE. After the shock passage, electron intensities at Cassini decayed on day 343 and evolved similarly (with the same decay rate) as those observed by Ulysses and earlier by ACE. It is possible that the energetic particle reservoir for this specific event was formed behind the traveling ICME, and therefore the arrival of energetic particles during the decay phase of the event at the different spacecraft was determined by the effects of this traveling structure.

It is also true that in some other cases transient structures (i.e., ICMEs) are not directly observed by the spacecraft, even though an energetic particle reservoir is formed (McKibben *et al.*, 2003). For example, during the Bastille Day 2000 event

(Figure 5.9), Ulysses did not observe the passage of the associated parent ICME. However, the Bastille Day 2000 event occurred during a period of intense solar activity when multiple CMEs were ejected from different longitudes, expanded through a large volume of the inner heliosphere (Smith *et al.*, 2001), and Ulysses observed prior to the event the passage of an SIR and a magnetic cloud (Sanderson, 2004). Both conditions seem appropriate for the creation of enhanced turbulent magnetic barriers beyond Ulysses necessary for the formation of heliospheric energetic particle reservoirs.

5.7 INFLUENCE OF INTERPLANETARY STRUCTURES ON SEP PROPAGATION

When the inner heliosphere is free of transient solar wind structures (such as SIRs, CIRs, or ICMEs), particle intensities measured in the interplanetary medium are modulated by the transport processes undergone by the particles as they travel from their sources to the spacecraft. The effects of pitch-angle scattering and adiabatic deceleration make isolated SEP events—seen at 1 AU as separated entities originated by two different solar events—merge into a single large SEP event at large heliocentric distances (Lario *et al.*, 2000b). McCarthy and O'Gallagher (1976) showed that energetic particle anisotropies decrease with heliocentric radial distance. In the absence of CME-driven shock effects, typical SEP events at 1 AU usually show anisotropy–time profiles that exhibit a sharp increase at the onset of the event followed by a gradual decrease. However, SEP events at large distances show a more rapid decrease in the anisotropy profiles (McCarthy and O'Gallagher, 1976).

Transient structures formed between the particle sources and the observer are also able to channel, confine and/or re-accelerate energetic particles, and thus modify the characteristics of the SEP events measured beyond 1 AU. Plasma structures rooted in the solar corona have been proven to be propagation channels for electrons of solar origin (Buttighoffer, 1998). Particle propagation within these structures is characterized by large mean-free paths and nearly scatter-free transport. Propagation channels have been observed to distances of nearly 5 AU in the ecliptic plane and are characterized by low-level magnetic field fluctuations (Buttighoffer, 1998; Buttighoffer *et al.*, 1999). Occasionally these propagation channels may be embedded within a CIR and allow for scatter-free particle propagation to high latitudes (Maia *et al.*, 1998). Injection of solar energetic electrons within ICMEs allows us to characterize the magnetic field topology of ICMEs. Rapid development of bidirectional electron flows indicates that ICMEs are flux loops rooted at the Sun (Malandraki *et al.*, 2001), whereas in some other cases ICMEs may present partial opening of field lines (Bothmer *et al.*, 1996; Malandraki, Sarris, and Tsiropoula, 2003).

Intervening plasma structures formed between the Sun and the spacecraft may be able to confine energetic particles and thus mitigate and/or delay the particle intensity increases at large heliocentric distances. Energetic particle enhancements are only observed once these transient structures move past the spacecraft (Lario *et al.*, 2004b). For example, Figure 5.17 shows 1.8–4.8 MeV ion anisotropy coefficients (five top panels), solar wind speed (sixth panel), and magnetic field magnitude and orientation (three bottom panels) during the events of January 2005. Ulysses was located at $R = 5.27$ and $\Lambda = 16°S$. Whereas 1 AU observations showed a sequence of SEP events from day 15 to 21, with the event with the hardest spectra observed on day 20 (Mewaldt *et al.*, 2005), Ulysses only observed a relatively large increase with onset overlapped with the passage of a relatively small CIR on day 22. The delay of the SEP event at these energies at Ulysses was most probably due to the effects of this CIR. Inserts in the top panel show pitch-angle distributions as measured by the LEMS120 telescope of HI-SCALE. With the exception of a short interval on day 25 (insert c), ion anisotropies during the rising phase of the event were small, with isotropic (insert a) or slightly anti-sunward (insert b) distributions. A large compound structure (most probably formed by the merging of multiple ICMEs) was observed by Ulysses from day 30 to 40 (gray-shaded area in Figure 5.17). Ion anisotropies throughout this interval were small and slightly sunward (insert e) or isotropic (insert f). Note that transverse anisotropies (A_{11}/A_0 and B_{11}/A_0) were practically zero throughout this event, indicating no net flow of particles across the magnetic field. The complexity of this SEP event embedded within CIRs and with the formation of compound ICMEs is characteristic of the SEP events observed by Ulysses during its solar-maximum orbit.

Particle anisotropies transverse to the local magnetic field different from zero have only been reported on two previous occasions throughout the Ulysses mission (Zhang *et al.*, 2003; Zhang, Jokipii, and McKibben, 2003). Both cases were associated with the passage of ICMEs (Sanderson, 2004). The interpretation of these transverse anisotropies as a result of particle motion perpendicular to the field lines is probably unlikely as the magnetic field inside ICME is very quiet and theories of transverse diffusion require the presence of field irregularities. Other possible explanations point out the existence of bidirectional field-aligned flows within the ICMEs together with a gradient intensity within the ICME (Roelof and Lario, 2004).

The confinement of energetic particles within ICMEs contributes also to shape the time–intensity profiles of SEP events. Figure 5.18 shows energetic particle, solar wind, and magnetic field observations during the passage of an overexpanding ICME at high heliolatitudes when Ulysses was immersed in fast solar wind streams during the solar-minimum (left) and solar-maximum (right) orbits (Bothmer *et al.*, 1995; Lario *et al.*, 2004a). The highest intensities were observed in association with the passage of the ICMEs and not with the shocks. This contrasts with the typical in-ecliptic 1 AU observations where peak intensities are observed in association with the passage of interplanetary shocks and particle intensity decay inside the ICME (Richardson, 1997). Although it is only a matter of contrast, particle intensities were higher inside the ICME than outside. ICMEs expanding into high-speed solar wind streams are not able to drive strong shocks that efficiently accelerate energetic particles. Particle confinement within the ICMEs is responsible for the slower decay of the intra-ICME intensities with respect to those outside the ICMEs (Lario *et al.*, 2004a).

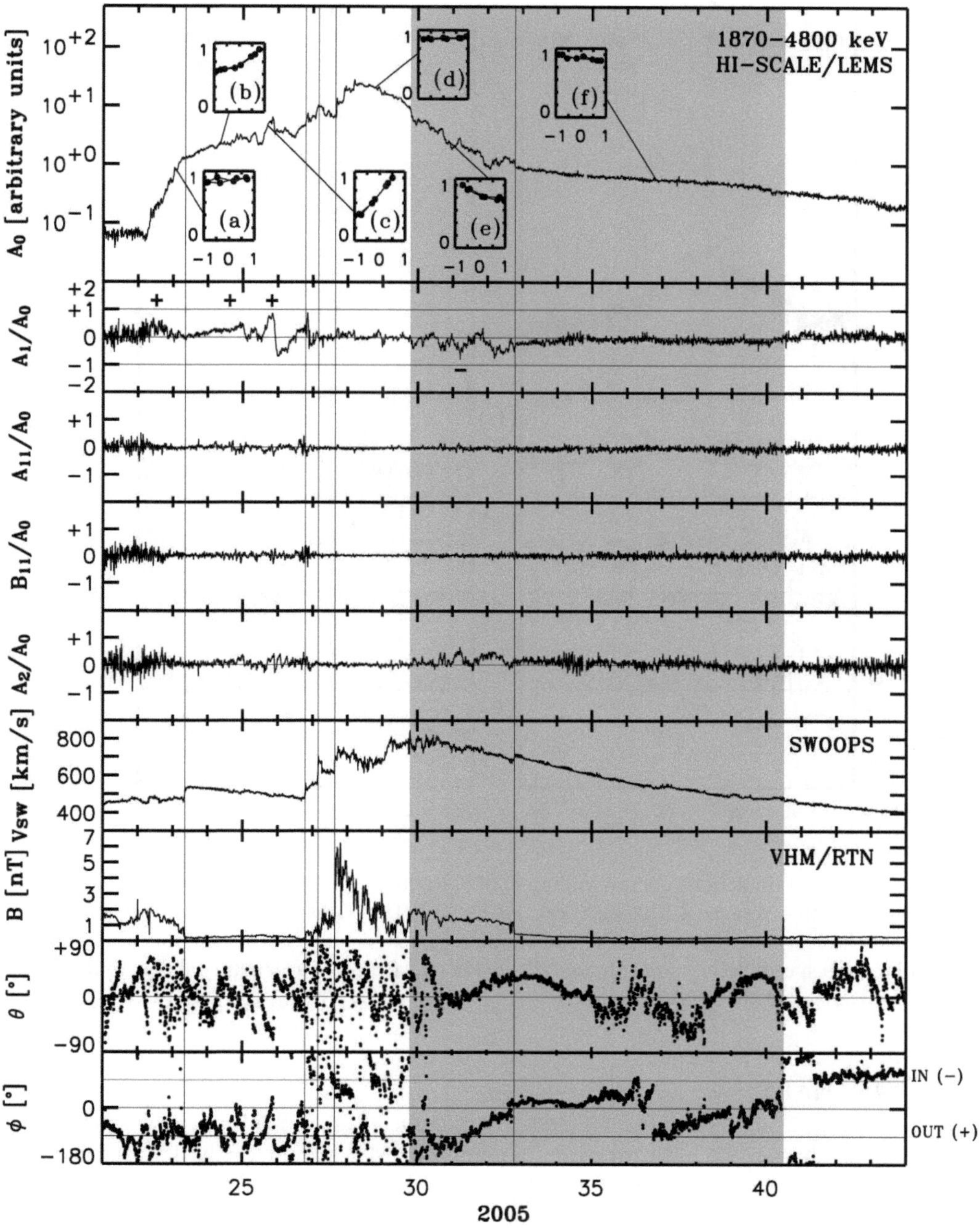

Figure 5.17. The same as Figure 5.2 but for the SEP event in January 2005. Solid vertical lines mark the arrival of interplanetary shocks. The gray-shaded area marks the passage of a compound solar wind stream formed by multiple ICMEs. Inserts in the top panel show pitch-angle distributions at selected times. Symbols +, –, and B in the second panel indicate anti-sunward, sunward, and bidirectional flows, respectively.

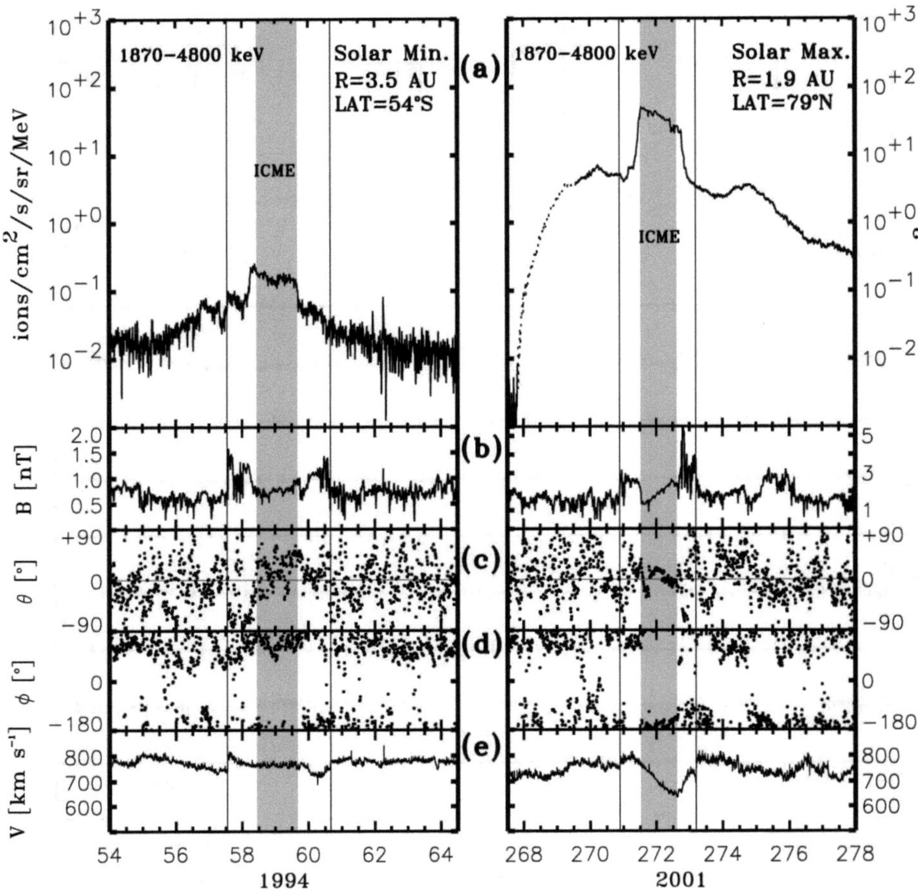

Figure 5.18. Left. In-ecliptic overexpanding ICME. Right. Polar overexpanding ICME. Top to bottom. (a) Spin-averaged 1,870–4,800 keV HI-SCALE/LEMS120 ion intensity, (b) magnetic field magnitude, (c) magnetic field RTN polar angle, (d) magnetic field RTN azimuth angle, and (e) solar wind speed. Solid vertical lines indicate the passage of interplanetary shocks and gray bars the passage of the ICMEs.

5.8 SUMMARY

1 In this chapter we have presented a comprehensive review of the Ulysses energetic particle observations throughout its solar-minimum and solar-maximum orbits. The transition from solar minimum to solar maximum is observed in the particle intensities, time–intensity profiles, and energetic particle abundances. The relatively simple structure of the heliosphere during solar minimum, dominated by CIR events and a relatively flat HCS, is replaced by a more much complex solar wind and magnetic field configuration at solar maximum with numerous transient events and highly inclined HCS.

2 A major surprise from the solar-minimum orbit observations is that the periodic recurrent enhancements of low-energy particles accelerated by the shocks bounding the CIRs persisted to the highest latitudes reached by Ulysses, even though the interaction regions were confined to latitudes less than about 35°. Interpretation of this particle enhancement was made in terms of the connection of the spacecraft to the CIR shocks and motions of the footpoints of field lines that connect to Ulysses.

3 The solar-maximum heliosphere showed elevated particle intensities at all latitudes that resulted from both the increasing level of solar activity and the dynamic evolution of the heliospheric structure. Multi-spacecraft energetic particle observations during solar maximum have shown two important pieces of information that lead to different interpretations of the particle transport processes in the heliosphere. The first observation is that large SEP events are simultaneously observed by spacecraft widely separated in longitude and latitude. The second observation is that particle fluxes measured in the late phase of large SEP events by widely separated spacecraft often present equal intensities (to within a small factor ~2–3) and evolve similarly in time (reservoirs).

4 In the absence of interplanetary structures, energetic particle anisotropies are field-aligned, indicating that there is no net flow of particles across the field lines. The observation of large SEP events during the solar-maximum north polar pass, when Ulysses was immersed in the fast solar wind and observed only a single magnetic field polarity, limits the possible excursion of field lines from low to high latitudes. Solar observations show that global coronal activity may be responsible for populating the inner heliosphere with energetic particles during major SEP events at both low and high latitudes and hence the concurrent observations of these events by widely separated spacecraft. Finally, the observation of energetic particle reservoirs suggests an efficient distribution of particles in the inner heliosphere that may occur in the downstream region of traveling interplanetary structures and by an efficient mechanism of cross-field diffusion.

5.9 ACKNOWLEDGMENTS

We thank the investigators of the COSPIN, EPAC, and HI-SCALE teams for their studies and research developed over the Ulysses mission. Special thanks to C. G. Maclennan, O. E. Malandraki, S. Hoang, A. Vourlidas, G. Stenborg, C. Tranquille, and R. G. Marsden for their help and for providing data used in this article. D. L. was supported by NASA research grants NAG5-13487 and NAG5-6113.

5.10 REFERENCES

Akimov, V. V., Ambroz, P., Belov, A. V., Berlicki, A., Chertok, I. M., Karlicky, M., Kurt, V. G., Leikov, N. G., Litvinenko, Yu. E., Magun, A. *et al.* (1996), Evidence for prolonged acceleration based on a detailed analysis of the long-duration solar gamma-ray flare of June 15 1991, *Solar Phys.*, **166**, 107.

Balogh, A., Beek, T. J., Forsyth, R. J., Hedgecock, P. C., Marquedant, R. J., Smith, E. J., Southwood, D. J., and Tsurutani, B. T. (1992), The magnetic field investigation on the Ulysses mission: Instru- mentation and preliminary results, *Astron. Astrophys. Suppl. Ser.*, **92**, 221–236.

Bame, S. J., McComas, D. J., Barraclough, B. L., Phillips, J. L., Sofaly, K. J., Chavez, J. C., Goldstein, B. E., and Sakurai, R. K. (1992), The Ulysses solar wind plasma experiment, *Astron. Astrophys. Suppl. Ser.*, **92**, 237–266.

Bame, S. J., Goldstein, B. E., Gosling, J. T., Harvey, J. W., McComas, D. J., Neugebauer, M., and Phillips, J. L. (1993), Ulysses observations of a recurrent high speed solar wind stream and the heliomagnetic streamer belt, *Geophys. Res. Lett.*, **20**, 2323–2326.

Biesecker, D. A. (1996), A search for ^{3}He enhancements in SEP and CIR-associated events with Ulysses/HI-SCALE, *Astron. Astrophys.*, **316**, 487.

Bothmer, V., Marsden, R. G., Sanderson, T. R., Trattner, K. J., Wenzel, K.-P., Balogh, A., Forsyth, R. J., and Goldstein, B. E. (1995), The Ulysses south polar pass: Transient fluxes of energetic ions, Geophys. Res. Lett., **22**, 3369–3372.

Bothmer, V., Desai, M. I., Marsden, R. G., Sanderson, T. R., Trattner, K. J., Wenzel, K.-P., Gosling, J. T., Balogh, A., Forsyth, R. J., and Goldstein, B. E. (1996), Ulysses observations of open and closed magnetic field lines within a coronal mass ejection, *Astron. Astrophys.*, **316**, 493–498.

Bougeret, J.-L., Kaiser, M. L., Kellogg, P. J., Manning, R., Goetz, K., Monson, S. J., Monge, N., Friel, L., Meetre, C. A., Perche, C. *et al.* (1995), WAVES: The radio and plasma wave investigation on the Wind spacecraft, *Space Sci. Rev.*, **71**, 231–265.

Brueckner, G. E., Howard, R. A., Koomen, M. J., Korendyke, C. M., Miichels, D. J., Moses, J. D., Socker, D. G., Dere, K. P., Lamy, P. L., Llebaria, A. *et al.* (1995), The large angle spectroscopic coronagraph (LASCO), *Solar Phys.*, **162**, 357–402.

Buttighoffer, A. (1998), Solar electron beams associated with radio type III bursts: Propagation channels observed by Ulysses between 1 and 4 AU, *Astron. Astrophys.*, **335**, 295–302.

Buttighoffer, A., Pick, M., Raviart, A., Hoang, S., Lin, R. P., Simnett, G. M., Lanzerotti, L. J., and Bothmer, V. (1996), Joint Ulysses and Wind observations of a particle event in April 1995, *Astron. Astrophys.*, **316**, 499–505.

Buttighoffer, A., Lanzerotti, L. J., Thomson, D. J., Maclennan, C. G., and Forsyth, R. J. (1999), Spectral analysis of the magnetic field inside particle propagation channels detected by Ulysses, *Astron. Astrophys.*, **351**, 385–392.

Cane, H. V., Erickson, W. C., and Prestage, N. P. (2002), Solar flares, type III radio bursts, coronal mass ejections and energetic particles, *J. Geophys. Res.*, **107**, A10, SSH14-1, doi:10.1029/2001JA000320.

Cliver, E. W., and H. V. Cane (2003), The last word, *EOS Trans. AGU*, **83**, 61–68.

Dalla, S., Balogh, A., Krucker, S., Posner, A., Müller-Mellin, R., Anglin, J. D., Hofer, M. Y., Marsden, R. G., Sanderson, T. R., Heber, B. *et al.* (2003), Delay in solar energetic particle onsets at high heliographic latitudes, *Annales Geophys.*, **21**, 1367–1375.

Delaboudinière, J.-P., Artzner, G. E., Brunaud, J., Gabriel, A. H., Hochedez, J. F., Millier, F., Song, X. Y., Au, B., Dere, K. P., Howard, R. A. *et al.* (1995), EIT: Extreme ultraviolet imaging telescope for the SOHO mission, *Solar Phys.*, **162**, 291–312.

Desai, M. I., Marsden, R. G., Sanderson, T. R., Lario, D., Roelof, E. C., Simnett, G. M., Gosling, J. T., Balogh, A., and Forsyth, R. J. (1999), Energy spectra of 50-keV to 20-MeV protons accelerated at corotating interaction regions at Ulysses, *J. Geophys. Res.*, **104**, 6705–6720.

Dougherty, M. K., Kellock, S., Southwood, D. J., Balogh, A., Smith, E. J., Tsurutani, B. T., Gerlach, B., Glassmeier, K.-H., Gleim, F., Russell, C. T. *et al.* (2004), The Cassini magnetic field investigation, *Space Sci. Rev.*, **114**, 331.

Fisk, L. A. (1996), Motion of the footpoints of the heliospheric magnetic field lines at the Sun: Implications for the recurrent energetic particle events at high heliographic latitudes, *J. Geophys. Res.*, **101**, 15547.

Forsyth, R. J., Balogh, A., Smith, E. J., Murphy, N., and McComas, D. J. (1995), The underlying magnetic field direction in Ulysses observations of the southern polar heliosphere, *Geophys. Res. Lett.*, **22**, 3321–3324.

Franz, M., Keppler, E., Krupp, N., Reuss, M. K., and Blake, J. B. (1995), The elemental composition in energetic particle events at high heliospheric latitudes, *Space Sci. Rev.*, **72**, 339–342.

Franz, M., Burgess, D., Keppler, E., Reuss, M. K., and Blake, J. B. (1997), Energetic particle anisotropies and remote magnetic connection at high solar latitudes, *Adv. Space Res.*, **26**, 855–858.

Franz, M., Keppler, E., Lauth, U., Reuss, M. K., Mason, G. M., and Mazur, J. E. (1999), Energetic particle abundances at CIR shocks, *Geophys. Res. Lett.*, **26**, 17–20.

Giacalone, J. (2002), The global transport of energetic particles in the heliospheric magnetic field, *EOS Trans. AGU*, **83**(47), Fall Meeting Suppl., Abstract SH72B-02.

Giacalone, J., and J. R. Jokipii (1997), Spatial variation of accelerated pickup ions at corotating interaction regions, *Geophys. Res. Lett.*, *24*, 1723–1726.

Giacalone, J., J. R. Jokipii, and M. Zhang (2001), The propagation of solar-energetic particles to high heliographic latitudes, *EOS Trans. AGU*, **82**(19), Spring Meeting Suppl., Abstract SH61B-06.

Gloeckler, G., Geiss, J., Roelof, E. C., Fisk, L. A., Ipavich, F. M., Ogilvie, K. W., Lanzerotti, L. J., von Steiger, R., and Wilken, B. (1994), Acceleration of interstellar pickup ions in the disturbed solar wind observed on Ulysses, *J. Geophys. Res.*, **99**, 17637–17643.

Gold, R. E., Krimigis, S. M., Hawkins, S. E., III, Haggerty, D. K., Lohr, D. A., Fiore, E., Armstrong, T. P., Holland, G., and Lanzerotti, L. J. (1998), Electron, Proton, Alpha Monitor on the Advanced Composition Explorer spacecraft, *Space Sci. Rev.*, **86**, 541.

Gosling, J. T., Bame, S. J., McComas, D. J., Phillips, J. L., Pizzo, V. J., Goldstein, B. E., and Neugebauer, M. (1995), Solar wind corotating stream interaction regions out of the ecliptic plane: Ulysses, *Space Sci. Rev.*, **72**, 99.

Hoang, S., Maksimovic, M., Bougeret, J.-L., Reiner, M. J., and Kaiser, M. L. (1998), Wind–Ulysses source location of radio emissions associated with the January 1997 Coronal Mass Ejection, *Geophys. Res. Lett.*, **25**, 2497–2500.

Hofer, M. Y., Marsden, R. G., Sanderson, T. R., and Tranquille, C. (2001), Energetic particle composition measurements at high heliographic latitudes around the solar activity maximum, in *Solar and Galactic Composition* (R. F. Wimmer-Schweingruber, ed.), AIP Conf. Proc., Vol. 598, pp. 189–193.

Hofer, M. Y., Marsden, R. G., Sanderson, T. R., and Tranquille, C. (2003a), Composition measurements over the solar poles close to solar maximum—Ulysses COSPIN/LET observations, in *Solar Wind Ten* (M. Velli *et al.*, eds.), AIP Conf. Proc., Vol. 679, pp. 183–186.

Hofer, M. Y., Marsden, R. G., Sanderson, T. R., Tranquille, C., and Forsyth, R. J. (2003b), Transition to solar minimum at high solar latitudes: Energetic particles from corotating interaction regions, *Geophys. Res. Lett.*, **30**, 8034, doi:10.1029/2003GL017138.

Hofer, M. Y., Marsden, R. G., Sanderson, T. R., and Tranquille, C. (2003c), From the Sun's south to the north pole—Ulysses COSPIN/LET composition measurements at solar maximum, *Ann. Geophys.*, **21**, 1383–1391.

Intriligator, D. S., Jokipii, J. R., Horbury, T. S., Intriligator, J. M., Forsyth, R. J., Kunow, H., Wibberenz, G., and Gosling, J. T. (2001), Processes associated with particle transport in corotating interaction regions and near stream interfaces, *J. Geophys. Res.*, **106**, 10625–10634.

Keppler, E. (1998a), The acceleration of charged particles in corotating interaction regions (CIR)—A review with particular emphasis on the Ulysses mission, *Surveys in Geophys.*, **19**, 211.

Keppler, E. (1998b), What causes the variations of the peak intensity of CIR accelerated energetic ion fluxes?, *Ann. Geophys.*, **16**, 1552–1556.

Keppler, E., Blake, J. B., Hovestadt, D., Korth, A., Quenby, J., Umlauft, G., and Woch, J. (1992), The Ulysses energetic particle composition experiment EPAC, *Astron. Astrophys. Suppl. Ser.*, **92**, 317–331.

Kerdraon, A., and J. Delouis (1997), *Coronal physics from radio and space observations* (G. Trottet, ed.), pp. 192, Harvard-Smithsonian Center for Astrophysics.

Kissmann, R., Fichtner, H., and Ferreira, S. E. S. (2004), The influence of CIRs on energetic electron flux at 1 AU, *Astron. Astrophys.*, **419**, 357–363.

Kóta, J., and J. R. Jokipii (1995), Corotating variations of cosmic rays near the south heliospheric pole, *Science*, **268**, 1024.

Krimigis, S. M., Mitchell, D. G., Hamilton, D. C., Livi, S., Dandouras, J., Jaskulek, S., Armstrong, T. P., Boldt, J. D., Cheng, A. F., Gloeckler, G. *et al.* (2004), Magnetosphere Imaging Instrument (MIMI) on the Cassini mission to Saturn/Titan, *Space Sci. Rev.*, **114**, 233–329.

Krucker, S., Larson, D. E., Lin, R. P., and Thompson, B. J. (1999) On the origin of impulsive electron events observed at 1 AU, *Astrophys. J.*, **519**, 864–875.

Kundu, M. R., and R. G. Stone (1984), Observations of solar radio bursts from meter to kilometer wavelengths, *Adv. Space Res.*, **4**, 261–270.

Lanzerotti, L. J., and T. R. Sanderson (2001), Energetic particles in the heliosphere, in *The Heliosphere near Solar Minimum: The Ulysses Perspective* (A. Balogh, R. G. Marsden, and E. J. Smith, eds.), pp. 259–286, Springer/Praxis, Chichester, UK.

Lanzerotti, L. J., Gold, R. E., Anderson, K. A., Armstrong, T. P., Lin, R. P., Krimigis, S. M., Pick, M., Roelof, E. C., Sarris, E. T., and Simnett, G. M. (1992), Heliosphere instrument for spectra, composition and anisotropy at low energies, *Astron. Astrophys. Suppl. Ser.*, **92**, 349–364.

Lario, D., and R. B. Decker (2002), The energetic storm particle event of October 20, 1989, *Geophys. Res. Lett.*, **29**, doi:10.1029/2001GL014017.

Lario, D., and G. M. Simnett (2004), Solar energetic particle variations, in *Solar Variability and Its Effects on Climate* (J. M. Pap and P. Fox, eds.), Geophysical Monograph 141, American Geophysical Union, pp. 195–216, doi:10.1029/141GM14.

Lario, D., Marsden, R. G., Sanderson, T. R., Maksimovic, M., Sanahuja, B., Balogh, A., Forsyth, R. J., Lin, R. P., and Gosling, J. (2000a), Energetic proton observations at 1 and 5 AU, 1: January–September 1997, *J. Geophys. Res.*, **105**, 18235–18250.

Lario, D., Marsden, R. G., Sanderon, T. R., Maksimovic, M., Sanahuja, B., Plunkett, S. P., Balogh, A., Forsyth, R. J., Lin, R. P., and Gosling J. T. (2000b), Energetic proton observations at 1 and 5 AU, 2: Rising phase of the solar cycle 23, *J. Geophys. Res.*, **105**, 18251.

Lario, D., Roelof, E. C., Forsyth, R. J., and Gosling, J. T. (2001a), 26-day analysis of energetic ion observations at high and low heliolatitudes: Ulysses and ACE, *Space Sci. Rev.*, **97**, 249–252.

Lario, D., Maclennan, C. G., Roelof, E. C., Gosling, J. T., Ho, G. C., and Hawkins, S. E., III (2001b), High-latitude Ulysses observations of the H/He intensity ratio under solar minimum and solar maximum conditions, in *Solar and Galactic Composition* (R. F. Wimmer-Schweingruber, ed.), AIP Conf. Proc., Vol. 598, pp. 183–188.

Lario, D., Haggerty, D. K., Roelof, E. C., Tappin, S. J., Forsyth, R. J., and Gosling, J. T. (2001c) Joint Ulysses and ACE observations of a magnetic cloud and the associated solar energetic particle event, *Space Sci. Rev.*, **97**, 277–280.

Lario, D., Roelof, E. C., Decker, R. B., Ho, G. C., Maclennan, C. G., and Gosling, J. T. (2003a), Solar cycle variations of the energetic H/He intensity ratio at high heliolatitudes and in the ecliptic plane, *Annales Geophys.*, **21**, 1229–1243.

Lario, D., Roelof, E. C., Decker, R. B., Ho, G. C., Maclennan, C. G., and Gosling, J. T. (2003b), Energetic H/He intensity ratio under solar maximum and solar minimum conditions: Ulysses observations, *Adv. Space Res.*, **32**, 585–590.

Lario, D., Roelof, E. C., Decker, R. B., and Reisenfeld, D. B. (2003c), Solar maximum low-energy particle observations at heliographic latitudes above 75 degrees, *Adv. Space Res.*, **32**, 579–584.

Lario, D., Decker, R. B., Roelof, E. C., Reisenfeld, D. B., and Sanderson, T. R. (2004a), Low-energy particle response to CMEs during the Ulysses solar maximum northern polar passage, *J. Geophys. Res.*, **109**, A01107, doi:10.1029/2003JA010071.

Lario, D., Livi, S., Roelof, E. C., Decker, R. B., Krimigis, S. M., and Dougherty, M. K. (2004b), Heliospheric energetic particle observations by the Cassini spacecraft: Correlation with 1 AU observations, *J. Geophys. Res.*, **109**, A09S02, doi:10.1029/2003JA010107.

Lario, D., Decker, R. B., Livi, S., Krimigis, S. M., Roelof, E. C., Russell, C. T., and Fry, C. D. (2005), Heliospheric energetic particle observations during the October–November 2003 events, *J. Geophys. Res.*, **110**, A09S11, doi: 10.1029/2004JA010940.

Laxton, N. F. (1997), Ulysses MeV ion fluxes and anisotropies at corotating interaction regions, *EOS Trans. AGU*, **78**(46), F531, Abstract SH11A-11.

Lehtinen, N. J., Pohjolainen, S., Karlický, M., Aurass, H., and Otruba, W. (2005), Non-thermal processes associated with rising structures and waves during a halo type CME, *Astron. Astrophys.*, **442**, 1049–1058.

Maclennan, C. G., and L. J. Lanzerotti (1995), Elemental abundances in corotating interaction regions at high solar latitudes, *Space Sci. Rev.*, **72**, 297–302.

Maclennan, C. G., and L. J. Lanzerotti (1998), Low energy anomalous ions at northern heliolatitudes, *Geophys. Res. Lett.*, **25**, 3473.

Maclennan, C. G., Lanzerotti, L. J., and Gold, R. E. (2003), Low energy charged particles in the high latitude heliosphere: Comparing solar maximum and solar minimum, *Geophys. Res. Lett.*, **30**, 8033, doi:10.1029/2003GL017080.

Maclennan, C. G., Lanzerotti, L. J., and Roelof, E. C. (2001), Populating an inner heliosphere reservoir (<5 AU) with electrons and heavy ions, *Proc. 27th International Cosmic Ray Conf.*, p. 3265.

Maia, D., and M. Pick (2004), Revisiting the origin of impulsive electron events: Coronal magnetic restructuring, *Astrophys. J.*, **609**, 1082–1097.

Maia, D., Malandraki, O., Pick, M., Sarris, E. T., Kasotakis, G., Lanzerotti, L. J., MacLennan, C. G., and Trochoutsos, P. C. (1998), Particle propagation channel detected at 4.7 AU inside a corotating interaction region, *J. Geophys. Res.*, **103**, 9545–9552.

Maia, D., Vourlidas, A., Pick, M., Howard, R., Schwenn, R., and Magalhães, A. (1999), Radio signatures of a fast coronal mass ejection development on November 6 1997, *J. Geophys. Res.*, **104**, 12507–12514.

Maia, D., Pick, M., Hawkins, S. E., III, Fomichev, V. V., and Jiřička, K. (2001), 14 July 2000, a near-global coronal event and its association with energetic particle electron events detected in the interplanetary medium, *Solar Phys.*, **204**, 199–214.

Malandraki, O. E., Sarris, E. T., and Tsiropoula, G. (2003), Magnetic topology of coronal mass ejection events out of the ecliptic: Ulysses/HI-SCALE energetic particle observations, *Ann. Geophys.*, **21**, 1249– 1256.

Malandraki, O. E., Sarris, E. T., Lanzerotti, L. J., Maclennan, C. G., Pick, M., and Tsiropoula, G. (2001), Tracing the magnetic field topology of coronal mass ejection events by Ulysses/HI-SCALE energetic particle observations in and out of the ecliptic, *Space Sci. Rev.*, **97**, 263–268.

Mann, G., Classen, H. T., Keppler, E., and Roelof, E. C. (2004), On electron acceleration at CIR related shock waves, *Astron. Astrophys.*, **391**, 749–756.

Marsden, R. G. (2004), Ulysses at solar maximum, in *The Sun and the Heliosphere as an Integrated System* (G. Poletto and S. T. Suess, eds.), pp. 91–112, Kluwer Academic, Dordrecht, The Netherlands.

Marsden, R. G., Sanderson, T. R., Tranquille, C., Trattner, K. J., Anttila, A., and Torsti, J. (1999), On the gradients of ACR oxygen at intermediate heliocentric distances: Ulysses/SOHO results, *Adv. Space Res.*, **23**, 531–534.

Mason, G. M. (2000), Composition and energy spectra of ions accelerated in corotating interaction regions, in *Acceleration and Transport of Energetic Particles Observed in the Heliosphere: ACE 2000 Symposium* (R. A. Mewaldt *et al.*, eds.), AIP Conf. Proc., Vol. 528, pp. 234–241.

Mason, G. M., and T. R. Sanderson (1999), CIR associated energetic particles in the inner and middle heliosphere, *Space Sci. Rev.*, **89**, 77–90.

Mason, G. M., von Steiger, R., Decker, R. B., Desai, M. I., Dwyer, J. R., Fisk, L. A., Gloeckler, G., Gosling, J. T., Hilchenbach, M., Kallenbach, R. *et al.* (1999), Origin, injection, and acceleration of CIR particles: Observations, *Space Sci. Rev.*, **89**, 327–367.

McCarthy, J., and J. J. O'Gallagher (1976), The radial variation of solar flare proton anisotropies observed in deep space on Pioneer 10 and 11, *Geophys. Res. Lett.*, **3**, 53–56.

McComas, D. J., Gosling, J. T., and Skoug, R. M. (2000), Ulysses observations of the irregularly structured mid-latitude solar wind during the approach to solar maximum, *Geophys. Res. Lett.*, **27**, 2437–2440.

McComas, D. J., Elliott, H. A., and von Steiger, R. (2002), Solar wind from high-latitude coronal holes at solar maximum, *Geophys. Res. Lett.*, **29**, doi:10.1029/2001GL014164.

McComas, D. J., Bame, S. J., Barker, P., Feldman, W. C., Phillips, J. L., Riley, P., and Griffee, J. W. (1998), Solar Wind Electron Proton Alpha Monitor (SWEPAM) for the Advance Composition Explorer, *Space Sci. Rev.*, **86**, 563.

McComas, D. J., Elliott, H. A., Gosling, J. T., Reisenfeld, D. B., Skoug, R. M., Goldstein, B. E., Neugebauer, M., and Balogh, A. (2002), Ulysses' second fast-latitude scan: Compexity near solar maximum and the reformation of polar coronal holes, *Geophys. Res. Lett.*, **29**, doi: 10.1029/2001GL013940.

McKibben, R. B. (1972), Azimuthal propagation of low-energy solar flare protons as observed from spacecraft very widely separated in solar azimuth, *J. Geophys. Res.*, **77**, 3957.

McKibben, R. B., Jokipii, J. R., Burger, R. A., Heber, B., Kóta, J., McDonald, F. B., Paizis, C., Potgieter, M. S., and Richardson, I. G. (1999), Modulation of cosmic rays and anomalous components by CIRs, *Space Sci. Rev.*, **89**, 307–326.

McKibben, R. B., and the Cospin Collaboration (2001), Ulysses COSPIN observations of the energy and charge dependence of the propagation of solar energetic particles to the Sun's south polar regions, *Proc. 27th International Cosmic Ray Conf.*, p. 3281.

McKibben, R. B., Connell, J. J., Lopate, C., Zhang, M., Anglin, J. D., Balogh, A., Dalla, S., Sanderson, T. R., Marsden, R. G., Hofer, M. Y. *et al.* (2003), Ulysses COSPIN observations of cosmic rays and solar energetic particles from the south pole to the north pole of the Sun during solar maximum, *Ann. Geophys.*, **21**, 1217.

McKibben, R. B., Anglin, J. D., Connell, J. J., Dalla, S., Heber, B., Kunow, H., Lopate, C., Marsden, R. G., Sanderson, T. R., and Zhang, M. (2005), Energetic particle observations from the Ulysses COSPIN instruments obtained during the October–November 2003 events, *J. Geophys. Res.*, **110**, A09S19, doi:10.1029/2005JA011049.

Mewaldt, R. A., Looper, M. D., Cohen, C. M. S., Mason, G. M., Haggerty, D. K., Desai, M. I., Labrador, A. W., Leske, R. A., and Mazur, J. E. (2005), Solar-particle energy spectra during the large events of October–November 2003 and January 2005, *Proc. 27th International Cosmic Ray Conf.*, **1**, p. 111.

Pick, M., and D. Maia (2005), Origin of complex Type III-L events and electron acceleration, *Adv. Space Res.*, **35**, 1876.

Pick, M., Lanzerotti, L. J., Buttighoffer, A., Sarris, E. T., Armstrong, T. P., Simnett, G. M., Roelof, E. C., and Kerdraon, A. (1995a), The propagation of sub-MeV solar electrons to heliolatitudes above 50°S, *Geophys. Res. Lett.*, **22**, 3373–3376.

Pick, M., Lanzerotti, L. J., Buttighofer, A., Hoang, S., and Forsyth, R. J. (1995b), Detection of a solar particle event at a heliolatitude of 73.8°S, *Geophys. Res. Lett.*, **22**, 3377–3380.

Pick, M., Maia, D., Kerdraon, A., Howard, R., Brueckner, G. E., Michels, D. J., Paswaters, S., Schwenn, R., Lamy, P., Llebaria, A. *et al.* (1998), Joint Nançay Radioheliograph and LASCO observations of coronal mass ejections—II. The 9 July 1996 event, *Solar Phys.*, **181**, 455–468.

Pohjolainen, S., Maia, D., Pick, M., Vilmer, N., Khan, J. I., Otruba, W., Warmuth, A., Benz, A., Alissandrakis, C., and Thompson, B. J. (2001), On-the-disk development of the halo coronal mass ejection on 1998 May 2, *Astrophys. J.*, **556**, 421–431.

Reames, D. V. (1999), Particle acceleration at the Sun and in the heliosphere, *Space Sci. Rev.*, **90**, 413–491.

Reames, D. V., L. M. Barbier, and C. K. Ng (1996), The spatial distribution of particles accelerated by coronal mass ejection shocks, *Astrophys. J.*, **466**, 473.

Reiner, M. J., Kaiser, M. L., Plunkett, S. P., Prestage, N. P., and Manning, R. (2000), Radio tracking of a white light coronal mass ejection from solar corona to interplanetary medium, *Astrophys. J.*, **529**, L53–L56.

Reuss, M. K., Fränz, M., and Keppler, E. (1996), Recurrent variations of anomalous oxygen in association with corotating interaction regions, *Ann. Geophys.*, **14**, 585.

Reuss, M. K., Keppler, E., Franz, M., Witte, M., Krupp, N., Wilken, B., Balogh, A., Forsyth, R. J., Moussas, X., Polygiannakis, J. M. *et al.* (1995), A particle event at 5 AU and 20 southern latitude from measurements with the EPAC instrument on Ulysses, *Space Sci. Rev.*, **72**, 343–346.

Reisenfeld, D. B., Gosling, J. T., Forsyth, R. J., Riley, P., and St. Cyr, O. C. (2003a) Properties of high-latitude CME-driven disturbances during Ulysses second northern polar passage, *Geophys. Res. Lett.*, **30**(19), 8031, doi:10.1029/2003GL017155.

Reisenfeld, D. B., Gosling, J. T., Steinberg, J. T., Riley, P., Forsyth, R. J., and St. Cyr, O. C. (2003b), CMEs at high northern latitudes during solar maximum: Ulysses and SOHO correlated observations, in *Solar Wind Ten* (M. Velli *et al.*, eds.), AIP Conf. Proc., Vol. 679, p. 210.

Richardson, I. G. (1997), Using energetic particles to probe the magnetic topology of ejecta, in *Coronal Mass Ejections* (N. Crooker *et al.*, eds.), pp. 189–198, Geophys. Monogr. Ser. 99.

Richardson, I. G., Barbier, L. M., Reames, D. V., and von Rosenvinge, T. T. (1993), Corotating MeV/amu ion enhancements at 1 AU or less from 1978 to 1986, *J. Geophys. Res.*, **98**, 13–32.

Roelof, E. C., and D. Lario (2004), Transverse anisotropies of 40–90 MeV solar energetic protons: A re-interpretation, *EOS Trans. AGU*, **85**(17), Jt. Assem. Suppl., Abstract SH24A-04.

Roelof, E. C., D. Lario, and D. K. Haggerty (2003), Role of CME-associated magnetic field increases in the solar particle chain, *EOS Trans. AGU*, **84**(46), Fall Meeting Suppl., Abstract SM41A-03.

Roelof, E. C., Simnett, G. M., and Armstrong, T. P. (1995), IMF connection for energetic protons observed at Ulysses via mid-latitude solar wind rarefaction regions, *Space Sci. Rev.*, **72**, 309–314.

Roelof, E. C., Simnett, G. M., and Tappin, S. J. (1996), The regular structure of shock-accelerated 40–100 keV electrons in the high-latitude heliosphere, *Astron. Astrophys.*, **316**, 481–486.

Roelof, E. C., Gold, R. E., Simnett, G. M., Tappin, S. J., Armstrong, T. P., and Lanzerotti, L. J. (1992), Low-energy solar electron and ions observed at Ulysses February–April 1991: The inner heliosphere as particle reservoir, *Geophys. Res. Lett.*, **19**, 1243.

Roelof, E. C., Simnett, G. M., Decker, R. B., Lanzerotti, L. J., Maclennan, C. G., Armstrong, T. P., and Gold, R. E. (1997), Reappearance of recurrent low-energy particle events at Ulysses/HI-SCALE in the northern heliosphere, *J. Geophys. Res.*, **102**, 11251–11262.

Sanderson, T. R. (2004), Propagation of energetic particles to high latitudes, in *The Sun and the Heliosphere as an Integrated System* (G. Poletto and S. T. Suess, eds.), pp. 113–145, Kluwer Academic, Dordrecht, The Netherlands.

Sanderson, T. R., Bothmer, V., Marsden, R. G., Trattner, K. J., Wenzel, K.-P., Balogh, A., Forsyth, R. J., and Goldstein, B. E. (1995), The Ulysses south polar pass: Energetic ion observations, *Geophys. Res. Lett.*, **22**, 3357.

Sanderson, T. R., Lario, D., Maksimovic, M., Marsden, R. G., Tranquille, C., Balogh, A., Forsyth, R. J., and Goldstein, B. E. (1999), Current sheet control of recurrent particle increases at 4–5 AU, *Geophys. Res. Lett.*, **26**, 1785–1788.

Sanderson, T. R., Marsden, R. G., Tranquille, C., Dalla, S., Forsyth, R. J., Gosling, J. T., and McKibben, R. B. (2003), Propagation of energetic particles in the high-latitude high-speed solar wind, *Geophys. Res. Lett.*, **30**, ULY 10-1, doi:10.1029/2003GL017306.

Scholer, M., Mann, G., Chalov, S., Desai, M. I., Fisk, L. A., Jokipii, J. R., Kallenbach, R., Keppler, E., Kóta, J., Kunow, H. *et al.* (1999), Origin, injection and acceleration of CIR particles: Theory, *Space Sci. Rev.*, **89**, 369–399.

Schrijver, C. J., and M. L. Derosa (2003), Photospheric and heliospheric magnetic field, *Solar Phys.*, **212**, 165–200.

Simnett, G. M. (2001), Energetic particle characteristics in the high-latitude heliosphere near solar maximum, *Space Sci. Rev.*, **97**, 231–242.

Simnett, G. M., and E. C. Roelof (1995), Reverse shock acceleration of electrons and protons at mid-heliolatitudes from 5.3–3.8 AU, *Space Sci. Rev.*, **72**, 303.

Simnett, G. M., Sayle, K. A., and Roelof, E. C. (1995), Differences between the 0.35–1.0 MeV/nucleon H/He ratio in solar and corotating events at high heliolatitude, *Geophys. Res. Lett.*, **22**, 3365–3368.

Simnett, G. M., Sayle, K., Roelof, E. C., and Tappin, S. J. (1994), Corotating particle enhancements out of the ecliptic plane, *Geophys. Res. Lett.*, **21**, 1561–1564.

Smith, C. W. (2003), The geometry of turbulent magnetic fluctuations at high heliographic latitudes, *Solar Wind Ten: Proc. 10th International Solar Wind Conf.*, CP679, pp. 413–416.

Smith, C. W., L'Heureux, J., Ness, N. F., Acuña, M. H., Burlaga, L. F., and Scheifele, J. (1998), The ACE Magnetic Fields Experiment, *Space Sci. Rev.*, **86**, 613.

Smith, C. W., Ness, N. F., Burlaga, L. F., Skoug, R. M., McComas, D. J., Zurbuchen, T. H., Gloeckler, G., Haggerty, D. K., Gold, R. E., Desai, M. I. *et al.* (2001), ACE observations of the Bastille Day 2000 interplanetary disturbances, *Solar Phys.*, **204**, 227.

Simpson, J. A., Anglin, J. D., Balogh, A., Bercovitch, M., Bouman, J. M., Budzinski, E. E., Burrows, J. R., Carvell, R., Connell, J. J., Ducros, R. *et al.* (1992), The Ulysses cosmic ray and solar particle investigation, *Astron. Astrophys. Suppl. Ser.*, **92**, 365–400.

Tranquille, C., Marsden, R. G., and Sanderson, T. R. (2001), Ulysses measurements of the solar cycle variation of 2–40 MeV/n ions in the inner heliosphere, in *Solar and Galactic Composition* (R. F. Wimmer-Schweingruber, ed.), AIP Conf. Proc., Vol. 598, pp. 195–200.

Tranquille, C., Marsden, R. G., Sanderson, T. R., and Hofer, M. Y. (2003), A survey of ^{3}He enhancements at 2–20 MeV/nucleon: Ulysses COSPIN/LET observations, *Ann. Geophys.*, **21**, 1245–1248.

Treumann, R. A., and T. Terasawa (2001), Electron acceleration in the heliosphere, *Space Sci. Rev.*, **99**, 135–150.

Trottet, G. (1986), Relative timing of hard X-rays and radio emissions during different phases of solar flares. Consequences for the electron acceleration, *Solar Phys.*, **104**, 145–163.

Vourlidas, A.., Pick, M., Hoang, S., and Démoulin, P. (2007), Identification of a peculiar radio source in the aftermath of large coronal mass ejection events, *Astrophys. J. Lett.*, **656**, L105–L108.

Wang, Y.-M., Pick, M., and Mason, G. M. (2006), Coronal holes, jets and the origin of ^{3}He-rich particle events, *Astrophys. J.*, **639**, 495–509.

Wimmer-Schweingruber, R. F., von Steiger, R., and Paerli, R. (1997), Solar wind stream interfaces in corotating interaction regions: SWICS/Ulysses results, *J. Geophys. Res.*, **102**, 17407–17417.

Zhang, M., Jokipii, J. R., and McKibben, R. B. (2003) Perpendicular transport of solar energetic particles in heliospheric magnetic fields, *Astrophys. J.*, **595**, 493.

Zhang, M., McKibben, R. B., Lopate, C., Jokipii, J. R., Giacalone, J., Kallenrode, M.-B., and Rassoul, H. K. (2003), Ulysses observations of solar energetic particles from the 14 July 2000 event at high heliographic latitudes, *J. Geophys. Res.*, **108**, 1154, doi:10.1029/2002JA009531.

6

Galactic and anomalous cosmic rays through the solar cycle: New insights from Ulysses

B. Heber and M. S. Potgieter

6.1 INTRODUCTION

Cosmic ray research began in 1912 when Victor Hess measured the intensity of the then unknown ionizing radiation with an electroscope in a balloon up to an altitude of about 5,000 m. He discovered that this very penetrating radiation, later called cosmic rays, was coming from outside the atmosphere (for a historic review, see Simpson, 2001).The systematic experimental study of cosmic rays began in the 1930s, using ground-based and balloon-borne ionization chambers. In the 1950s it expanded on a much larger scale with neutron monitors, coordinated world-wide during the International Geophysical Year (IGY) in 1957 (Simpson, 2000). When Parker (1958) described the solar wind, the theoretical research of cosmic rays began, stimulated by the beginning of *in situ* space observations that have led over four decades to important space missions, including the Ulysses mission to high heliolatitudes.

6.1.1 Particle populations in the heliosphere

Within the heliosphere, energetic charged particles of different origin can be identified. In the previous chapters solar energetic particles and particles accelerated by interplanetary shock waves have been discussed. In what follows we will only discuss galactic and anomalous cosmic rays, Jovian electrons, and their propagation and modulation in the heliosphere.

Galactic cosmic rays

Galactic cosmic rays (GCRs) consist of energetic electrons and nuclei which are a direct sample of material from far beyond the solar system. They are accelerated by shock waves in the galaxy from, for example, supernova remnants, pulsars, or active galactic nuclei. The remarkable feature of cosmic rays is their energy spectra,

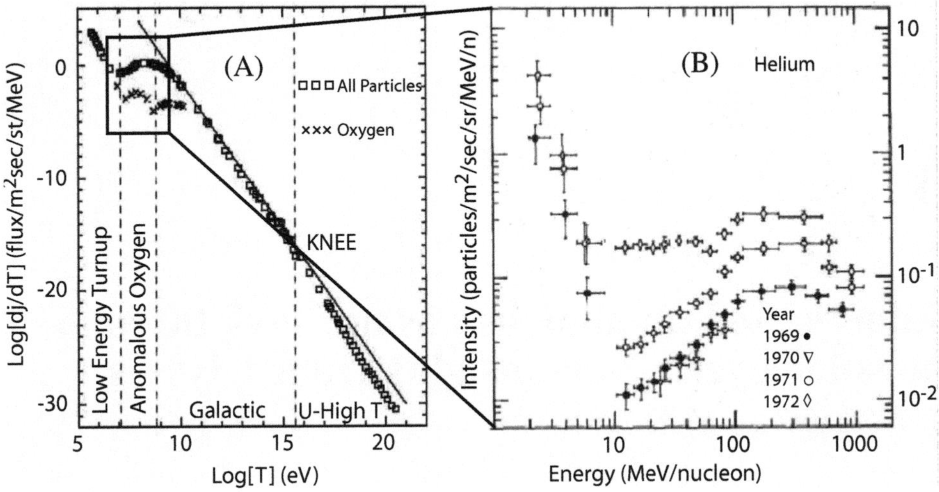

Figure 6.1. (A) Energy spectrum of cosmic rays measured in the vicinity of Earth, over some 10 orders of magnitude, showing a relatively featureless power-law distribution (Jokipii, 1989). However, at the lowest indicated energies the effects of solar modulation become evident for galactic and anomalous particles. (B) Helium energy spectra showing the peculiar variation between 10 to 50 MeV/nucleon of anomalous cosmic ray intensities with the solar cycle (Garcia-Munoz, Mason, and Simpson, 1973).

displayed in Figure 6.1. From $\sim 10^6$ eV to 10^{20} eV these spectra, over some 10 orders of magnitude variation in intensity, show a relatively featureless power-law distribution but with at least two "breaks" in the power-law.

GCRs consist primarily of hydrogen nuclei ($\sim$92%) and He nuclei ($\sim$7%). The heavier nuclei (1%) provide information about cosmic-ray origin through their elemental and isotopic composition (Wiedenbeck, 2000). At energies below a few GeV the influence of solar modulation becomes important.

6.1.2 Cosmic ray modulation

Measurements by various particle detectors have shown that the intensity varies on different timescales. Figure 6.2 shows the time profile of >3 GV galactic cosmic rays, as measured by the Climax neutron monitor (*http://ulysses.uchicago.edu/Neutron Monitor*), and the monthly smoothed sunspot number. The short-term variations observed at Earth and at various spacecraft are mostly correlated with disturbances originating at the Sun—for example, coronal mass ejections (Cane, 2000) and the interaction of solar wind streams with different speeds forming corotating interaction regions beyond the Earth's orbit (Richardson, 2004). On longer timescales the cosmic ray flux varies in anti-correlation with the 11-year and 22-year solar activity cycle. Thus, cosmic rays entering the region surrounding the Sun are increasingly modulated as they traverse that volume of space dominated by the Sun, called the heliosphere.

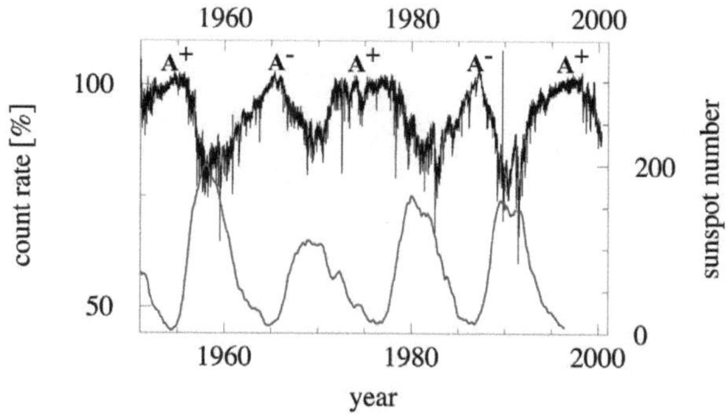

Figure 6.2. Time profile of >3 GV GCRs, as measured by the Climax neutron monitor (upper curve, *http://ulysses.uchicago.edu/NeutronMonitor*), and the monthly smoothed sunspot number (NSSDC), showing the anti-correlation of GCR intensities with the solar cycle. Marked by A^+ (A^-) is the polarity epoch of the solar magnetic field (from Heber, 2001).

Anomalous cosmic rays

Anomalous cosmic rays (ACRs) were discovered in the 1970s when Garcia-Munoz, Mason, and Simpson (1973) found an unexpected shape of the helium spectrum below ~100 MeV/n (see Figure 6.1 and fig. 1 in Moraal, 2001). Fisk, Kozlovsky, and Ramaty (1974) postulated the following mechanism as a source for these particles. The principal ideas were further developed by Vasyliunas and Siscoe (1976), discussed in detail by Moraal (2001) and le Roux (2001), and are summarized in Figure 6.3. Neutral interstellar atoms enter the heliosphere and are ionized by the interaction with the solar wind and/or solar radiation and are picked up by the solar wind. Pickup ions are convected out to the heliospheric termination shock and are accelerated to cosmic ray energies. The process of shock acceleration has been theoretical, as described by Pesses, Eichler, and Jokipii (1981) and Lee and Fisk (1982). Interstellar neutral helium and the hydrogen and helium pickup ions were measured with instruments onboard the AMPTE (Möbius *et al.*, 1985) and the Ulysses spacecraft (Witte *et al.*, 1993; Gloeckler *et al.*, 1993).

The ACR component is different from GCRs in a number of respects:

1. ACRs are mostly singly charged, while GCRs are fully stripped atoms.
2. ACRs should reflect the elemental and isotopic composition of pickup ions and therefore of the local interstellar neutrals, while the GCR composition is modified during their propagation within the galaxy.
3. The maximum energy of ACRs should be restricted to several hundred MeV, whereas GCRs are accelerated to much higher energies by presumably much larger shocks.

For details of the current paradigm see the recent review by Fichtner (2001). Note that the *in situ* measurements by Voyager 1 at the heliospheric termination shock are posing new questions to the ACR paradigm (Potgieter, 2006).

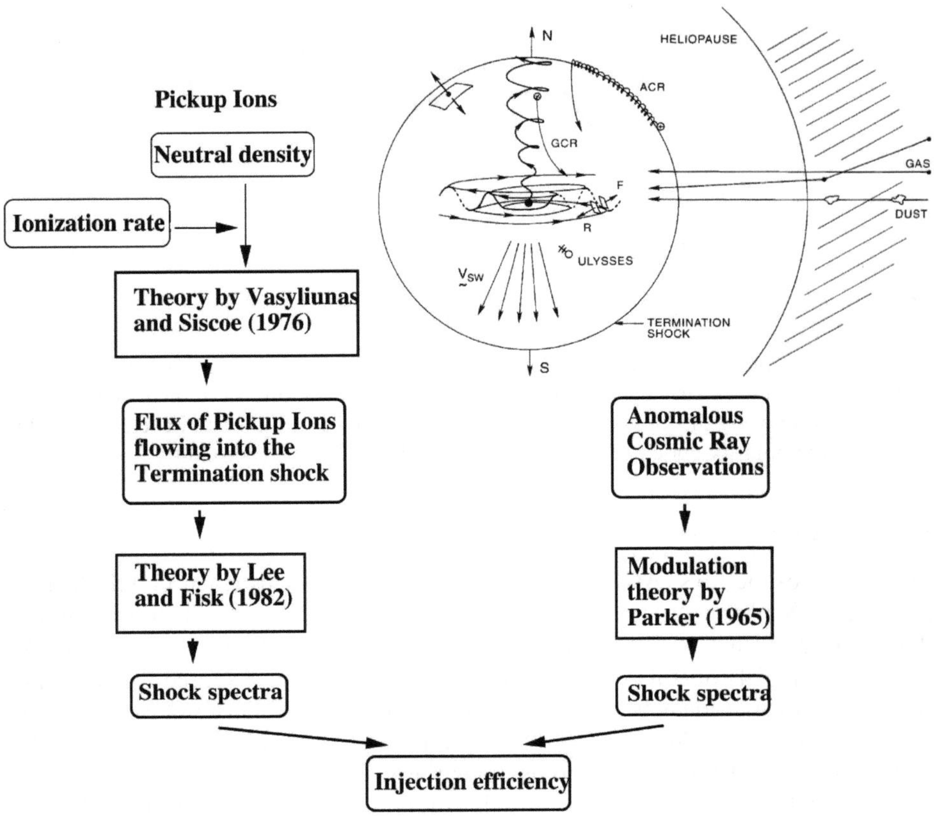

Figure 6.3. The interaction of the solar wind with the local interstellar medium defines the heliosphere (upper right panel). Pickup ions are generated from interstellar neutrals by ionization. These pickup ions are accelerated at the heliospheric termination shock to become ACRs (from Heber and Cummings, 2001).

Jovian electrons

Historically, it became clear that Jupiter was a continuous source of MeV electrons in the solar system when Pioneer 10 came within 1 AU of the planet (Teegarden *et al.*, 1974; Simpson *et al.*, 1974). Figure 6.4 from Pyle and Simpson (1977) displays the trajectories of the planets Earth (E), Jupiter (J), and Saturn (S) and both Pioneer 10 and 11. The solid and dashed lines in part (b) show the counting rate of 3–6 MeV electrons and its distance r dependent with respect to Jupiter, respectively. The $\sim 1/r$ dependence can be explained by diffusion from a continously emitting point source (Conlon, 1978; Pyle and Simpson, 1977) which is limited to less than 1 AU behind the planet. In addition, Jovian electron studies resulted in the first strong observational evidence for a diffusive transport of electrons perpendicular to the mean heliospheric magnetic field (Chenette, Conlon, and Simpson, 1974). Teegarden *et al.* (1974) further identified Jupiter as the source of "quiet time" electron increases previously observed

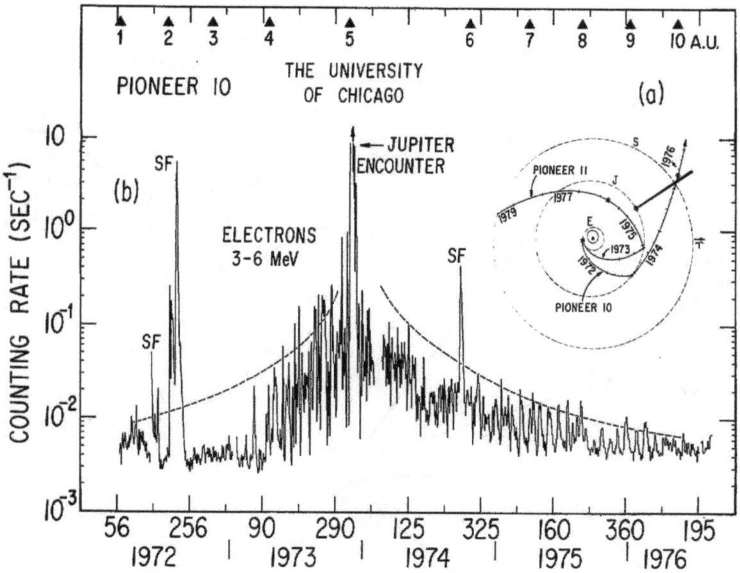

Figure 6.4. (a) Trajectories of the Earth (E), Jupiter (J), and Saturn (S) together with those of Pioneer 10 and 11. (b) The solid and dashed lines display the daily averages of the 3–6 MeV electron counting rate from 1972 to 1976 compared with the $1/r$ dependence from Jupiter, respectively (Pyle and Simpson, 1977).

at 1 AU (McDonald, Cline, and Simnett, 1972; L'Heureux, Fan, and Meyer, 1972). This variability is caused mainly by varying heliospheric conditions—for example, by corotating interaction regions (Conlon and Simpson, 1977; Conlon, 1978; Rastoin, 1995; Kissmann, Fichtner, and Ferreira, 2004).

6.2 SELECTED COSMIC RAY OBSERVATIONS

In order to put Ulysses observations of galactic and anomalous cosmic rays into context, we first briefly review measurements made by space probes at or near 1 AU and by the Pioneer and Voyager spacecraft in the outer heliosphere. Figure 6.5 (adapted from Heber and Cummings, 2001) displays in part (a) the trajectory of different planets, the two Voyagers, Pioneers, and the Ulysses spacecraft. Note that the two Voyagers and Pioneer 11 are heading towards the nose of the heliosphere, while Pioneer 10 is moving in the direction of the tail. The dashed lines are the projections of the spacecraft trajectories onto the ecliptic plane. In part (b) the heliographic latitude as a function of heliocentric distance is shown for the two Voyagers and Ulysses only. The inner heliosphere paths are omitted for the Voyagers.

Pioneer 10 and Pioneer 11 launched in 1972 and 1973 ran out of power in 2003 and 1995 at a distance of about 80 AU and 44.7 AU, respectively. The instruments

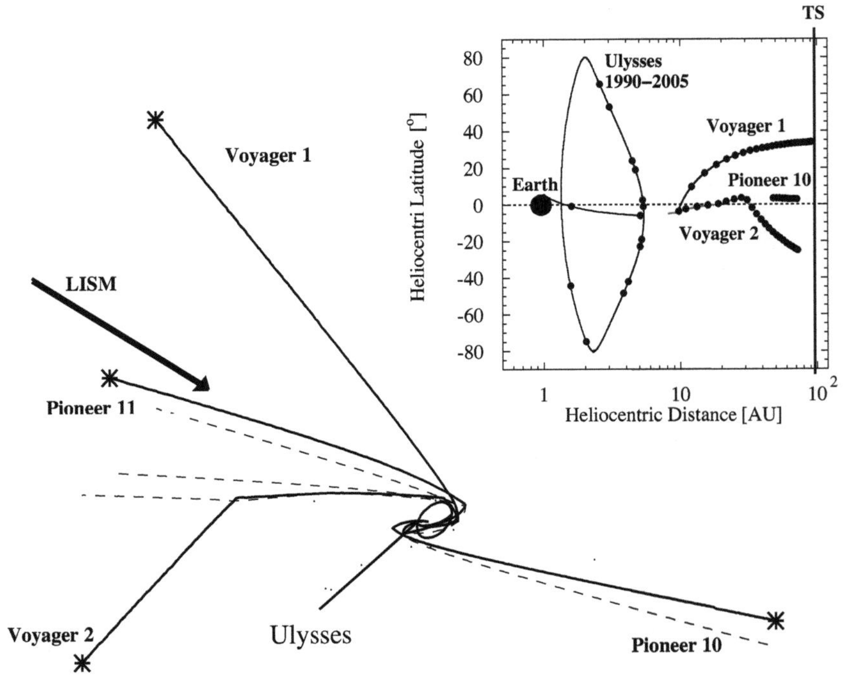

Figure 6.5. Ulysses, Pioneer, and Voyager trajectories are displayed in the left part. The heliographic latitude as a function of radial distance is shown in the right part. Marked by the shaded dot is the region close to Earth where a fleet of spacecraft is located, such as the Advanced Composition Explorer, the Solar and Heliospheric Observatory, the WIND and IMP spacecraft. Distances are plotted on a logarithmic scale (adapted from Heber and Cummings, 2001).

aboard the Pioneer spacecraft provided important and very useful information on cosmic ray nuclei and electrons (McKibben *et al.*, 1973; Lopate, 1991). The two Voyager spacecraft launched in 1978 left the outer planets in the 1980s to begin their mission to the termination shock, heliosheath, and interstellar space. Today, Voyager 1 is the most distant man-made object and is beyond 100 AU from the Sun. Marked by the innermost circle is the region close to Earth, where a fleet of spacecraft, such as the Advanced Composition Explorer (ACE), the Solar and Heliospheric Observatory (SOHO), the WIND and IMP spacecraft, has been exploring the inner heliosphere using advanced instrumentation. From Figure 6.5 it is evident that the decade of the 1990s was unique in investigating radial and latitudinal gradients in the inner as well as in the outer heliosphere.

6.2.1 Observations close to Earth

At energies below a few GeV/nucleon the influence of solar modulation on the galactic cosmic ray energy spectra becomes important. Figure 6.6 displays the varia-

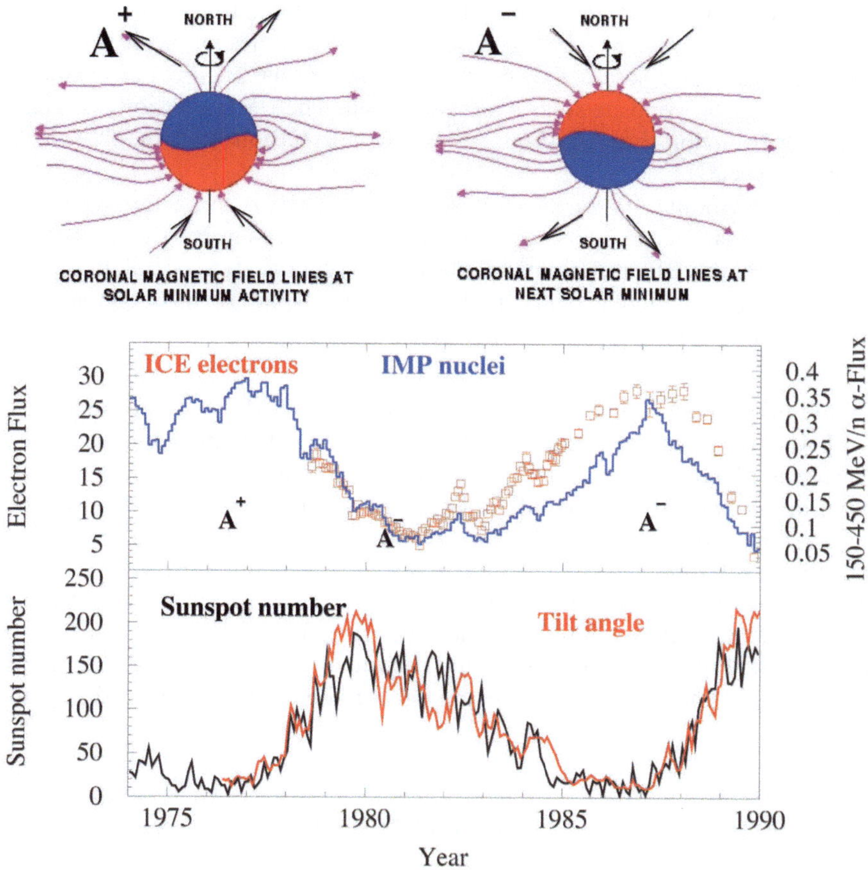

Figure 6.6. Solar modulation of galactic cosmic rays of both charge signs, monthly sunspot number, and tilt angle α of the heliospheric current sheet. Marked by A^+ (A^-) are times when the solar magnetic field is directed inward (outward) from the Sun in the northern polar and outward (inward) in the southern polar region, as sketched in the top panel.

tion of GeV particles with the solar cycle but for particles of opposite charge. It displays in the lower panel the monthly sunspot number (black line) and the evolution of the maximum latitudinal extension of the heliospheric current sheet (tilt angle, red line). The upper panel gives the cosmic ray variation close to Earth of galactic cosmic ray helium and electrons, measured by the IMP and ICE spacecraft. When comparing the neutron monitor measurements with the helium intensity–time profile it is obvious that both profiles are very similar. From Figure 6.6 three characteristic features of the cosmic ray intensity history are evident:

1. Both helium and electrons vary in anti-correlation with the 11-year solar activity cycle, leading to the conclusion that galactic cosmic rays are modulated as they traverse the heliosphere.

2. In the 1960s and 1980s (A^-), when the solar magnetic field was pointing towards the Sun in the northern hemisphere, the time profiles of positively charged particles peaked, whereas they were more or less flat in the 1970s and 1990s (A^+) during the opposite solar magnetic epoch. The electrons, however, had the opposite behavior showing clearly the close correlation with the 22-year solar magnetic cycle.
3. Cosmic ray modulation during increased solar activity is characterized by several large steps that are easily recognized from observations at Earth and beyond, as shown in Figure 6.6, and discussed above. These large steps correlate with long-lasting intense magnetic fields in the outer heliosphere, called global merged interaction regions (Burlaga, Perko, and Pirraglia, 1993).

Figure 6.7 displays quiet time counting rates of >70 MeV protons as measured aboard IMP 8 (dark curve) and Voyager 2 (light curve). By comparing the intensity–time histories of both spacecraft it is evident that the amplitude of the solar cycle variation depends on the spacecraft radial position. The minimum intensity in the early 1990s was about 0.15 c/s and 0.3 c/s at Earth and at about 40 AU, respectively. The maximum intensities of 0.5 c/s at Earth and 0.7 c/s at about 70 AU were reached in 1997 and late 1998. The delay Δt of the onset in modulation in the outer heliosphere corresponds approximately to the time the solar wind needs to travel from 1 AU to 70 AU. Figure 6.7 also shows that the radial gradient depends on the phase of the solar cycle. Indeed, there has been a debate about whether the radial gradient

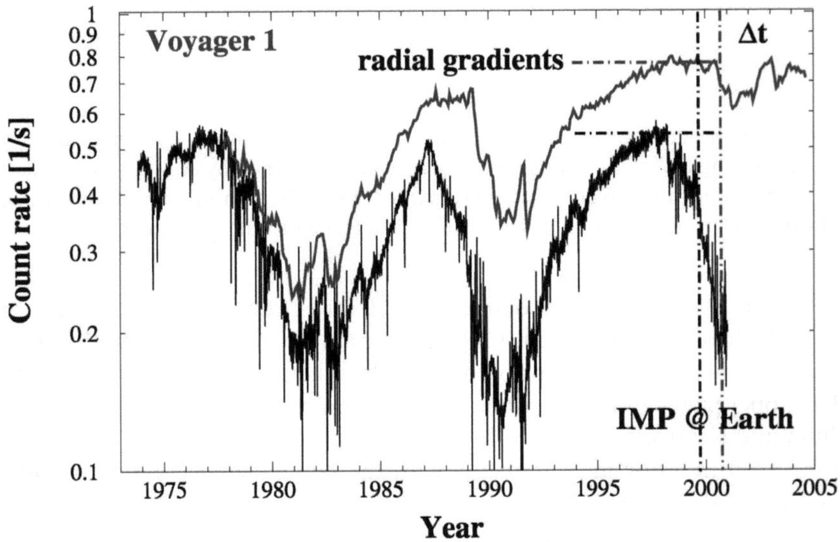

Figure 6.7. As an illustration of the positive radial gradient, the count rate of >70 MeV protons as measured by the Goddard Spaceflight Center instrument onboard Voyager 2, is compared with the University of Chicago instrument onboard IMP 8. Obviously, the intensity is always higher in the outer heliosphere.

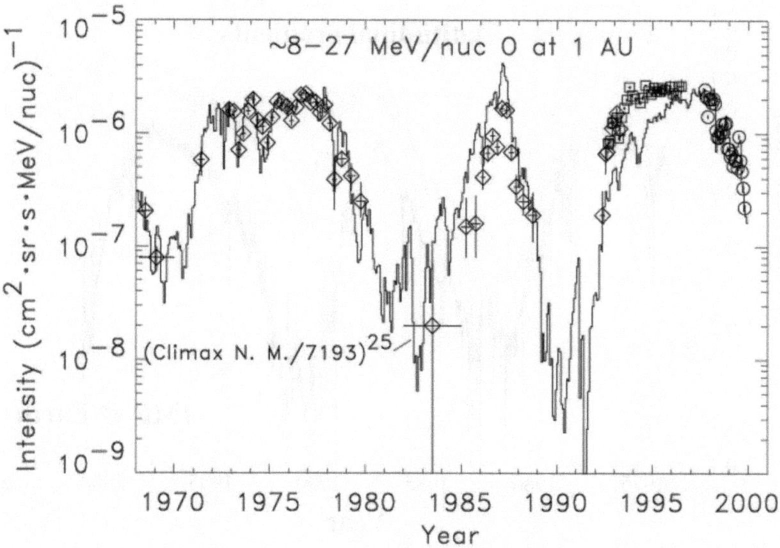

Figure 6.8. Comparison of the time profile of ACR oxygen (symbols) with GCRs (solid line) at Earth. While the high-energy GCRs are only modulated by a few percent, ACR oxygen varies by more than 2 orders of magnitude (Leske *et al.*, 2000).

should be calculated by taking the measurements at the same times or using the intensities measured in the outer heliosphere at a time $t + \Delta t$ (Heber *et al.*, 1993).

Figure 6.8 displays the time profile of ACR oxygen and GCRs above 3 GV. While the high-energy GCRs are only modulated by a few percent, ACR oxygen varies by more than 2 orders of magnitude. In contrast to GCRs, the generation process of ACRs might also be solar cycle dependent (e.g., Fichtner, 2001). Since ACRs are difficult to measure during high solar activity in the inner heliosphere, results mainly during the last solar minimum are reviewed.

In Figure 6.9 observations from IMP and Voyager 1 are shown together with Voyager 2 measurements. For most time periods the intensity measured at Voyager 1 is larger than at Voyager 2 and at Earth, indicating measurable positive radial gradients also in the outer heliosphere. Of special interest is the 1980s' solar minimum: during a long time period the intensity measured at Voyager 2 exceeded the one observed at Voyager 1, although Voyager 1 is farther out in the heliosphere. At that time Voyager 2 was close to the ecliptic and Voyager 1 at 30°N. Thus, Cummings, Stone, and Webber (1987) showed that this was the first direct measurement of a (negative) latitudinal gradient.

6.2.2 The transport equation

The transport of cosmic rays in the heliosphere is described by Parker's (1965) transport equation. If $f(\mathbf{r}, P, t)$ is the cosmic ray distribution function with respect

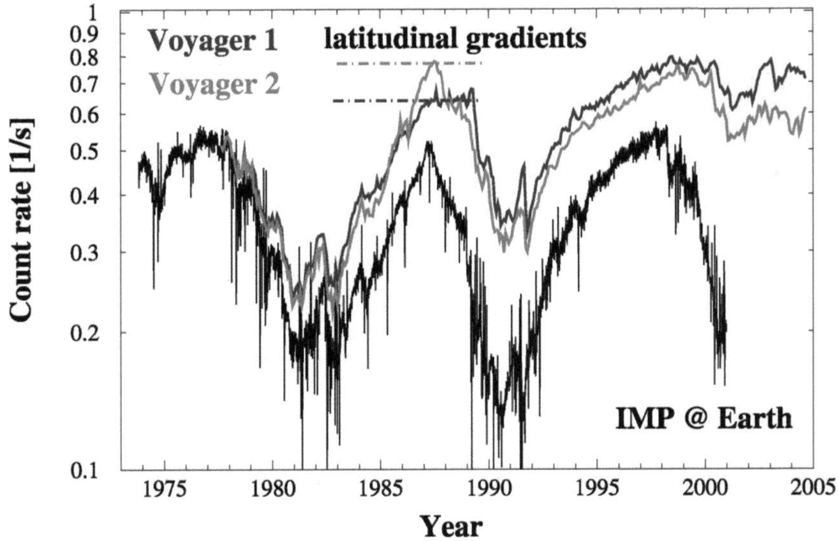

Figure 6.9. In comparison with Figure 6.7, the count rate of >70 MeV protons onboard Voyager 1 is inserted. Although Voyager 1 was farther away from the Sun than Voyager 2 in 1987, the intensity is higher at Voyager 2. Since Voyager 1 was at about 30°N and Voyager 2 still close to the ecliptic, the figure illustrates the existence of a negative latitudinal gradient from 1985 to 1987.

to particle rigidity P, then the cosmic ray variation with time t and position $\mathbf{r}$ is given by:

$$\frac{\partial f}{\partial t} = -\Big(\underbrace{V}_{a} + \underbrace{\langle \mathbf{v}_D \rangle}_{d}\Big) \cdot \nabla f + \underbrace{\nabla \cdot (\kappa_{(s)} \cdot \nabla f)}_{c} + \underbrace{\frac{1}{3}(\nabla \cdot \mathbf{V}) \frac{\partial f}{\partial \ln P}}_{b} + \underbrace{Q}_{e}, \quad (6.1)$$

where terms on the right-hand side represent the following mechanisms:

a. An outward convection caused by the radially directed solar wind velocity $\mathbf{V}$. During solar minimum activity this radial flow has a large latitudinal dependence. Beyond the termination shock $\mathbf{V}$ becomes increasingly latitudinal- and azimuthal-dependent.
b. Adiabatic energy changes depending on the sign of the divergence of $\mathbf{V}$. Inside the termination shock and towards the Sun, adiabatic energy losses become increasingly important. Beyond the termination shock, adiabatic heating may occur especially in the direction that the heliosphere is moving. Diffusive shock acceleration is implicity described by this term in the transport equation.
c. Diffusion caused by turbulent irregularities in the background heliospheric magnetic field. The symmetric part of the diffusion tensor $\boldsymbol{\kappa}_{(s)}$ consists of a diffusion coefficient parallel to the background magnetic field ($\boldsymbol{\kappa}_{\parallel}$) and a perpendicular diffusion coefficient for the radial ($\boldsymbol{\kappa}_{\perp r}$) and polar direction ($\boldsymbol{\kappa}_{\perp\theta}$), respectively, as displayed in Figure 6.10. It follows that the values of the three diffusion coeffi-

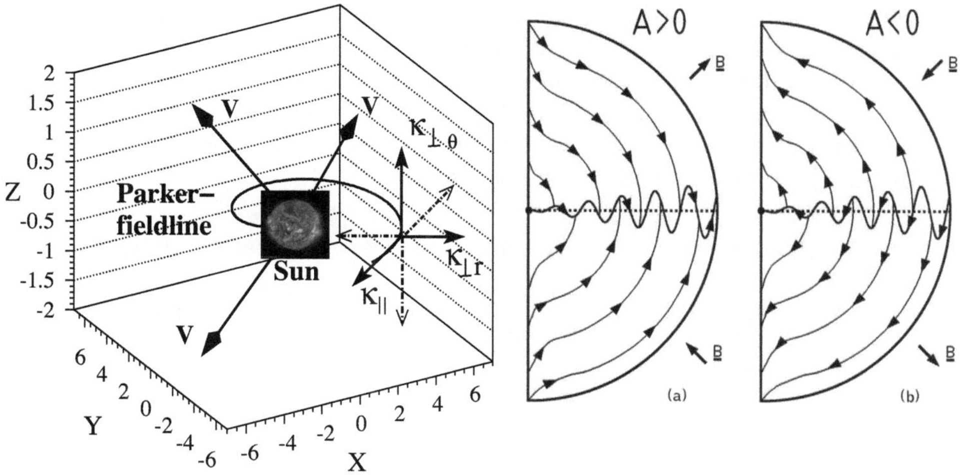

Figure 6.10. The different elements of the diffusion tensor with respect to the Parker spiral (left). The arrows *V* indicate the radially expanding solar wind velocity. The global drift pattern of positively charged particles in an A>0 and A<0 solar magnetic epoch, together with a wavy current sheet, are shown in the right panels.

cients depend on time (i.e., solar activity), on one's position in the heliosphere, and the energy (rigidity) of the cosmic rays. This means that for a full three-dimensional and time-dependent treatment, the transport equation has to be solved in five "numerical" dimensions.

d. Gradient, curvature, and current sheet drifts in the global heliospheric magnetic field.The averaged guiding center drift velocity for a near isotropic galactic cosmic ray distribution is given by:

$$\langle \mathbf{v}_D \rangle = \frac{Pv}{3} \nabla \times \frac{\mathbf{B}}{B^2},$$

v is the particle speed, and B is the magnitude of the background heliospheric magnetic field **B**. With the solar magnetic field directed outward from the Sun in the northern polar region and inward in the southern polar region, as displayed in Figure 6.10 (right-hand side), positively charged particles are expected to drift into the inner heliosphere primarily over the solar poles and out along the heliospheric current sheet. This period is known as the A>0 magnetic polarity epoch. The drift pattern of negatively charged particles is then in the opposite direction so that the intensity at Earth of these particles strongly depends on the latitudinal excursion of the heliospheric current sheet, whereas the intensity of positively charged particles varies significantly less (Potgieter and le Roux, 1992). The situation reverses in an A<0 magnetic cycle, shown in Figure 6.10b.

e. The last term represents any additional sources—for example, for anomalous cosmic ray particles being accelerated at the solar wind termination shock, or a source of Jovian electrons which at energies of a few MeV can make a major contribution to the particle distribution in the inner heliosphere relatively close to the ecliptic (Ferreira *et al.*, 2001a). The location of the Jovian magnetosphere with respect to the mean heliospheric magnetic field provides therefore ideal test particles to study heliospheric particle propagation. A term describing diffusion in momentum space that produces stochastic acceleration is usually neglected for long-term cosmic ray modulation studies.

Rewriting Equation (6.1) in heliocentric spherical coordinates (r, θ, ϕ) gives:

$$\begin{aligned}
\frac{\partial f}{\partial t} &= \left[\frac{1}{r^2}\frac{\partial}{\partial r}(r^2 K_{rr}) + \frac{1}{r\sin\theta}\frac{\partial}{\partial\theta}(K_{\theta r}\sin\theta) + \frac{1}{r\sin\theta}\frac{\partial K_{\phi r}}{\partial\phi} - V\right]\frac{\partial f}{\partial r} \\
&+ \left[\frac{1}{r^2}\frac{\partial}{\partial r}(rK_{r\theta}) + \frac{1}{r^2\sin\theta}\frac{\partial}{\partial\theta}(K_{\theta\theta}\sin\theta) + \frac{1}{r^2\sin\theta}\frac{\partial K_{\phi\theta}}{\partial\phi}\right]\frac{\partial f}{\partial\theta} \\
&+ \left[\frac{1}{r^2\sin\theta}\frac{\partial}{\partial r}(rK_{r\phi}) + \frac{1}{r^2\sin\theta}\frac{\partial K_{\theta\phi}}{\partial\theta} + \frac{1}{r^2\sin^2\theta}\frac{\partial K_{\phi\phi}}{\partial\phi}\right]\frac{\partial f}{\partial\phi} \\
&+ K_{rr}\frac{\partial^2 f}{\partial r^2} + \frac{K_{\theta\theta}}{r^2}\frac{\partial^2 f}{\partial\theta^2} + \frac{K_{\phi\phi}}{r^2\sin^2\theta}\frac{\partial^2 f}{\partial\phi^2} + \frac{2K_{r\phi}}{r\sin\theta}\frac{\partial^2 f}{\partial r\,\partial\phi} \\
&+ \frac{1}{3r^2}\frac{\partial}{\partial r}(r^2 V)\frac{\partial f}{\partial \ln p} + Q_{source}(r, \theta, \phi, p, t).
\end{aligned} \tag{6.2}$$

The diffusion tensor is then written as:

$$\begin{bmatrix} K_{rr} & K_{r\theta} & K_{r\phi} \\ K_{\theta r} & K_{\theta\theta} & K_{\theta\phi} \\ K_{\phi r} & K_{\phi\theta} & K_{\phi\phi} \end{bmatrix} = \begin{bmatrix} \kappa_{\parallel}\cos^2\psi + \kappa_{\perp r}\sin^2\psi & -\kappa_A\sin\psi & (\kappa_{\perp r} - \kappa_{\parallel})\cos\psi\sin\psi \\ \kappa_A\sin\psi & \kappa_{\perp\theta} & \kappa_A\cos\psi \\ (\kappa_{\perp r} - \kappa_{\parallel})\cos\psi\sin\psi & -\kappa_A\cos\psi & \kappa_{\parallel}\sin^2\psi + \kappa\cos^2\psi \end{bmatrix}, \tag{6.3}$$

with ψ the spiral angle of the magnetic field with respect to the radial direction.

The components of the gradient and curvature drift velocity are:

$$\left.\begin{aligned}
\langle \mathbf{v}_d \rangle_r &= -\frac{\mathrm{A}}{r\sin\theta}\frac{\partial}{\partial\theta}(\sin\theta K_{\theta r}), \\
\langle \mathbf{v}_d \rangle_\theta &= -\frac{\mathrm{A}}{r}\left[\frac{1}{\sin\theta}\frac{\partial}{\partial\phi}(K_{\phi\theta}) + \frac{\partial}{\partial r}(rK_{r\theta})\right], \\
\langle \mathbf{v}_d \rangle_\phi &= -\frac{\mathrm{A}}{r}\frac{\partial}{\partial\theta}(K_{\theta\phi}),
\end{aligned}\right\} \tag{6.4}$$

with $\mathrm{A} = sign(qB)$ determining the drift direction of particles with charge q in a magnetic field B, shown in Figure 6.10.

The present understanding of the mechanisms of global modulation in the heliosphere, as described above, is considered essentially correct. However, the main obstacle in solving Equation (6.1) is insufficient knowledge of spatial rigidity and especially the temporal dependence of diffusion coefficients and drifts, including the underlying features of the magnetic field turbulence, and structure at high heliolatitudes, the size and geometry of the heliosphere (e.g., where is the heliopause located?), and the values of the local interstellar spectra for the different cosmic ray species. In what follows the various parameters of importance to galactic and anomalous cosmic ray modulation in the heliosphere are discussed.

6.2.3 The diffusion tensor

The spatial and rigidity dependence of the elements of the diffusion tensor are not well-known. Serious efforts are therefore being made to improve the situation following three approaches. (1) Determining the diffusion coefficients fundamentally from basic micro-physics (diffusion and turbulence theory). (2) Partly based on fundamental theory but constrained by cosmic ray observations (e.g., Burger, van Niekerk, and Potgieter, 2001). (3) Primarily based on compatibility studies (e.g., Ferreira and Potgieter, 2004) between state-of-the-art modulation models and a large set of cosmic ray observations. The last two approaches have contributed significantly in limiting the values of the various diffusion coefficients, as a result of the comprehensive numerical models that have been developed and applied over the past 20 years (as discussed below), and also the excellent cosmic ray observations from a unique combination of spacecraft in the heliosphere. The first approach is more difficult, but progress is being made to come to an *ab initio* formulation (Bieber, 2003) of cosmic ray modulation in which the diffusion coefficients are calculated from basic diffusion (scattering) theories and from the underlaying fluctuating parameters based on plasma and turbulence theories using known features of the solar wind and the heliospheric magnetic field (Burger, 2000). These approaches must eventually be tested against cosmic ray observations made at Earth and on spacecraft.

Diffusion theory involves several turbulence parameters so that one needs to understand how solar wind turbulence evolves throughout the heliosphere, also at high heliolatitudes. Even in the simplest formulation, this would involve specification of the turbulence energy density and a correlation scale length. While *in situ* observations inside 1 AU from the Sun can be used as boundary conditions, an understanding of the process throughout the heliosphere is required (McKibben, 2005). As discussed by Parhi *et al.* (2001) and Bieber (2003), developing such an *ab initio* formulation faces some major challenges: (1) A satisfactory theory of diffusion parallel and perpendicular (radial and latitudinal) to the large-scale magnetic field. Theoretical formulations of diffusion coefficients by, for example, Bieber and Matthaeus (1997) and numerical simulations by, for example, Giacalone and Jokipii (1999) are not yet fully compatible and converging. (2) Perpendicular diffusion in a two-component slab/two-dimensional turbulence depends critically on an "outer/ultra scale" about which little observational information exists, even in the ecliptic plane. (3) The radial and latitudinal variation of the parallel and perpendicular diffusion coefficients depends on the corresponding variation of the correlation length which is also poorly understood. (4) An advanced formal description of realistic global gradient and curvature drifts over a complete 11-year cycle from first principles is still to be developed. However, significant progress has been made on all these facets, a few examples are Matthaeus *et al.* (2003), Dröge (2005), Shalchi and Schlickeiser (2004), le Roux *et al.* (2005), Shalchi *et al.* (2006), and Giacalone, Jokipii, and Matthaeus (2006).

6.2.4 Solar wind, magnetic field, and the current sheet

Apart from the diffusion coefficients all cosmic ray transport models also require knowledge of the global structure and geometry of the heliosphere, the heliospheric magnetic field, the current sheet, and the solar wind velocity. Observations by the Pioneer, Voyager, Ulysses and other spacecraft have contributed significantly to understanding the spatial dependence and time evolution of these features. A major contribution was the confirmation that **V** is not uniform over all latitudes but that it can be divided into fast and slow solar wind regions during solar-minimum conditions (McComas *et al.*, 2000).The latitude-dependent radial solar wind speed inside the termination shock can be approximated for modeling purposes by

$$V(\theta) = V_0\left(1.5 \mp 0.5 \tanh\left[16.0\left(\theta - \frac{\pi}{2} \pm \varphi\right)\right]\right), \tag{6.5}$$

with $V_0 = 400$ km/s, $\varphi = \alpha + 15\pi/180$, and with all angles in radians, for northern and southern hemispheres—top and bottom signs in Equation (6.5), respectively; α is the angle between the Sun's rotation and magnetic axes known as the current sheet tilt angle which changes significantly with solar activity. The role of φ is to determine at which polar angle V starts to increase from 400 km/s towards 800 km/s during solar- minimum conditions (Moeketsi *et al.*, 2005). For solar-maximum modulation conditions it is usually simply assumed that $V(\theta) = V_0$. Beyond the termination shock, at radial distances approaching the heliopause, **V** obtains an additional strong latitudinal component.

Apart from the convection caused by the solar wind, the divergence of **V** is equally important because it describes the adiabatic energy changes of cosmic rays. If it is positive—the case in most of the heliosphere—cosmic ray ions experience large energy losses resulting in a characteristic spectral shape below a few hundred MeV in the inner heliosphere. At the termination shock it is negative and beyond the shock it may vary between positive and negative, with interesting effects for anomalous cosmic rays when it is negative, such as an increasing intensity beyond the termination shock (Lange, Fichtner, and Kissmann, 2006).

One of the most fundamental properties of the heliosphere is that its magnetic field is convected outward with the solar wind causing the heliosphere to be magnetodynamically embedded in the interstellar medium. The magnetic field features determine to a very large extent the transport of energetic particles. In order to properly understand modulation, especially at large heliolatitudes, the geometry, structure, and properties of the magnetic field must be known. With the observation of recurrent cosmic ray variations at high heliolatitudes without corresponding variations in the magnetic field, it became evident that the Parker (1958) description of the heliospheric magnetic field is an oversimplification, particularly at high latitudes. The magnetic field equations are usually modified to account for deviations of the Parker field at high latitudes. Jokipii and Kóta (1989) argued that since the radial field lines at the poles are in a state of unstable equilibrium, the smallest perturbation may cause the "collapsing" of the field line. The solar surface has a granular turbulent character that changes with time and solar latitude. The "footpoints" of the polar field lines wander randomly, creating transverse components, causing deviations from the smooth Parker geometry. The net effect is highly irregular and compressed field lines so that the magnitude of the mean magnetic field at the poles is greater than in the smooth magnetic field of a pure Parker spiral. Qualitatively, such a modification is supported by measurements made of the magnetic field in the polar regions of the heliosphere by Ulysses (Balogh *et al.*, 1995). For a recent treatment of these issues, see also Giacalone, Jokipii, and Matthaeus (2006).

Fisk (1996) pointed out that a different correction needs to be made to the Parker spiral model for the simple reason that the Sun does not rotate rigidly but differentially, with the solar poles rotating ~20% slower than the solar equator. The interplay between the differential rotation of the footpoints of the field lines in the photosphere of the Sun, and the subsequent non-radial (superradial) expansion of the field lines with the solar wind from coronal holes, can result in excursions of the field lines with heliographic latitude, illustrated in Figure 6.11. This effect accounts for observations from the Ulysses spacecraft of recurrent energetic particle events at higher latitudes. The magnetic field lines at high latitudes can be connected directly to corotating interaction regions in the solar wind at lower latitudes. When the footpoint trajectories on the source surface can be approximated by circles offset from the solar rotation axis with an angle β_A, an analytical expression for this field can be obtained as given by Zurbuchen, Schwadron, and Fisk (1997).

A field with a meridional component leads to a more complicated form of the transport equation than for a Parker-type field. It is inherently three-dimensional and time-dependent so that the increase in the number of mixed derivatives results in the numerical codes that are used to solve the transport equation easily becoming unstable (Jokipii and Kóta, 2000). It is unlikely that this type of field can persist with increasing solar activity. The properties of these hybrid fields have been studied extensively (e.g., Burger and Hitge, 2004), but because of its inherent complexity this type of field is not yet fully incorporated as a standard approach in numerical modulation models. Although the Jokipii–Kóta modification is to some extent unsatisfactory, it is still well-motivated and the most convenient to apply. For a review, see Burger (2005).

A major corotating structure in the heliosphere is the current sheet which divides the heliospheric magnetic field into hemispheres of opposite polarity. Every ~11 years the solar magnetic field changes sign across this current sheet. It has a wavy structure and is rooted in the coronal magnetic field, well-correlated to solar activity. The waviness originates because the magnetic axis of the Sun is tilted relative to the rotational axis, approximated by using the tilt angle α. During high levels of activity, the observed tilt angle increases to as much as $\alpha \approx 75°$, beyond that it becomes undetermined during times of extreme solar activity. During times of low solar activity the axis of the magnetic equator and the heliographic equator become nearly aligned, causing a relative small waviness, $\alpha \approx 5°$ to $10°$. The wavy structure of the

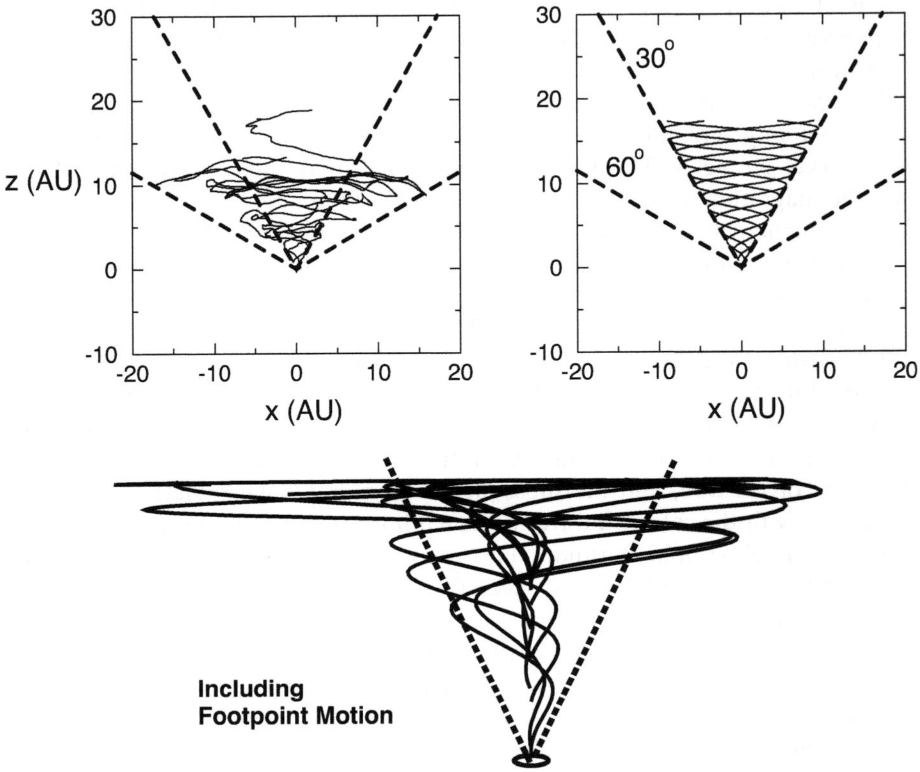

Figure 6.11. Illustration of the magnetic field lines as projected out into the heliosphere for the stochastically modified heliospheric magnetic field (Giacalone and Jokipii, 1999, upper left panel), the Parker heliospheric magnetic field (upper, right panel), and the modified heliospheric magnetic field using the footpoint motion, as suggested by Fisk (1996) (adapted from Fisk and Jokipii, 1999).

sheet is carried outwards by the solar wind (Forsyth, Balogh, and Smith, 2002). For periods of high levels of solar activity the dipole-like appearance of the Sun's magnetic field changes into more complex configurations. The wavy structure of the current sheet plays an important role in cosmic ray modulation as pointed out by Thomas and Smith (1981). For a review on the features and the importance of the wavy current sheet, see Smith (2001).

6.2.5 Size and geometry of the heliosphere

The relevant spatial regions of the heliosphere (the cosmic ray modulation domain or volume) are: (1) The region confined by the heliospheric termination shock. (2) The inner heliosheath between the termination shock and the heliopause. (3) The heliopause and the outer heliosheath. (4) The bow shock, and (5) then the local interstellar medium.

Until recently, the heliosphere was assumed to be spherical in most modulation models with an "outer boundary" at radial distances beyond ~100 AU, although it has not always been discussed clearly in the literature what it is that this "outer boundary" physically corresponds to. Presently, it is considered to be the highly asymmetrical heliopause (well-defined in the heliospheric nose direction but ill-defined in the tail direction), implying that it is the region where cosmic ray modulation actually starts, although it cannot be excluded that modulation at energies less than a few hundred MeV may occur beyond the heliopause. Assuming the termination shock to be spherical is still reasonable (Lange, Fichtner, and Kissmann, 2006).

Studying the role of the termination shock and that of the heliosheath in cosmic ray modulation with numerical models has become most relevant since Voyager 1 crossed the termination shock on 16 December 2004 (e.g., Burlaga *et al.*, 2005). Voyager 1 and 2 (and Pioneer 10 and 11) observations over 22 years and now out to ~100 AU have also shown markedly different behavior for minimum modulation conditions between the radial intensity profiles for periods of opposite magnetic polarities and that most of the residual modulation for these periods took place beyond where the termination shock was found. Models indicate that a heliosheath of several tens of AU should have a noticeable effect on the modulation of low-energy galactic and anomalous cosmic rays, and may act as an almost steady modulation front or barrier (e.g., Langner, Potgieter, and Webber, 2003).

The typically assumed heliocentric distances in the upwind (nose) direction are 80–100 AU for the termination shock (it varies with solar activity), 150–200 AU for the heliopause, and 300–400 AU for the bow shock. These distances are much larger in the downwind direction; the termination shock is probably at 150 AU. For detailed discussions, see, for example, Scherer, Fichtner, and Stawicki (2002); Zank and Müller (2003); Fahr (2004); Borrmann and Fichtner (2005); and Malama, Izmodenov, and Chalov (2006).

6.2.6 Termination shock and anomalous cosmic rays

The supersonic solar wind must merge with the local interstellar medium that surrounds the heliosphere. It first makes a transition from a supersonic into a subsonic flow at the heliospheric termination shock, in order for the solar wind ram pressure to match the interstellar thermal pressure. A shock is created because the internal wave speed suddenly becomes larger than the plasma propagation speed. The termination shock was first suggested by Parker (1961). For recent reviews, see Zank (1999), Fichtner (2001), and le Roux (2001).

At the lowest level of complexity the termination shock is expected to be a fast-mode MHD shock that is attempting to propagate sunward against the solar wind flow. It is therefore a reverse shock, so that the upstream side is closest to the Sun and the downstream side is farther from the Sun. Accordingly, the solar wind plasma should be compressed, heated, deflected, and slowed across the shock, while the magnetic field should increase. At a more complex level, the various charged particle populations may have sufficient energy density to modify the termination shock from being a primarily MHD shock to being a cosmic ray modified shock. This might affect the detailed shock structure, including the compression ratio of the plasma density across the shock and the location of the shock and the heliopause. The termination shock plays a crucial role in studies of the anomalous cosmic ray component but since it is a rather weak shock it is less important for the modulation of galactic cosmic rays.

The discovery of anomalous helium by Garcia-Munoz, Mason, and Simpson (1973) has provided a powerful new tool with which the heliosphere has been probed. Soon thereafter anomalous oxygen (Hovestadt *et al.*, 1973), nitrogen (McDonald *et al.*, 1974), and other species were observed. Fisk, Kozlovsky, and Ramaty (1974) recognized that all these elements have high first-ionization potentials and proposed that these elements enter the heliosphere as interstellar neutrals because of the movement of the heliosphere through interstellar space. They penetrate deeply into the heliosphere before they become singly ionized by charge exchange with the solar wind ions, electron collisions, or photo-ionization. These singly ionized atoms are then picked up (therefore called pickup ions) by the solar wind and convected outwards to the termination shock where they can be accelerated in principle to higher energies. These accelerated particles then diffuse back into the heliosphere to form the anomalous component of cosmic rays. They are modulated by the same processes as the galactic component. Möbius *et al.* (1985) obtained the first conclusive evidence of the solar wind picking up singly ionized interstellar helium (He^+). These aspects were reviewed from a theoretical and experimental point of view by McKibben (1998), Klecker (1999), Heber (2001), and Chalov (2005).

Pesses, Eichler, and Jokipii (1981) proposed that the termination shock was the place where the pickup ions could be accelerated to sufficiently high energies to be classified as anomalous particles, mainly through diffusive shock acceleration. Diffusive shock acceleration resulting from an infinite plane shock always gives rise to a power-law spectrum that depends only on the compression ratio of the shock. In practice, shocks are seldom plane or stationary. The power-law can only be achieved up to such a value of energy as there is time for the particles to reach that level. This still remains the most plausible explanation for the source of anomalous cosmic rays but it has become controversial since Voyager 1 observed different features for these accelerated particles that cannot easily be explained using only diffusive shock acceleration, so that alternative mechanisms have been proposed (e.g., by Fisk, Gloeckler, and Zurbuchen, 2006) or that the population of accelerated particles are more complex (e.g., by le Roux, Fichtner, and Zank, 2000).

6.2.7 Local interstellar spectra

In order to study the transport of cosmic rays in the heliosphere and to find proper diffusion coefficients it is important that the local interstellar spectra of the different particle species are known with adequate accuracy. For this, galactic propagation models are needed and significant progress has been made in computing galactic spectra for all cosmic ray species during the past decade (Moskalenko *et al.*, 2002). This work has to be extended to calculate (very) local interstellar spectra. Presently, galactic spectra are simply used as local interstellar spectra.

In the inner heliosphere, along the Ulysses trajectory, the modulated ion intensities are dominated by adiabatic energy losses below a few hundred MeV. For anti-protons these effects are less pronounced because their galactic spectrum is predicted to be much lower at low energies than for protons (Langner and Potgieter, 2004). Galactic electrons and positrons (with a completely different spectral shape), in contrast to cosmic ray ions, do not experience very large adiabatic energy losses and also fewer drifts (Potgieter, 1996). However, in the inner heliosphere electrons are completely dominated up to about 30 MeV by Jovian electrons and out to 10 AU in the ecliptic regions (Ferreira *et al.*, 2001a). In the outer heliosphere, electrons and positrons below 100 MeV should also experience relatively large modulation in the heliosheath so that—as for cosmic ray ions—the local interstellar spectra below a few hundred MeV may not be observed until a spacecraft crosses the heliopause into the local interstellar medium (Ferreira, Potgieter, and Webber, 2004).

6.2.8 Cosmic ray modulation models

In the late 1960s, the convection–diffusion model and the force field approximation were developed (Gleeson and Axford, 1967). The latter is still in use today and appears to be rather robust when applied to observations on Earth where adiabatic cooling is very large. For a recent appreciation of this approach and its limitations, see Caballero-Lopez and Moraal (2004).

Significant progress has been made over the past three decades in solving the transport equation numerically with increasing sophistication and complexity. Fisk (1976, 1979) developed the first numerical model of the transport equation, assuming a steady-state and spherical symmetry (spatially one-dimensional, 1-D). He then included a polar angle dependence to form an axisymmetric (spatially two-dimensional, 2-D) steady-state model, the first important step in the theoretical study of cosmic ray modulation at high heliolatitudes. Jokipii and Kopriva (1979) and Moraal, Gleeson, and Webb (1979) took the second step when they separately developed 2-D steady-state models including gradient and curvature drifts with a flat current sheet.The first 2-D models to emulate the waviness of the current sheet were developed by Potgieter and Moraal (1985) and Burger and Potgieter (1989), later improved by Hattingh and Burger (1995). These models emphasized the importance of global particle drifts and how the modulation at high latitudes is changed.

The first 1-D time-dependent model was developed by Perko and Fisk (1983), later extended to 2-D to include drifts and other off-ecliptic aspects by le Roux and Potgieter (1995) enabling the study of long-term cosmic ray modulation effects and the effect of outward-propagating merged interaction regions. Kóta and Jokipii (1991) developed a model that could be used to study corotating interaction regions which proved to be very useful in understanding recurrent modulation at high heliolatitudes. Another important step in modulation modeling came with the inclusion of the solar wind termination shock in models (Jokipii, 1986) that have given most plausible explanations to several observed features of the anomalous component. Various models addressing anomalous particle modulation and acceleration were independently developed by Potgieter and Moraal (1988), Potgieter (1998), le Roux, Potgieter, and Ptuskin (1996), Steenberg and Moraal (1996), and Langner, Potgieter, and Webber (2003), to mention only a few. The practical utilization of a 3-D time-dependent termination shock and modulation model is still beyond the capacity of desktop computers. Self-consistent, mostly hydrodynamic models of the heliosphere and the heliospheric interface with the interstellar medium have also been constructed (e.g., Zank and Müller, 2003; Izmodenov *et al.*, 2003; Scherer and Fahr, 2003; and references therein). However, these models cannot be directly applied to cosmic ray modulation studies and must be used in conjunction with transport models in order to obtain cosmic ray spectra and gradients at all latitudes (Scherer and Ferreira, 2005).

6.2.9 Modeling the 11-year and 22-year cycles

A major issue with time-dependent modeling is what to assume for the time dependence of the diffusion coefficients, which is significantly more difficult to do from first principles than their energy or spatial dependence. A basic departure point (required to make progress) for the time dependence of the transport parameters to describe global long-term modulation is that propagating barriers (solar wind and magnetic field structures inhibiting the easy access of cosmic rays) are formed (and later dissipate) in the heliosphere during the 11-year activity cycle. The concept was first implemented in a model by Perko and Fisk (1983) and later extended by Potgieter and le Roux (1989). This is especially applicable to the phase of the solar activity cycle before and after solar-maximum conditions when large steps in the particle intensities have been observed. In fact, a wide range of interaction regions occur in the heliosphere, the largest being called global merged interaction regions (GMIRs) introduced by Burlaga and Ness (1993) and Burlaga, Perko, and Pirraglia (1993). They observed that a clear relation exists between cosmic ray decreases (recoveries) and the time-dependent decrease (recovery) of the magnetic field magnitude and extent local to the observation point. The paradigm on which these modulation barriers is based is that interaction (and rarefaction) regions form with increasing radial distance from the Sun. This happens when two different solar wind speed regions become radially aligned to form an interaction region when the fast one runs into the slower one, resulting in compression fronts with forward and backward shocks. When these relatively narrow interaction regions are extended and wrap almost around the Sun they are called corotating interaction regions (CIRs). Between 8 AU and 10 AU these CIRs begin to spread, merge, and interact to form merged interaction regions (MIRs).

Perko and Burlaga (1990) introduced MIRs in modeling as outward-propagating regions of enhanced magnetic field magnitude relative to the background field which then cause a localized region of decreased diffusion coefficients, acting in the process as diffusion barriers but also as drift barriers to the incoming cosmic rays. The latter was extensively modeled by Potgieter and le Roux (1994). Beyond 20 AU the MIRs merge to form GMIRs, which can become large in extent and capable of causing the large step-like changes in cosmic rays. Potgieter *et al.* (1993) found that the periods during which GMIRs affect long-term modulation depend on their rate of occurrence, the radius of the heliosphere, the speed with which they propagate, their spatial extent (and amplitude), especially their latitudinal extent (to disturb drifts), and the background modulation conditions (diffusion coefficients) they encounter. Drifts, on the other hand, dominate the solar-minimum modulation periods up to 4 years so that during an 11-year cycle a transition must occur (depending on how solar activity develops) from a period dominated by drifts to a period dominated by these propagating structures.

Equally important to long-term cosmic ray modulation are gradient, curvature, and current sheet drifts as confirmed by comprehensive modeling done by Potgieter *et al.* (1993) and le Roux and Potgieter (1995). They showed that it was possible to simulate, to the first order, a complete 22-year modulation cycle by including a combination of drifts, with time-dependent tilt angles, and GMIRs in a time-dependent modulation model. For reviews of their work, see Potgieter (1997) and references therein.

For recent contributions and appreciation of this process, see Zank and Müller (2003), Ferreira and Scherer (2006), and Florinski and Zank (2006).

6.2.10 The compound modeling approach to long-term modulation

A subsequent step in modeling long-term modulation came when Cane *et al.* (1999) and Wibberenz, Richardson, and Cane (2002) pointed out that the step decreases observed at Earth could not be primarily caused by GMIRs because they occurred well before any GMIRs could form beyond 10–20 AU. Instead, they suggested that time-dependent global changes in the heliospheric magnetic field over an 11-year cycle might be responsible for long-term modulation. Following the work of le Roux and Potgieter (1995), relating this approach to changes in the diffusion coefficients, Ferreira and Potgieter (2004) combined these changes with time-dependent drifts to simulate long-term in the inner heliosphere. They called it the compound modeling approach. It was assumed that all the diffusion coefficients change time dependently $\propto B(t)^{-n}$, with $B(t)$ the observed magnetic field at Earth and n a function of rigidity and the current sheet tilt angle. The latter provides, from a cosmic ray perspective, a very realistic proxy for solar activity. These changes are then propagated outwards at solar wind speed to form propagating modulation barriers throughout the heliosphere, changing with the solar cycle. With $n = 1$ and $B(t)$ changing by an observed factor of 2 over a solar cycle, this approach resulted in a variation of the diffusion coefficients by a factor of 2 only, which is good enough to simulate the 11-year modulation for neutron monitor cosmic ray observations at Earth, but not for lower rigidities. In order to reproduce spacecraft observations at energies below a few GeV, $n(P, t)$ must depend on time (solar activity) and rigidity.

Ferreira and Potgieter (2004) confirmed that using the current sheet tilt angle as the only time-dependent modulation parameter resulted in compatibility with solar-minimum observations but not for intermediate to solar-maximum conditions. The computed modulation amplitude was too small, illustrating that wavy current sheet drifts alone cannot be responsible for the modulation of galactic cosmic rays over a complete 11-year cycle. Using the compound approach resolved this problem. Applied at Earth and along the Ulysses trajectory, this approach is remarkably successful over a period of 22 years; for example, when compared with 1.2 GV electron and helium observations at Earth, it produces the correct modulation amplitude and most of the modulation steps. Some of the simulated steps did not have the correct magnitude and phase, indicating that refinement of this approach is still needed, allowing for some merging of the propagating structures. However, solar-maximum modulation could be largely reproduced for different cosmic ray species using this relatively simple concept, while maintaining all the other major modulation features during solar minimum, such as the flatter modulation profile for electrons (helium) in 1987 (1997), but a sharper profile in 1997 (1987). Important, especially from a Ulysses point of view, is that this modeling approach also produces the observed charge sign dependent modulation from minimum to maximum solar activity. Charge sign dependent modulation is one of the important features of cosmic ray modulation because it is the most direct indication of gradient, curvature, and current sheet drifts in the heliosphere, as is discussed below.

The compound approach also involves two other important features. First, to account for the latitude dependence of cosmic ray protons and the lack thereof for electrons along the Ulysses trajectory over the 22-year cycle, a significant increased perpendicular diffusion towards the polar regions is required, mainly to reduce the large latitudinal effects caused by unmodified drifts. This is in addition to the time dependence of the diffusion coefficients. Second, during periods of large solar activity, drifts must be reduced additionally to better describe the observed electron-to-He-intensity ratio at Earth and the electron-to-proton ratio along the Ulysses trajectory during the period when the magnetic field polarity reverses (see Figure 6.24).

Ndiitwani *et al.* (2005) and Ferreira and Scherer (2006) compared model results with Voyager 2 observations and found that the compound approach could also account to a large extent for cosmic ray modulation in the outer heliosphere but merging of neighboring propagating barriers seems still necessary to realistically simulate the really large steps as occurred in 1981, 1983, and 1991. The question is how much merging occurs and what happens with these propagation barriers in the heliosheath and beyond?

6.3 COSMIC RAY DISTRIBUTION AT SOLAR MINIMA

In order to understand solar and heliospheric modulation it is vital to reproduce the spatial distribution and the energy spectra of cosmic rays in the three-dimensional heliosphere around solar-minimum periods.

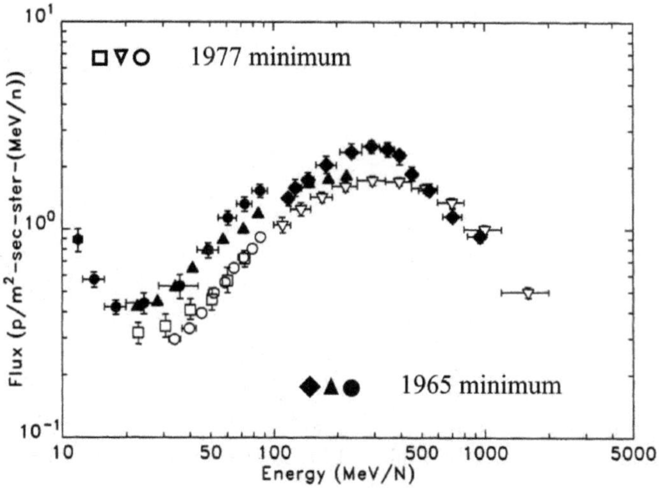

Figure 6.12. Galactic cosmic ray proton spectra (a), as measured by IMP during the 1960s' and 1970s' solar minima (Beatty, Garcia-Munoz, and Simpson, 1985).

Energy spectra and radial gradients

Figure 6.12 displays the cosmic ray spectra for protons for the 1965 and 1977 solar minimum, respectively. A closer inspection of the spectrum shows that (1) the energy spectra follows an E^1 law at several 10 MeV. It is also obvious that (2) the intensities are higher in the A>0 than in the A<0 epoch at energies below 700 MeV and (3) *vice versa* at energies above 700 MeV (Reinecke and Potgieter, 1994).

Fujii and McDonald (2005) determined the radial distribution of anomalous cosmic rays during the A<0 and the A>0 epochs. Figure 6.13 shows their result. While the intensity increases from $\sim 10^{-2}$ c/(s sr Mev/n cm^2) at 1 AU to $5 \cdot 10^{-1}$ at ~ 40 AU in an A<0 solar magnetic epoch it reaches only values of $1 \cdot 10^{-1}$ during the A>0 epoch. Thus the radial gradient depends on the phase of the solar cycle.

Charge states of ACRs

Above energies of a few MeV the charge-state composition can only be determined by using the Earth's magnetic field as a filter, which requires a spacecraft in a high inclination or polar Earth orbit such as the Solar, Anomalous, and Magnetospheric Particle Explorer (SAMPEX, Baker *et al.*, 1993). Klecker *et al.* (1998) showed that at higher energies the charge-state distribution differs significantly from that expected for oxygen. While ACR oxygen has been found to be singly charged at energies of about 10 MeV the 20–28 MeV/n oxygen distribution could only fit if the authors assumed that O^+ and O^{++} contribute approximately equally. Klecker *et al.* (1998) found that the ratio of O^+ to all charge states is decreasing with increasing energy and by analyzing N and Ne, Klecker *et al.* (1997) concluded that the total energy better

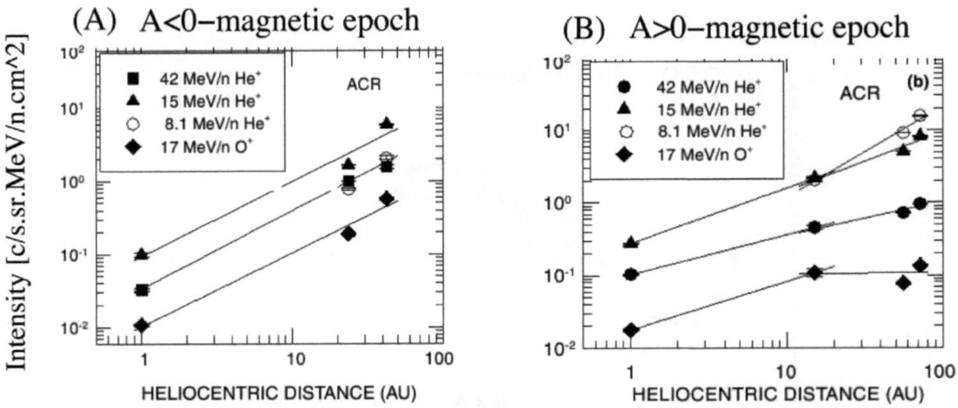

Figure 6.13. Radial intensity distributions of anomalous cosmic rays for the A<0 (A) and A>0 epochs (B) have been determined by Fujii and McDonald (2005). The radial gradient is larger during an A<0 than during an A>0 epoch.

organizes the charge-state composition. For total energies above ~350 MeV, more than half of the ACRs are multiply charged. The occurrence of multiply charged ACRs is due to electron stripping and can be explained within the current paradigm if typical timescales are of the order of 1 year or more (Mewaldt *et al.*,1996). This is in agreement with theoretical considerations (Potgieter and Moraal, 1988; Jokipii, 1996).

6.3.1 Ulysses observations at solar minimum

Ulysses observations during solar minimum have been discussed in several reviews (e.g., McKibben, 1998, 2001; Heber and Potgieter, 2000; Heber and Marsden, 2001; Heber and Potgieter, 2006). They are ideally suited to investigate particle propagation in the inner three-dimensional heliosphere. In Figures 6.15 and 6.16 we summarize first the observations for solar minimum.

The local interstellar proton spectrum

Figure 6.14(A) displays the Ulysses and Earth trajectory during the minimum fast-latitude scans. Marked by shading are the Ulysses polar passes, which are those periods during which the spacecraft is above 70° heliographic latitude in either hemisphere. An important prediction from drift-dominated modulation models is the expectation that protons will have large positive latitudinal gradients in an A>0 solar magnetic epoch (Potgieter, 1998; Heber and Marsden, 2001). Especially the intensity of protons below several 100 MeV should have increased by an order of magnitude. Electrons, on the other hand, were expected to show negative latitudinal gradients. Figure 6.14 illustrates this in part (B), where the expected proton spectra at 1 AU in the ecliptic and at 80° latitude are displayed together with the local inter-

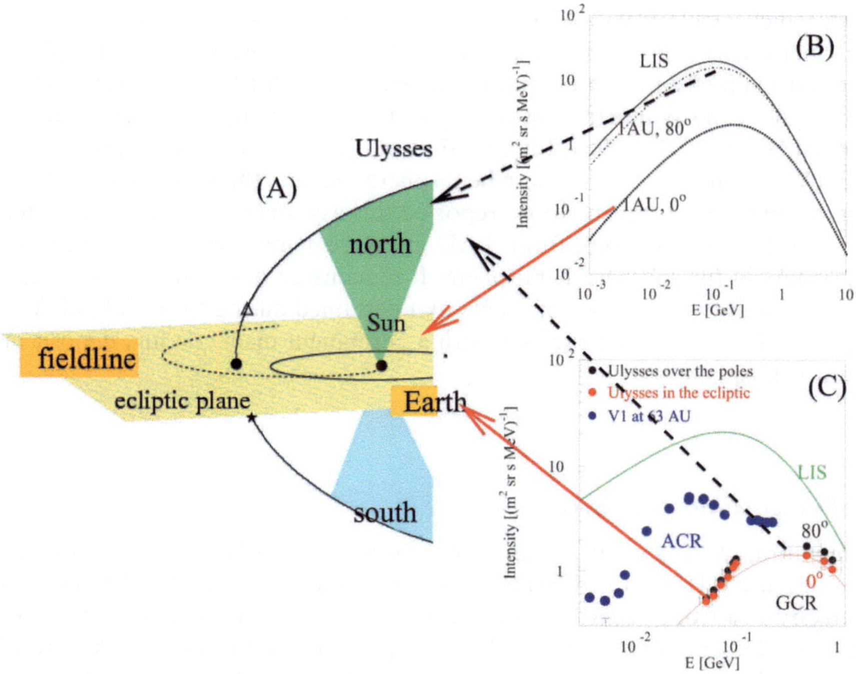

Figure 6.14. Panel (A) shows the Ulysses and Earth orbit during the first fast-latitude scan in 1994/1995. In panel (B) and (C) the expected and measured proton spectra in the ecliptic and over the poles are displayed, respectively (Heber and Potgieter, 2000; Heber *et al.*, 1996).

stellar spectrum (LIS). The model parameters have been chosen such that the 1 AU spectrum fits typical ecliptic 1 AU solar-minimum spectra. At energies below several 100 MeV an increase by an order of magnitude was expected and the LIS should become almost unmodulated at polar latitudes.The Ulysses observations during solar minimum are given in panel (C) together with the Voyager observations at 63 AU. The red symbols and line correspond to the Ulysses observations and the calculation for the heliographic equator, respectively. The black symbols are Ulysses measurements above 70°. In contrast to expectations the measured spectrum over the poles is still lower than the Voyager measurements and highly modulated. Thus, Ulysses did not measure the LIS during the minimum of solar cycle 22—with positive charged particles drifting inwards at polar regions—and led Heber *et al.* (1996a) to the conclusion that it is impossible to determine LIS in the inner heliosphere. Therefore, the LIS will only be measurable by a space probe, like Interstellar Probe (Liewer *et al.*, 2000) or the Interstellar Heliopause Explorer—investigated by ESA in 2003 (Leipold *et al.*, 2003)—to be sent far beyond the heliospheric termination shock.

The north–south-asymmetry and its consequences

A real surprise of the Ulysses mission was the observation that the galactic cosmic ray flux was not symmetric with the heliographic equator. This is illustrated in Figure

6.14(C), which displays 9-day running averages of Ulysses-to-Earth ratios of 35–70 MeV per nucleon protons, helium, and >100 MeV protons as a polar plot. The data are normalized during the equator crossing in March 1995. A constant ratio of 1 means a spherically symmetric cosmic ray distribution. Simpson, Zhang, and Bame (1996) and Heber *et al.* (1996a) found a shift of ~7–10° of the minimum intensity of >100 MeV protons into the southern hemisphere. Neither the solar wind experiments nor the magnetic field investigations reported this asymmetry. Only 5 years later did magnetic field investigations from 1 AU measurements confirm a deficit of the magnetic flux in the southern hemisphere. It remains an open question whether this observation was an occurrence of events that pertained during the rapid pole-to-pole passage of Ulysses or was correlated with a permanent magnetic flux deposit in the southern heliosphere.

Latitudinal gradients of electrons

Determination of the spatial gradients of 2.5 GV electrons is less straightforward. It relies on Figure 6.15(B) (from Heber *et al.*, 1999) which displays the 2.5 GV proton-to-electron ratio from mid-1994 to the end of 1995. The solid curve represents the variation of the temporally detrended 2.5 GV proton count rates only. As a result of this curve—almost a perfect fit to the proton-to-electron ratio—one has to conclude that the contribution of electron latitudinal gradients to this ratio is negligible.

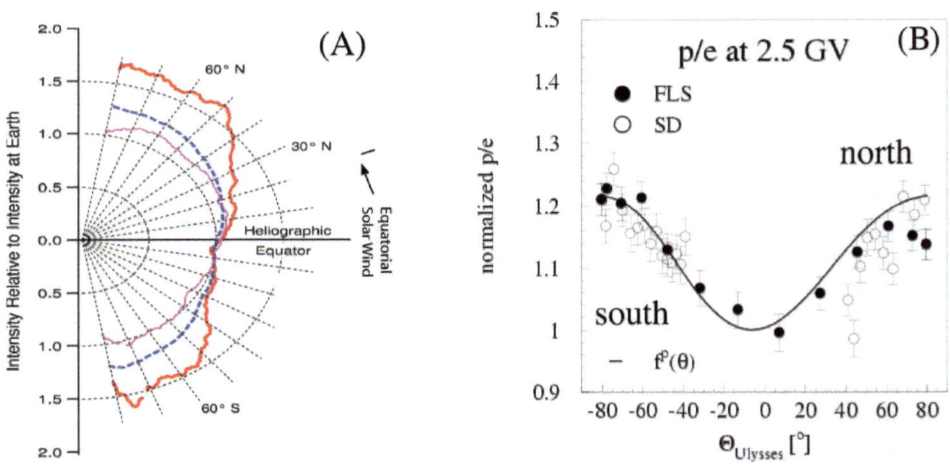

Figure 6.15. Panel (A) displays the daily-averaged Ulysses-to-Earth ratios for ~50 MeV and >125 MeV protons as well as anomalous helium by the purple, blue, and red curve, respectively. These ratios not only show the expected positive latitudinal but also an unexpected north–south asymmetry of galactic anomalous cosmic rays (McKibben *et al.*, 1996). Panel (B) shows the proton-to-electron ratio as a function of Ulysses heliographic latitude. The curve displays the variation due to protons only, indicating no latitudinal gradients for electrons (Heber *et al.*, 1999).

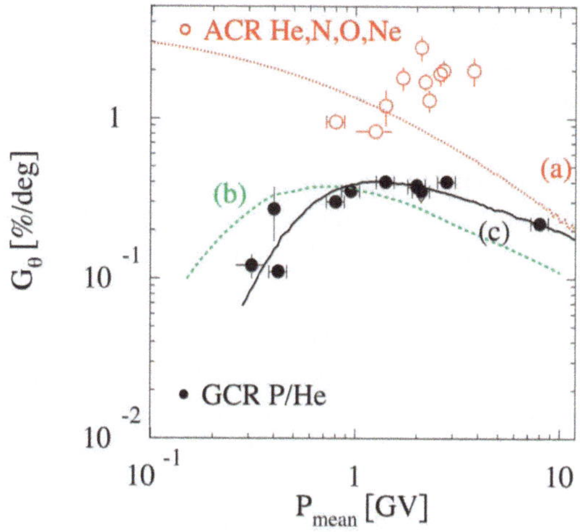

Figure 6.16. The latitudinal gradients as a function of particle rigidity are shown for anomalous (open symbols) and galactic cosmic rays (filled symbols). This figure is a compilation of Trattner *et al.* (1996), Heber *et al.* (1996a, 1999), and McKibben *et al.* (1996). In comparison with GCRs, ACRs have much larger gradients. The different curves superseded reflect the results of different model calculations for GCR protons.

The energy dependence of the latitudinal gradient

Figure 6.16 displays in part (A) Ulysses' and Earth's position during the fast-latitude scan in 1994 and 1995. Figure 6.16 displays in panel (B) the mean latitudinal gradient—that is, the ratio of the temporal detrended intensities above 70°N and 70°S and the corresponding intensity below 20°—as a function of particle rigidity during the fast-latitude scan in 1994 and 1995. From the figure it is evident that the latitudinal gradient G_θ for protons shows a local maximum of about 0.3%/degree between 1 and 2 GV. It should be mentioned that Heber *et al.* (1999) found nearly the same rigidity dependence using the data from the slow northern descent between 1995 and 1997.

The latitudinal gradients for different ACRs, as determined by Trattner *et al.* (1996) and Heber *et al.* (1996a, 1999), are much larger than the corresponding GCR ones and no maximum occurs in the investigated rigidity range. MacLennan and Lanzerotti (1998) determined the energy spectra of MeV oxygen, nitrogen, and neon. The resulting latitudinal gradients however are smaller than the one determined by Heber *et al.* (1999). In contrast to Heber *et al.* (1999) they did not determine quiet time periods.

Recurrent modulation

In the time interval extending from July 1992 to July 1994 Ulysses climbed from 10°S heliographic latitude up to over 70°S. In this time lapse solar-minimum conditions were gradually approached which in turn led to stable and long-lasting corotating interaction regions (CIRs). Paizis *et al.* (1997, 1999) and Zhang (1997) analysed recurrent cosmic ray decreases associated with ~30 registered CIRs. They studied the amplitude evolution of the 26-day recurrent cosmic ray decreases at different energies, derived its rigidity dependence, and found that

1. the amplitude has a maximum around 25°–30° and
2. the rigidity dependence of both the latitudinal gradient as well as the 26-day variation amplitude show a remarkable similarity.

Paizis *et al.* (1999) attributed the first point to a combined effect of two different causes: the effects of CIRs at low latitudes and the magnetic connection between low- and high-latitude regions. They also showed that energy changes can explain the similarity of the rigidity dependence of the gradients and the amplitude (Zhang, 1997).

An example of such a CIR event is displayed in Figure 6.17. The figure displays from top to bottom 6-hour averages of MeV protons, keV electrons, compositional signatures of the solar wind, galactic cosmic rays, magnetic field, and solar wind speed from 10 January 1993 to 9 February 1993. The CIR event can be identified unambigously by characteristic plasma and magnetic field data and thus allows the investigation of recurrent particle events and cosmic ray decreases (panel 4).

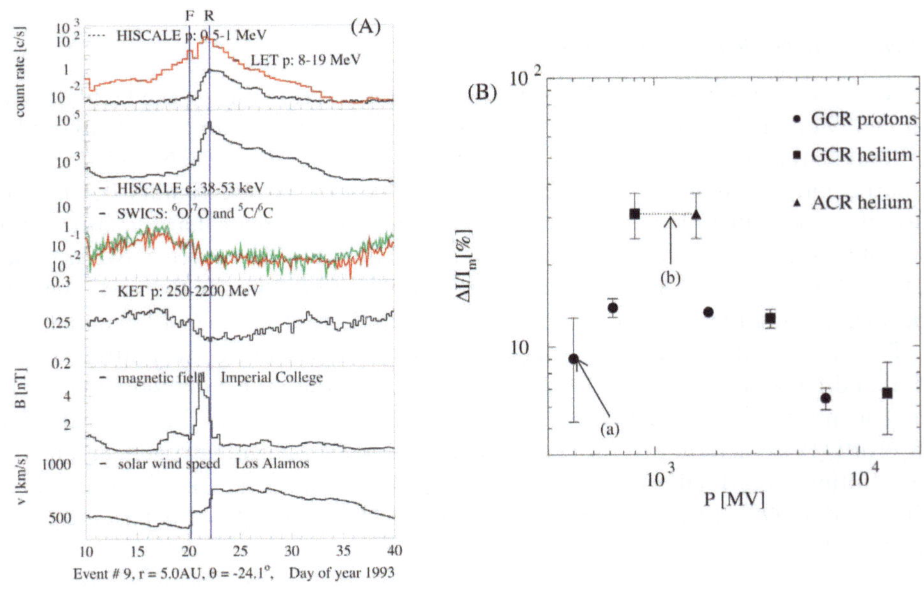

Figure 6.17. From top to bottom 6-hour averages of MeV protons, keV electrons, compositional signatures of the solar wind, galactic cosmic rays, magnetic field, and solar wind speed from 10 January 1993 to 9 February 1993. Galactic cosmic rays are modulated over short time intervals. Such recurrent modulation had already been reported in the 1960s. In panel (B) the amplitude of these short-term modulations is shown as a function of particle rigidity (Paizis *et al.*, 1999).

MeV electrons at high heliolatitudes

For energies below 300 MeV, cosmic ray electrons (and positrons) give a direct indication of the diffusion transport because they do not experience large adiabatic energy changes and their modulation is unaffected by global gradient and curvature drifts. Apart from galactic electrons, other dominant sources of electrons, especially in the energy range of 0.2–25.0 MeV and for radial distances <10 AU, are solar flares and the Jovian magnetosphere (Eraker, 1982, and references therein). Since Jupiter is a non-central source of electrons with respect to the heliospheric magnetic field, Jovian electrons provide a handy tool to investigate the particle propagation properties in the inner heliosphere. Jovian electron studies in the 1970s resulted in the first strong observational evidence for a diffusive transport of electrons perpendicular to the mean heliospheric magnetic field (Chenette, Conlon, and Simpson, 1974). Chenette *et al.* (1977) developed an analytic model to describe the propagation of Jovian electrons in the inner heliosphere. Applying the same model, Ferrando *et al.* (1999) could fit the Ulysses observations (red curves in Figure 6.18, left) from launch to the end of 1993, when Ulysses was still within 30° heliographic latitude. From that latitude on, the model calculation leads to too small intensities (green curve). As described in Section 7.4 a three-dimensional steady-state modulation code became available in the 1990s, so Ferreira *et al.* (2001a, b) used such a code to model the Ulysses observations including galactic cosmic rays as second source of energetic electrons. However, even this advanced model did not fit the Ulysses observations when using similar diffusion coefficients to those of Ferrando *et al.* (1999).

This initiated a detailed analysis of the diffusion tensor at low rigidities (Ferreira *et al.*, 2001a). The right part of Figure 6.18 (from Ferreira *et al.*, 2001b) shows the results of calculations using the large difference between the two perpendicular coefficients of the diffusion tensor. They used $\boldsymbol{\kappa}_{\perp,\vartheta} = 0.02 \cdot \boldsymbol{\kappa}_{\|}$ in the ecliptic and $\boldsymbol{\kappa}_{\perp,\vartheta} = 0.12 \cdot \boldsymbol{\kappa}_{\|}$ over the poles. In contrast $\boldsymbol{\kappa}_{\perp,r} = 0.01 \cdot \boldsymbol{\kappa}_{\|}$ everywhere. From the figure it is evident that a good agreement has been found between observations and model calculations. As a consequence of these calculations Jovian electrons are only dominant in the inner heliosphere close to the ecliptic. In addition, it was found by Burger, Potgieter, and Heber (2000) that the two elements of the diffusion tensor, perpendicular to the mean magnetic field, scale differently from each other.

There are two competing models, which are qualitatively able to explain such observation. The first by Jokipii and co-workers is based on stochastic processes whereas the one by Fisk and co-workers introduces systematic modification of the heliospheric magnetic field:

1. One explanation is that because of enhanced latitudinal diffusion the temporal variation at high heliolatitudes is determined by the interaction regions at low latitudes. The delay is then explained by the details of the diffusion process, especially perpendicular diffusion.
2. The second explanation relies on analysis of the Ulysses magnetic field data which showed that the polar magnetic field was dominated by strong variations. Systematic modifications of the standard Parker theory of the heliospheric

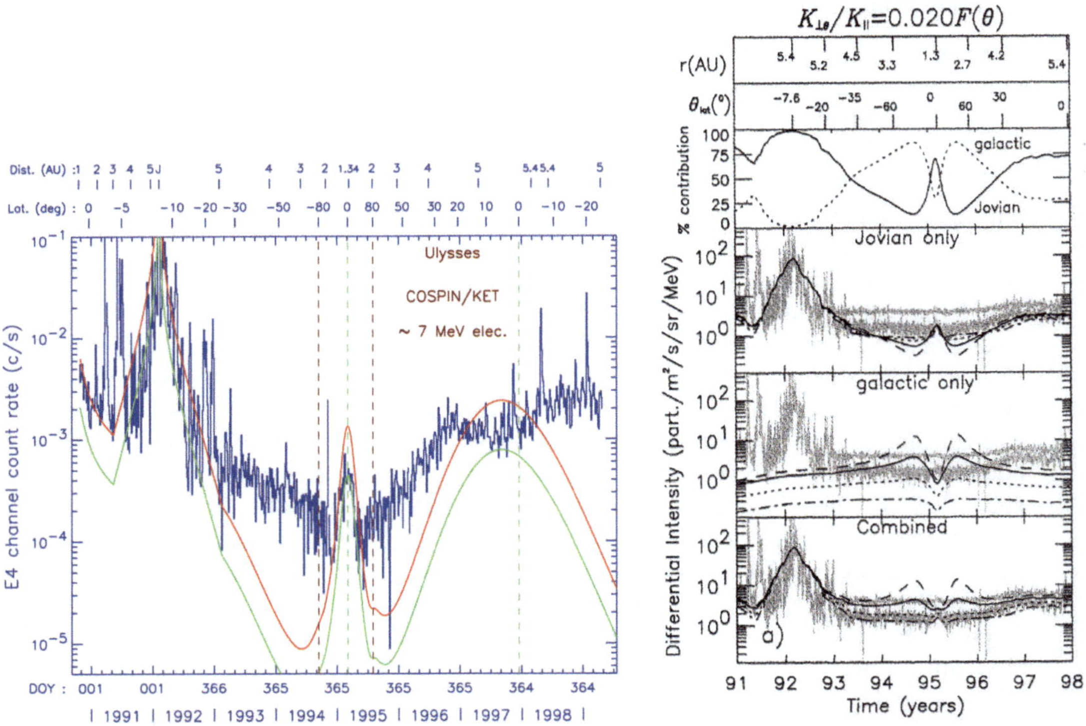

Figure 6.18. Left: Four-day averaged count rate of the Ulysses KET 3–10 MeV electron channel. The full lines are the predictions of the model by Conlon (1978) scaled to fit either the 1992 (red line) or the 1994 (green) observations close to the ecliptic. Right: From top to bottom: radial distance and heliographic latitude of Ulysses, relative contribution of galactic and Jovian electrons to the total flux for different parameter sets, intensity–time profile of 3–10 MeV electrons—as in the left part—with the contribution of Jovian-only electrons and galactic-only electrons, and the sum of both sources (Ferreira *et al.*, 2001b).

magnetic field would then cause a meridional field component along which particles may move easily to polar latitudes.

Since both proposals explain the observations equally well, only magnetic field measurements in the distant, high latitude heliosphere, where the systematic effects will be larger than the statistical variation, may prove which one is correct.

6.4 THE TRANSITION FROM SOLAR MINIMUM TO SOLAR MAXIMUM

As discussed before, cosmic ray modulation is caused by a number of physical processes, including spatial diffusion in the turbulent heliospheric magnetic field, convection and adiabatic deceleration in the expanding solar wind, and gradient and curvature drift in the large-scale magnetic fields. The strength and relative importance of these processes varies with the location in the heliosphere and with the 22-year solar cycle (Jokipii and Wibberenz, 1998). The global modulation picture changed significantly when the first high-latitude results were obtained by Ulysses during solar-minimum conditions (see Section 6.3). While solar energetic particles are discussed by Pick and Lario (2007), we will discuss here the observations and current interpretation of galactic cosmic rays and MeV electrons during the transition from solar minimum to maximum.

6.4.1 Galactic cosmic rays during the 1990–2000 A>0 solar magnetic cycle

Figure 6.19 shows the solar polar magnetic field strength as determined by Hoeksema (*http://quake.stanford.edu/~wso/*) for the southern and northern hemispheres (gray line). From the superimposed 20 nHz smoothed solar polar magnetic field strength in the northern and southern hemisphere it follows that the two hemispheres reversed their polarities around 1980, 1990, and 2000, followed by the heliospheric magnetic field reversal. In what follows we will concentrate on the importance of drifts and summarize the current understanding of the solar cycle dependence of the diffusion coefficients in the context of MeV electrons, for which drifts are of minor importance.

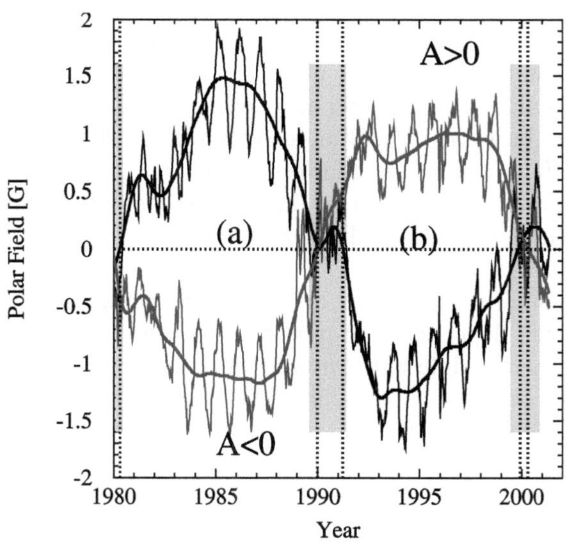

Figure 6.19. Solar polar magnetic field strength (from *http://quake.stanford.edu/~wso/*) for the southern (black) and northern hemisphere (gray). The smoothed curves display the 20 nHz low-pass filtered values. Marked by shading are time periods of the solar magnetic field reversal from 1989 to 1991 and from 1999 to 2000. (a) and (b) indicate time periods close to solar minimum investigated by Evenson (1998) and Heber *et al.* (1999).

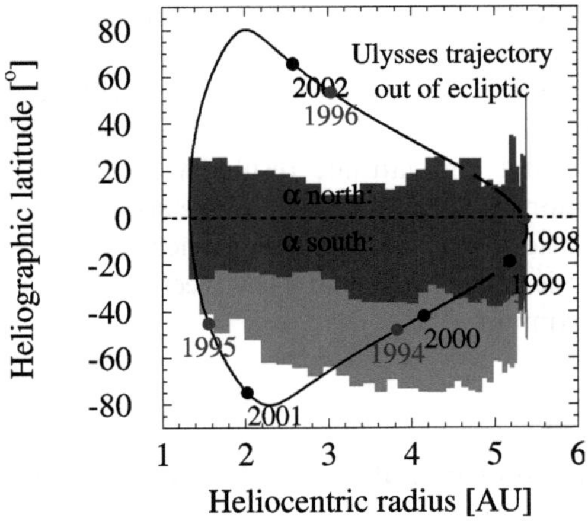

Figure 6.20. Ulysses trajectory from beginning of 1993 to 2002. Solid circles mark the start of each year. The dark and light histograms show the evolution of the maximum latitudinal extent α of the heliospheric current sheet as a function of time during the first and second orbit of Ulysses.

The variation of the tilt angle α together with the first and second out-of-ecliptic orbits of Ulysses are displayed in Figure 6.20 (from Heber *et al.*, 2002b). The dark and light histograms show the evolution of α during the first and second orbit. Inspection of Figure 6.20 shows that α was below 40° during most of the first orbit, but values above 60° have been observed for the second pass. Heber *et al.* (2002b) restricted the data analysis to the time period up to the end of 2000, when the inner heliosphere was dominated by an A>0 heliospheric magnetic field configuration. At first sight it seems to be a very difficult task to disentangle the effects of temporal and spatial variation. Fortunately, the changes caused by solar activity, latitude, and radial distance do not occur in phase, so that conclusions about the spatial gradients as well as about the differences in the behavior of electrons and protons could be drawn from the observations.

Good experimental indicators for drift effects in modulation are: (1) the difference in the latitudinal dependence of oppositely charged particles during the same polarity epoch and (2) the different temporal variation of differently charged cosmic rays caused by the variation of the heliospheric current sheet in a solar cycle (Potgieter *et al.*, 1997; Heber, 2001).

Gradients

Figure 6.21 (from Heber *et al.*, 2002b) displays with curves C_R and C_I the 26-day averaged quiet time count rates of above 250 MeV protons as measured by Ulysses and IMP. The green curve displays the Ulysses count rate, corrected for the radial movement of the spacecraft with a radial gradient of 2.2%/AU. The lower panel shows the count rate ratio (black) of the corrected Ulysses and IMP intensities. This ratio is consistent with one during the time period "C" in and from early 1997 to mid-

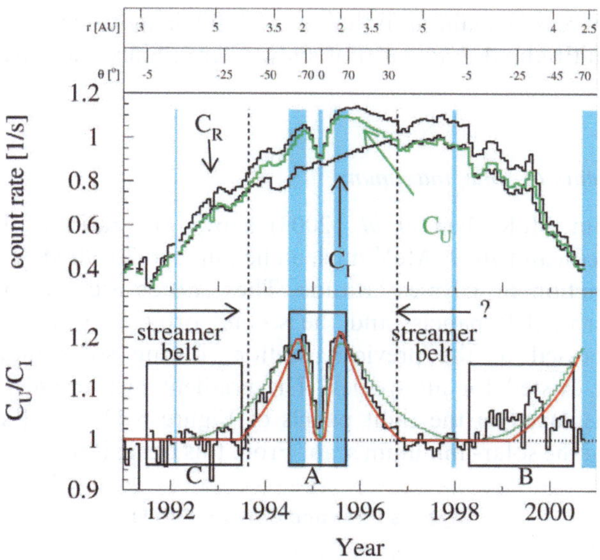

Figure 6.21. Upper panel: The 26-day averaged quiet time count rates C_R of Ulysses >250 MeV protons; IMP guard detector C_I from the GSFC instrument (Richardson, Cane, and Wibberenz, 1999) from 1991 to 2001.The green curve C_U displays the >250 MeV proton count rate, corrected for Ulysses radial variation by using a radial gradient of 2.2%/AU. Lower panel: Ratio C_U/C_I (black) in comparison with the expected variation by a Gaussian shape (see Heber *et al.*, 1996a). Note that the ratio is consistent with one within the streamer belt, corresponding to a zero latitudinal gradient. The red curve is a different representation of the previously observed latitudinal variation (Paizis *et al.*, 1995). For the explanation of periods "A", "B" and "C" see text. Ulysses' distance to the Sun and its heliographic latitude are shown in the top panel.

1998. Heber *et al.* (2002b) interpreted the constant ratio as a result of a vanishing latitudinal gradient.

The temporal variation of the ratio C_U/C_I from mid-1993 to late 1997 is unambiguously correlated with Ulysses' latitude.The green and red superimposed curves display the time profile which is expected from the observations during the fast-latitude scan. As a consequence, the data during period "A" are represented very well by the Gaussian shape of the proton latitudinal variation. However, the green curve is not an ideal fit to the data for other time periods. Because of that the alternative representation (red curve) of the proton latitudinal gradient is favored with a zero gradient in the low-latitude regions (streamer belt) and $G_\theta = 0.25\%/$ degree for latitudes above 25° (Heber *et al.*, 1996a). The increase of the ratio C_U/C_I for the period "B" starting in 1998, when Ulysses was still at low latitudes, was caused by an increase of the radial gradient. Such an increase is consistent with the analysis of Belov *et al.* (2001) and has been reported also during previous solar cycles (Chen and Bieber, 1993). After that period, when the spacecraft moved below

$-30°$, the ratio increased again until the end of 1999, in agreement with the previous estimates of the latitudinal gradient (not shown here). Since then the ratio has been constant.

Latitudinal gradients at solar maximum

Figure 6.22 (from McKibben *et al.*, 2003) shows the ratio of Ulysses-to-IMP 35–70 MeV/nucleon and 70–90 MeV/nucleon helium and 70–90 MeV and >100 MeV protons as a function of Ulysses latitude. These ratios are shown for the solar-minimum first orbit (left panels) and the second orbit, at solar maximum (right panels). As discussed in the previous section, during solar minimum a clear latitude gradient existed for all species. The gradient measurements during solar maximum are displayed in the right panels of Figure 6.22. The fluctuations were larger than during the solar-minimum scan. From this figure it follows that there is no

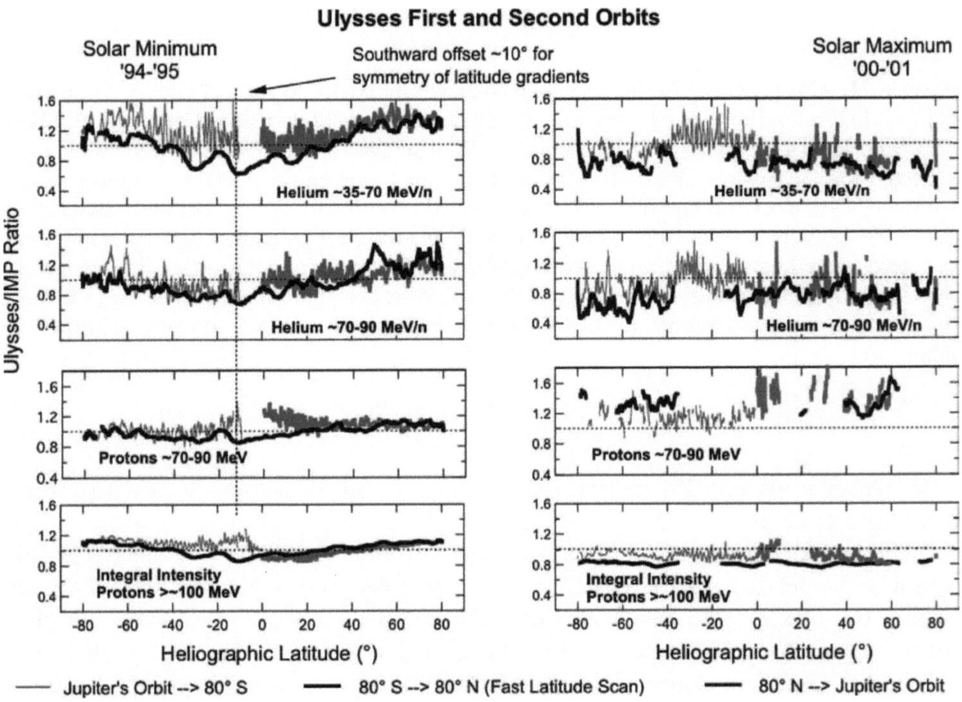

Figure 6.22. Ulysses-to-IMP-8-count-rate ratio as a function of latitude. Several energy ranges as noted in the panels are shown. Panels on the left contain observations from Ulysses' first (solar minimum) orbit, and panels on the right contain observations from the second (solar maximum) orbit. The dark lines identify observations taken during the fast-latitude scans, which provide the most definitive information concerning cosmic ray latitudinal gradients. Gray lines identify observations made during the climb to high southern latitude from aphelion (light line) and the return to low latitudes (heavy line) following the north polar pass (McKibben *et al.*, 2003).

evidence for a measurable gradient larger than the fluctuations (see also Heber *et al.*, 2003a).

Charge sign dependence

Heber *et al.* (2002b) analyzed the temporal variation of the electron-to-proton ratio for a location at 1 AU near the heliographic equator. For this purpose, they corrected the data for radial and latitudinal variations. While the radial gradient was assumed to be the same for protons and electrons (Chen and Bieber, 1993; Clem, Evenson, and Heber, 2002) the proton data have been corrected for the latitudinal gradient. Figure 6.23 shows in the upper and middle panel the directly measured and corrected e/p ratio along the Ulysses orbit. In Figure 6.23 a horizontal line at a reference level e/p = 0.92 has been drawn. Though this level is reached during short periods of time only, it ought to be representative for a medium range of tilt angles around 40–50° where the slope of the intensity versus tilt angle variation is roughly the same for both particle types (see Burger and Potgieter, 1999).

With respect to the level at a ratio of 0.92, the structures in the e/p ratio can be characterized as follows. The Λ-shape during period "A" seen in the e/p ratio measured along the Ulysses orbit (upper panel) is not found in the corrected one. The relatively constant value (Ferrando *et al.*, 1996) from mid-1993 to end-1994 has been replaced by a continuous increase. The e/p ratio lies systematically above 0.92 for a time period around solar minimum between about mid-1995 and mid-1998, see the

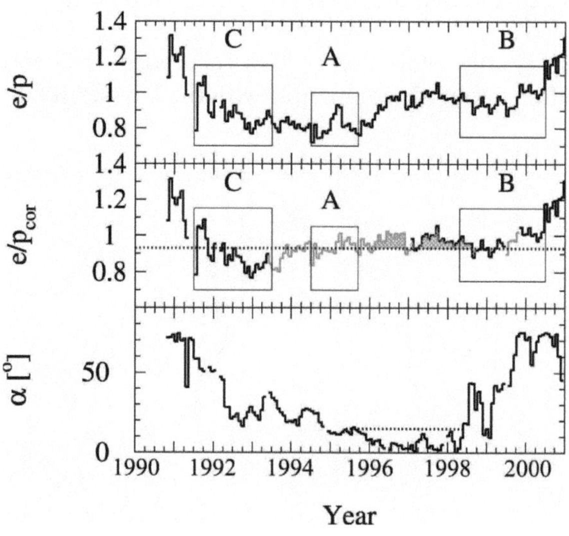

Figure 6.23. From top to bottom: Measured 26-day averaged 2.5 GV e/p ratio from launch to 2001 along the Ulysses orbit.The second panel displays the "heliographic equator equivalent" e/p ratios as described in the text. The lowest panel shows the evolution of the maximum latitudinal extent of the heliospheric current sheet α shifted by 5 solar rotations to later times. Ulysses' distance from the Sun and its heliographic latitude are shown at the top.

dashed part in the second panel. During this time the tilt angle is below a value of 15°, as indicated in the bottom panel. This increase during the occurrence of low-tilt angles near solar-minimum periods was found by Heber *et al.* (1999) and has been extensively discussed before. An increase of the e/p ratio with the transition to solar-maximum conditions commences around mid-1999. The e/p ratios are roughly the same in 1990/1991 and in 2000, both periods of solar maximum, indicating that charge sign dependent modulation is small. It is important to note that with increasing solar activity in cycle 23 the variation of the radial gradient does not occur at the same time as the latitudinal gradient and the e/p ratio. This means that variations of different cosmic ray transport parameters do not occur in phase. Qualitatively, the decrease of the latitudinal gradient at high latitudes with increasing solar activity might be coupled with the extension of slow solar wind regions to high latitudes.

Interpretation of the Ulysses measurements

As mentioned before, Wibberenz, Richardson, and Cane (2002) suggested that time-dependent global changes in the heliospheric magnetic field might be responsible for long-term modulation. This approach was the starting point for the compound approach, which combines the effects of the global changes in the HMF magnitude with drifts, therefore also time-dependent current sheet "tilt angles", in order to establish realistic time-dependent diffusion coefficients. Ndiitwani *et al.* (2005) used this approach in order to model the Ulysses observations. The first step was to investigate the importance of drifts over the 22-year solar magnetic cycle. For all their computations it is assumed that the HMF switched polarity (from A>0 to A<0) at 2000.2.

Figure 6.24 displays the total number of drifts needed in the model (red line), to compute a modulation amplitude compatible with the Ulysses/KET observations for

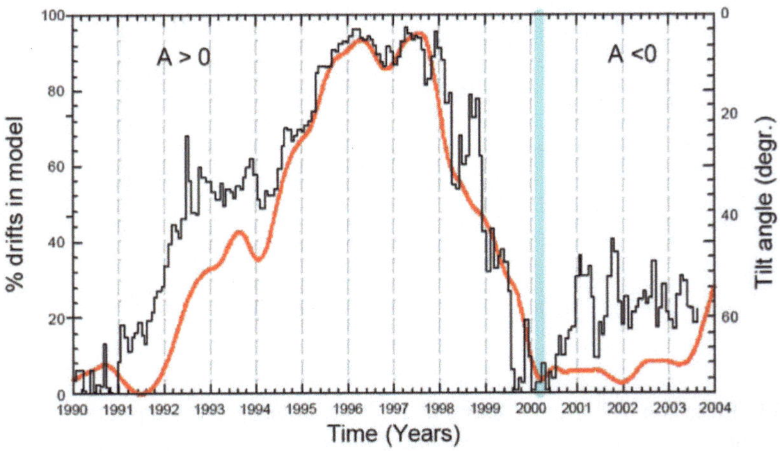

Figure 6.24. The percentage of drifts (red line) in the model that gives a realistic modulation for various stages of the solar cycle for both the 2.5 GV electron and protons. As a proxy for solar activity the tilt angles as used in the model are shown by the black line. Note the different scales.

different phases of the solar cycle. The percentage drift (on the left-hand side) is shown with respect to the varied tilt angle. It is important to note that the values for the tilt angle are shown from being small at the top to large at the bottom. From this figure a close relation of the percentage of drifts with the solar activity is evident for solar minimum. It is less obvious for solar-maximum activity. Large to no drifts are needed in order to model Ulysses observations for solar minimum, varying between 80% and 100% for at least 3 years, and for solar maximum, respectively. As soon as solar activity increases, drifts follow to suit the increase in modulation. During solar-maximum conditions drifts are reduced to less than 10% for most of this period (see also fig. 3 in Ndiitwani *et al.*, 2005).

Solar magnetic field reversal

Ndiitwani *et al.* (2005) showed how the electron-to-proton ratio can be used to bracket the time of the solar magnetic field reversal. In Figure 6.25 three model solutions are shown corresponding to three different specified times of the polarity reversal, from A>0 to A<0. The black line corresponds to 2000.2, the red line to 2000.4, and the green line to 2000.0. Magnetic field polarity reversal influences model computations, although only for approximately 1 year. This is the timescale needed for information to travel from the inner heliosphere to the heliopause. After 2000.4, these solutions converged again because the whole heliosphere is then filled with heliospheric magnetic fields corresponding to the A<0 polarity cycle. A choice between 2000.2 and 2000.4 is—from a cosmic ray modulation perspective—optimal.

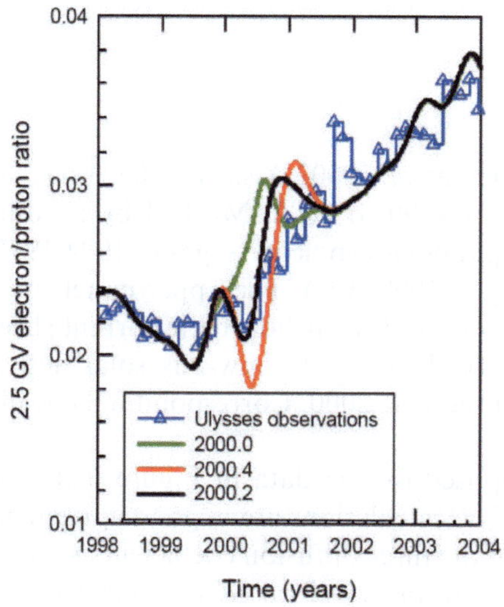

Figure 6.25. Computed and observed 2.5 GV e/p (not normalized) along the Ulysses trajectory. Three model solutions are shown corresponding to different scenarios of fixed polarity reversal times in the model. The black line corresponds to an HMF polarity reversal in 2000.2, the red line to 2000.4, and the green line to 2000.0.

For earlier values the computed e/p increases too fast, and for latter values the computed e/p decreases too much after 2000, which is not observed. However, in these modulation models this polarity reversal occurs at a specified time-step, on the order of a solar rotation, which may be contrasted with observations which predict a longer time period of several months (Jones, Balogh, and Smith, 2003).

Radial gradient

The charge sign or solar cycle dependence of the radial gradients is expected to be pronounced only at solar minimum. It was shown by Fujii and McDonald (2005) that the radial gradient for protons varies with polarity; it is larger in an $A<0$ than in an $A>0$ solar magnetic epoch at solar minimum. Clem, Evenson, and Heber (2002) reported that the radial gradient of cosmic ray electrons in the heliosphere at rigidities of 1.2 and 2.5 GV from 1 to 5 AU appears to be the same as those for positive particles of the same rigidity. As shown above the radial gradient of 2.5 GV, protons increased from 2.2%/AU to 3.5%/AU from solar minimum to maximum, respectively, making a time-dependent determination of the electron radial gradients mandatory. The computations by Ndiitwani *et al.* (2005) resulted in radial gradients of about 2.9%/AU for electrons and 1.9%/AU, for the $A>0$ polarity cycle in the equatorial region at solar minimum. The latter are in good agreement with values summarized above. They could also show that for solar maximum there is a general increase in the radial gradients for both particle species in the equatorial regions and that the values become similar for protons and electrons. This is primarily caused by the small levels of drifts present and explains the results found by Clem, Evenson, and Heber (2002).

6.4.2 MeV electrons

Figure 6.26 (from Heber *et al.*, 2002a) shows the 3-day averaged count rate of 3–10 MeV electrons from 1990 to 2002. Marked by shading are the Jovian flyby in 1992 (JE), the two rapid pole-to-pole passages in 1994/1995 and 2000/2001 (FLS), and the ecliptic crossing in 1998 (EC). The upper panel shows the evolution of the maximum latitudinal extent α of the heliospheric current sheet, which can be used as a proxy for solar activity. From 1998 onwards solar activity was increasing and reached its maximum in the year 2000. Correspondingly, α increased again to values above 70°.

The curve superimposed on the data in Figure 6.26 is the modeling result by Ferreira *et al.* (2001b). The calculations are in good agreement with the observations until 1998. The values of the three diffusion coefficients $\kappa_{\|}$, $\kappa_{\perp,r}$, and $\kappa_{\perp,\theta}$ organize the intensity distribution of Jovian as well as GCR electrons as a function of the spacecraft position relative to Jupiter (e.g., fig. 5 in Ferreira *et al.*, 2001b).

After 1998 the model starts to deviate significantly from the observations. Similar to the results presented in fig. 7 of Ferreira *et al.* (2001b) the relative contribution of the two sources varies systematically with latitude. For the model run selected in

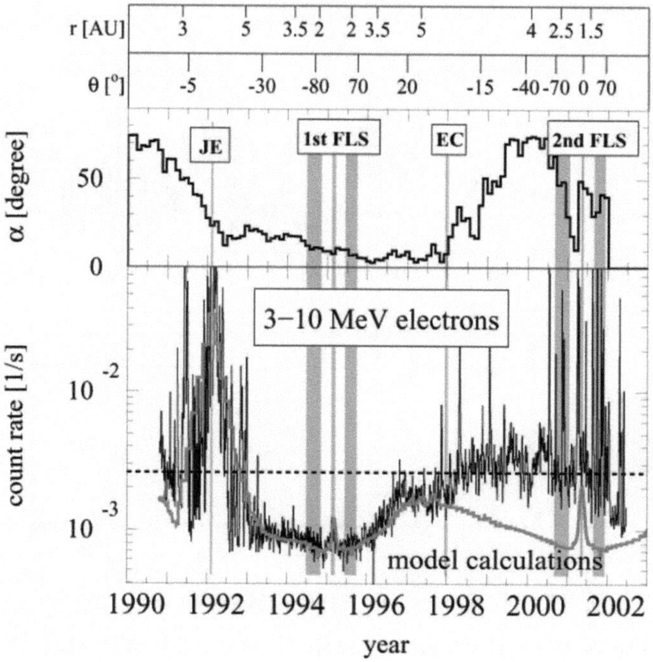

Figure 6.26. The 3-day averaged count rates of 3–10 MeV electrons from Heber *et al.* (2002a). The gray and dotted curves represent the results of a model run (Ferreira *et al.*, 2001a), and the observed count rate in 1991, respectively. At the top, Ulysses radial distance and heliographic latitude are shown. The upper panel shows the evolution of the maximum latitudinal extent α of the heliospheric current sheet, as determined by Hoeksema (*http://quake.Stanford.EDU:80/~wso/*).

Figure 6.26 the two contributions are practically the same in early 1999, whereas in late 2000 the galactic component was about a factor of 4 larger than the Jovian component. Although the observations at high southern heliographic latitudes are somehow masked by the number of solar particle events, the data during the 2000/2001 fast-latitude scan show only a weak dependence on latitude, excluding the possibility that a temporal increase of the electron intensity cancels the expected decrease with latitude. The strong longitudinal variation of the Jovian electrons is reflected in the peak around the ecliptic crossing in 2001, because at this time Ulysses had a very favorable magnetic connection with Jupiter. None of these observed features could be represented by these models. This discrepancy initiated the studies by Ferreira *et al.* (2003a), Ferreira, Potgieter, and Scherer (2004) and Moeketsi *et al.* (2005). They investigated the variation of $\kappa_\perp^\theta$ as a function of time and its consequences for the computed intensity time profiles. Ferreira *et al.* (2003b) showed that a reduction in the enhancement of $\kappa_\perp^\theta$ towards the poles from its solar-minimum value seems necessary, and that it should be correlated with changes in the observed latitudinal dependence of the solar wind speed in a self-consistent manner. The

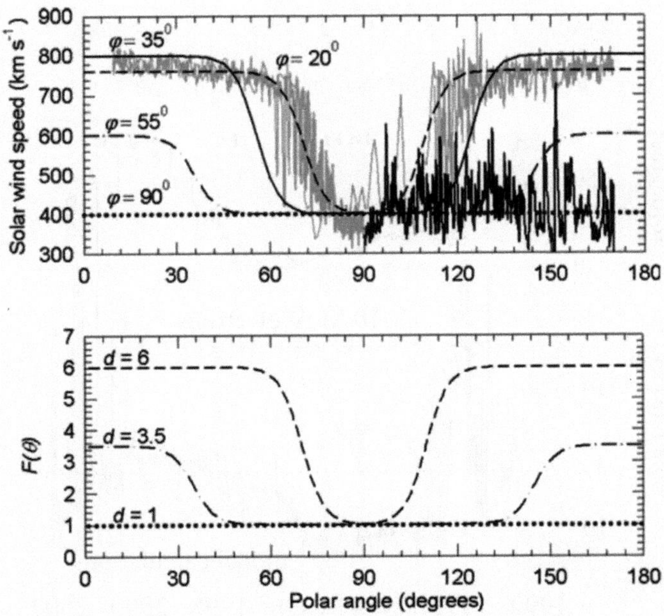

Figure 6.27. Measured and modeled solar wind speeds v (upper panel) and $F(\theta)$ (lower panel) as a function of polar angle and solar activity (from Moeketsi *et al.*, 2005).

variation of the solar wind speed v is displayed in the upper panel of Figure 6.27. While at solar minimum the speed increases from 400 km/s to 800 km/s, at about 20° latitude it stays around 400 km/s at solar maximum. Thus, there is no latitudinal dependence at solar maximum (McComas *et al.*, 2002). The $f(\theta)$ shown in the lower panel represents the latitude dependence of $\kappa_\perp^\theta$. The parameter d therefore reflects the time dependence of $\kappa_\perp^\theta$. Simulations are for $\varphi = 20°$ assumed to correspond to $d = 6$, a scenario for typical solar-minimum conditions; $\varphi = 55°$ corresponding to $d = 3.5$, a scenario assumed for intermediate solar activity conditions, and $\varphi = 90°$, assumed to correlate with $d = 1$ for extreme solar-maximum conditions.

Figure 6.28 (from Moeketsi *et al.*, 2005) displays in the first two panels the Ulysses trajectory parameters and in the lower panels the measured 3–10 MeV electron intensity–time profile together with calculations using the set of parameters found by Moeketsi *et al.* (2005). The lower two panels show the contribution of Jovian and galactic cosmic ray electrons to the model calculations. From the figure it is evident that the computed intensities are in good agreement with the calculations using $d = 3.5$. In addition, Henize, Ferreira, and Potgieter (2003) and Lange, Fichtner, and Kissmann (2006) also varied the Jovian electron source strength, increasing it with solar activity. From these analyses, it is evident that the Ulysses electron observations between 1998 and 2003 represent a picture of a very complicated interplay of several time-dependent modulation parameters for solar-maximum activity. Thus, the utilization of time-dependent three-dimensional modeling is mandatory.

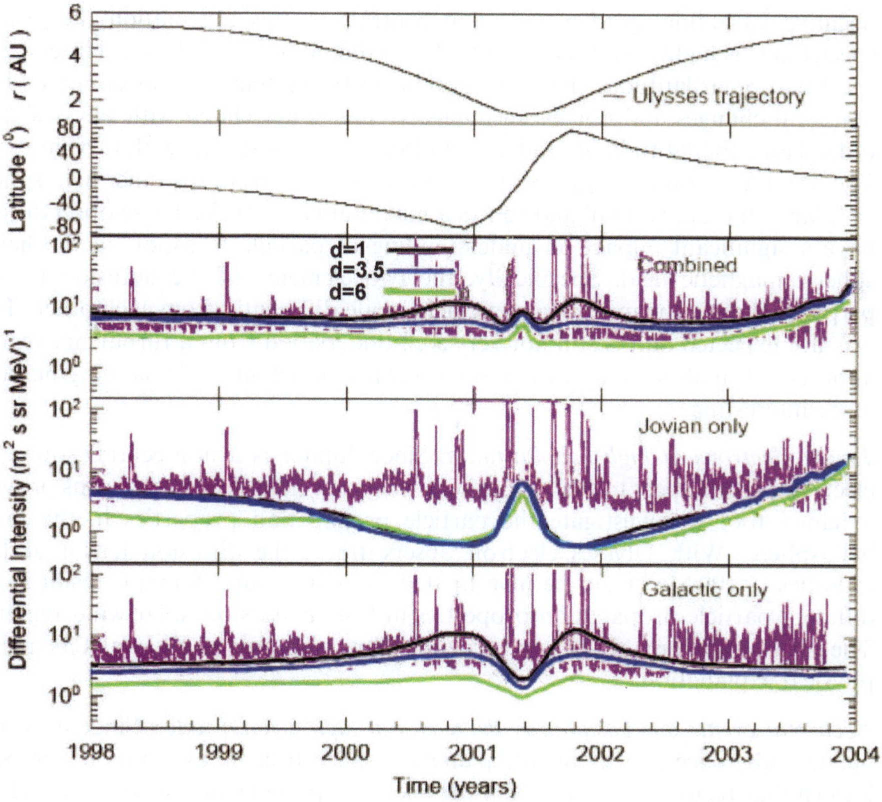

Figure 6.28. The top two panels show Ulysses' radial distance and heliographic latitude. The following three panels show the observed 3–10 MeV electron observations together with the computed 7 MeV combined Jovian and galactic electron, Jovian electron, and galactic electron intensities. Computations are shown for different *d*s, as shown in the previous figure (Moeketsi *et al.*, 2005).

6.5 SUMMARY

6.5.1 Solar minimum

Ulysses observations during solar minimum were ideally suited to complement the investigation of spatial structure and its consequences for our understanding of particle propagation in the inner heliosphere. The observations can be summarized as follows:

1 *Spatial gradients and the local interstellar spectra*: Of particular interest are cosmic ray intensities over the solar poles in an A>0 solar magnetic epoch. Protons below several 100 MeV would have large positive latitudinal gradients and their intensity should have increased by an order of magnitude, if local interstellar values are approached. Electrons, on the other hand, would show negative latitudinal gradients. But in contrast to this expectation the proton spectrum was highly modulated and Ulysses did not observe the local interstellar spectra at polar latitudes. The variation in electron intensities was dominated by temporal changes and not by an intensity change correlated with Ulysses' latitude. Thus, the latitudinal gradient of electrons is consistent with it being zero. The observed latitudinal gradient of cosmic ray protons in the inner heliosphere at solar minimum is small and shows a maximum at ~2 GV. These observations have a significant impact on understanding of particle transport in the heliospheric magnetic field. Specifically, the two elements of the diffusion tensor, perpendicular to the mean magnetic field, scale differently from each other. This was not expected and is still not very well-understood from a turbulence theory approach. It makes the modeling of cosmic rays much more demanding but also more interesting.

2 *Jovian electrons at high heliolatitudes*: Since Jupiter is a non-central source of electrons with respect to the heliospheric magnetic field, Jovian electrons provide a handy tool to investigate the particle propagation properties in the inner heliosphere. With Ulysses electron observations, the diffusion tensor at low rigidities (<100 MeV) can be investigated in great detail. A major result is the different particle propagation properties in fast *versus* slow solar wind regimes. The large anisotropy of the two perpendicular components still awaits its microphysical explanation.

3 *Recurrent events and cosmic ray decreases at high heliolatitudes*: Since measurements with space probes in interplanetary space became available it has been known that recurrent variations in the energetic particle intensities are observed in association with the occurrence of recurrent fast and slow solar wind streams. Normally, the passage of corotating interaction regions causes recurrent depressions in the cosmic ray flux and MeV/n nuclei and keV electron intensity increases centered around the forward and reverse shock. From latitudes above 40°S Ulysses was embedded in fast solar wind originating from the southern polar coronal hole. As expected, corotating forward and reverse shock waves disappeared. In contrast to what had been expected, recurrent particle increases and galactic cosmic ray decreases were observed up to polar latitudes. Even more surprising was the fact that the 40–65 keV electrons were delayed from the 0.5–1.0 MeV protons by up to 4 days. To explain this observation two competing proposals have been put forward:

(a) One explanation is that because of enhanced latitudinal diffusion the temporal variation at high heliolatitudes is determined by the interaction regions at low latitudes. The delay is then explained by the details of the diffusion process, especially perpendicular diffusion.

(b) The second explanation relies on the analysis of the Ulysses magnetic field data which showed that the polar magnetic field was dominated by strong variations. Systematic modifications of the standard Parker theory of the heliospheric magnetic field would then cause a meridional field component along which particles may move easily to polar latitudes.

Since both proposals explain the observations equally well, only magnetic field measurements in the distant, high-latitude heliosphere—where the systematic effects will be larger than the statistical variation—may prove which one is correct.

4 *The north–south-asymmetry and its consequences*: A real surprise of the Ulysses mission was the observation that the galactic cosmic ray flux was not symmetric with the heliographic equator. Neither the solar wind experiments nor the magnetic field investigations reported this asymmetry. Only 5 years later did magnetic field investigations from 1 AU measurements confirm a deficit of the magnetic flux in the southern hemisphere. It remains an open question whether this observation was an occurrence of events that pertained during the rapid pole-to-pole passage of Ulysses or is correlated with a permanent magnetic flux deposit in the southern heliosphere.

6.5.2 Solar maximum

The Ulysses orbit is ideally suited to investigate the variation in the latitudinal dependence of charged particles in the inner heliosphere at solar maximum and compare this with solar-minimum conditions. The heliospheric magnetic structure during solar maximum is characterized by a highly inclined current sheet that was distorted by coronal mass ejections at all latitudes. The following summarizes the observations around solar maximum:

1 *Gradients and charge sign dependence at solar maximum*: Gradient measurements during solar maximum exhibited fluctuations of the order of 2 and larger, so that no unambiguous evidence could be found for latitudinal cosmic ray gradients. In contrast to solar-minimum conditions, the cosmic ray distribution was almost spherically symmetric around solar maximum, consistent with model computations. From this the important conclusion was made that since the latitudinal gradients were positive at solar minimum, the total modulation is relatively higher at polar latitudes than in the ecliptic.

Model computations predicted the electron-to-proton ratio to have a W-shape in $A>0$ epochs and an M-shape during $A<0$ epochs, with the ratio always decreasing from large values to small values during solar maximum in an $A<0$ to $A>0$ transition, but increasing from an $A>0$ to $A<0$ transition. The ratio should however return to the same values during every reversal. This was confirmed by the Ulysses observations. It was concluded that less than 10% drifts were required at extreme solar maximum to explain the observations when the magnetic field reversed. The proton-to-electron-intensity ratios turned out to be an excellent indicator of when the magnetic field actually reversed. It was found that the reversal must have occurred between late 2000 and mid-2001.

2 *Particle events and the reservoir effect*: It was found for 35–70 MeV and 70–95 MeV protons that essentially all large events at Earth also produce com-

parable intensity increases at Ulysses during the period from Ulysses' south polar pass in 2000 to the north polar pass in 2001. The onset of the events at high latitudes being delayed and the delay being ordered by Ulysses' latitude emphasized the similarity of all studied events during the first days of this part of the Ulysses trajectory. Nearly equal particle intensities at Ulysses and close to Earth occurred after 3–4 days. After formation, this "reservoir" slowly dissipated as a combined result of normal modulation processes. A similar reservoir effect was observed in the 1970s. Without a Ulysses' fast-latitude scan, it would have been very difficult to distinguish between these high-latitude and low-latitude measurements. Thus, it was concluded that either an acceleration front for energetic particles in large events extends over a broad range in latitude and longitude, or that mechanisms exist to transport particles efficiently across the mean magnetic field close to the Sun.

6.5.3 Insights on particle propagation in a turbulent astrophysical plasma

The Ulysses mission to high heliolatitudes led to several insights concerning propagation and modulation theory—in particular, the relative importance of the various diffusion coefficients. It was concluded that in order to obtain agreement between current modulation models and Ulysses observations, enhancing $\boldsymbol{\kappa}_{\perp\theta}$ latitudinally and changing the rigidity dependence of $\boldsymbol{\kappa}_{\perp\theta}$ differently from $\boldsymbol{\kappa}_{\|}$ was essential. The latitudinal enhancement is related to the different solar wind regimes observed around solar minimum.

The observed electron-to-proton ratios (implicitly also containing the radial and latitudinal gradients) indicated that large particle drifts were occurring during solar minimum but diminished significantly toward solar maximum when less than 10% drifts were required in models to explain the observed values.

These combined observational and modeling studies initiated new projects concerning diffusion and turbulence theories which have become known as the *ab initio* approach to modulation theory and modeling, with exceptional progress being made the past few years.

6.5.4 Cosmic ray modulation surprises from Ulysses

From the Ulysses cosmic ray observations and corresponding modulation modeling it is evident that cosmic rays studies have led to new and surprising insights from large-scale phenomena (like the global magnetic field) to microphysical processes (like the wave particle interaction). In what follows we conclude by listing the most important cosmic ray modulation surprises according to the scale they operate in the heliosphere:

1 *Large scale*

(a) A north–south asymmetry in cosmic ray modulation with respect to the heliospheric equator.

(b) Small latitudinal gradients, implying that the local interstellar spectra cannot be observed in the inner polar regions of the heliosphere.
(c) Essentially no latitudinal gradients and little drifts at solar maximum.

2 *Intermediate scale*

(a) Recurrent particle events at high heliolatitudes without direct corresponding evidence in the solar wind and magnetic field.
(b) Jovian electrons at high heliolatitudes and the consequently implied effective latitudinal transport.

3 *Micro scale*

(a) The latitudinal enhancement of perpendicular diffusion in the polar direction.
(b) The difference in the rigidity dependence of latitudinal and radial perpendicular diffusion.

The above topics will become even more important, for comparative reasons, when the spacecraft performs its third rapid pole-to-pole passage in 2007 and 2008.

6.6 ACKNOWLEDGMENTS

This work was partially funded within the framework of the bi-lateral collaboration program between South Africa and Germany by the Deutsche Forschungsgemeinschaft (DFG) and the South African Research Foundation (NRF). The ULYSSES/KET project is supported under grant No. 50 ON 9105 by the German Bundesminister für Wirtschaft through the Deutsches Zentrum für Luft- und Raumfahrt. We gratefully acknowledge the support of the International Space Science Institute (ISSI) in Bern, Switzerland, where a large part of this review was written during visits in 2006 and 2007. It is our pleasure to thank Bruce McKibben and Steve Suess for reading and improving the paper substantially.

6.7 REFERENCES

Baker, D. N., Mason, G. M., Figueroa, O., Colon, O., Watzin, J. G., and Aleman, R. (1993), An overview of the solar, anomalous, and magnetospheric particle explorer (sampex) mission, *IEEE Transactions on Geosc. and Remote Sensing*, **31**, 531–541.

Balogh, A., Smith, E. J., Tsurutani, B. T., Southwood, D. J., Forsyth, R. J., and Horbury, T. S. (1995), The Heliospheric Magnetic Field over the South Polar Region of the Sun, *Science*, **268**, 1007–1010.

Beatty, J. J., Garcia-Munoz, M., and Simpson, J. A. (1985), The cosmic-ray spectra of H-1, H-2, and He-4 as a test of the origin of the hydrogen superfluxes at solar minimum modulation, *Astrophysical Journal*, **294**, 455–462.

Belov, A., Eroshenko, E., Heber, B., Yanke, V., Raviart, A., Rohrs, K., Müller-Mellin, R., Kunow, H., Wibberenz, G., and Paizis, C. (2001), Latitudinal and radial variation of

> 2 GeV/n protons and alpha-particles in the southern heliosphere at solar maximum: ULYSSES COSPIN/KET and neutron monitor network observations, in *Proc. 27th ICRC*, pp. 3996–3999.

Bieber, J. W. (2003), Transport of charged particles in the heliosphere: Theory, *Advances in Space Research*, **32**, 549–560.

Bieber, J. W., and Matthaeus, W. H. (1997), Perpendicular Diffusion and Drift at Intermediate Cosmic-Ray Energies, *Astrophysical Journal*, **485**, 655.

Borrmann, T., and Fichtner, H. (2005), On the dynamics of the heliosphere on intermediate and long time-scales, *Advances in Space Research*, **35**, 2091–2101.

Burger, R. A. (2000), Galactic Cosmic Rays in the Heliosphere, in *AIP Conf. Proc. 516: 26th International Cosmic Ray Conference*, ICRC XXVI, p. 83.

Burger, R. A. (2005), Cosmic-ray modulation and the heliospheric magnetic field, *Advances in Space Research*, **35**, 636–642.

Burger, R. A., and Hitge, M. (2004), The Effect of a Fisk-Type Heliospheric Magnetic Field on Cosmic-Ray Modulation, *Astrophysical Journal*, **617**, L73–L76.

Burger, R. A., and Potgieter, M. S. (1989), The calculation of neutral sheet drift in two-dimensional cosmic-ray modulation models, *Astrophysical Journal*, **339**, 501–511.

Burger, R. A., and Potgieter, M. S. (1999), The effect of large heliospheric current sheet tilt angles in numerical modulation models: A theoretical assessment, in *Proc. 26th Int. Cosmic Ray Conf.*, Vol. 7, pp. 1316.

Burger, R. A., Potgieter, M. S., and Heber, B. (2000), Rigidity dependence of cosmic ray proton latitudinal gradients measured by the Ulysses spacecraft: Implications for the diffusion tensor, *Journal of Geophysical Research*, **105**, 27447–27456.

Burger, R. A., van Niekerk, Y., and Potgieter, M. S. (2001), An Estimate of Drift Effects in Various Models of the Heliospheric Magnetic Field, *Space Science Reviews*, **97**, 331–335.

Burlaga, L. F., and Ness, N. F. (1993), Large-scale distant heliospheric magnetic field: Voyager 1 and 2 observations from 1986 through 1989, *Journal of Geophysical Research*, **98**, 17451–17460.

Burlaga, L. F., Perko, J., and Pirraglia, J. (1993), Cosmic-ray modulation, merged interaction regions, and multifractals, *Astrophysical Journal*, **407**, 347–358.

Burlaga, L. F., Ness, N. F., Acuña, M. H., Lepping, R. P., Connerney, J. E. P., Stone, E. C., and McDonald, F. B. (2005), Crossing the Termination Shock into the Heliosheath: Magnetic Fields, *Science*, **309**, 2027–2029.

Caballero-Lopez, R. A., and Moraal, H. (2004), Limitations of the force field equation to describe cosmic ray modulation, *Journal of Geophysical Research (Space Physics)*, **109**(A18), 1101.

Cane, H. V. (2000), Coronal Mass Ejections and Forbush Decreases, *Space Science Reviews*, **93**, 55–77.

Cane, H., Wibberenz, G., Richardson, I. G., and von Rosenvinge, T. T. (1999), Cosmic ray modulation and the solar magnetic field, *Geophysical Research Letters*, **26**, 565.

Chalov, S. V. (2005), Acceleration of interplanetary pick-up ions and anomalous cosmic rays, *Advances in Space Research*, **35**, 2106–2114.

Chen, J., and Bieber, J. W. (1993), Cosmic-Ray Anisotropies and Gradients in Three Dimensions, *Astrophysical Journal*, **405**, 375–389.

Chenette, D. L., Conlon, T. F., and Simpson, J. A. (1974), Bursts of relativistic electrons from Jupiter observed in interplanetary space with the time variation of the planetary rotation period, *Journal of Geophysical Research*, **79**, 3551.

Chenette, D. L., Conlon, T. F., Pyle, K. R., and Simpson, J. A. (1977), Observations of Jovian electrons at 1 AU throughout the 13 month Jovian synodic year, *Astrophysical Journal*, **215**, L95–L99.

Clem, J., Evenson, P., and Heber, B. (2002), Cosmic Electron Gradients in the Inner Heliosphere, *Geophysical Research Letters*, **29**, doi:10.1029/2002G.

Conlon, T. F. (1978), The interplanetary modulation and transport of Jovian electrons, *Journal of Geophysical Research*, **83**, 541–552.

Conlon, T. F., and Simpson, J. A. (1977), Modulation of Jovian electron intensity in interplanetary space by corotating interaction regions, *Astrophysical Journal*, **211**, L45–L49.

Cummings, A. C., Stone, E. C., and Webber, W. R. (1987), Latitudinal and radial gradients of anomalous and galactic cosmic rays in the outer heliosphere, *Geophysical Research Letters*, **14**, 174–177.

Dröge, W. (2005), Probing heliospheric diffusion coefficients with solar energetic particles, *Advances in Space Research*, **35**, 532–542.

Eraker, J. H. (1982), Origin of the low-energy relativistic interplanetary electrons, *Astrophys. J.*, **257**, 862.

Evenson, P. (1998), Cosmic Ray Electrons, *Space Science Reviews*, **83**, 63–73.

Fahr, H.-J. (2004), Global structure of the heliosphere and interaction with the local interstellar medium: Three decades of growing knowledge, *Advances in Space Research*, **34**, 3–13.

Ferrando, P., Raviart, A., Haasbroek, L. J., Potgieter, M. S., Dröge, W., Heber, B., Kunow, H., Müller-Mellin, R., Sierks, H., Wibberenz, G., and Paizis, C. (1996), Latitude variations of ~7 MeV and >300 MeV cosmic ray electron fluxes in the heliosphere: ULYSSES COSPIN/KET results and implications, *Astronomy & Astrophysics*, **316**(2), 528–537.

Ferrando, P., Raviart, A., Heber, B., Bothmer, V., Kunow, H., Müller-Mellin, R., and C., P. (1999), Observation of a 7 MeV electron super-flux at 5 AU by Ulysses, in *Proc. 26th Int. Cosmic Ray Conference*, Vol. 7, pp. 17–25.

Ferreira, S. E. S., and Potgieter, M. S. (2004), Long-Term Cosmic-Ray Modulation in the Heliosphere, *Astrophysical Journal*, **603**, 744–752.

Ferreira, S. E. S., Potgieter, M. S., and Moeketsi, D. M. (2003), Modulation effects of a changing solar wind speed on low-energy electrons, *Advances in Space Research*, **32**, 675–680.

Ferreira, S. E. S., Potgieter, M. S., and Scherer, K. (2004), Modulation of Cosmic-Ray Electrons in a Nonspherical and Irregular Heliosphere, *Astrophysical Journal*, **607**, 1014–1023.

Ferreira, S. E. S., Potgieter, M. S., and Webber, W. R. (2004), Modulation of low-energy cosmic ray electrons in the outer heliosphere, *Advances in Space Research*, **34**, 126–131.

Ferreira, S. E. S., and Scherer, K. (2006), Time Evolution of Galactic and Anomalous Cosmic-Ray Spectra in a Dynamic Heliosphere, *Astrophysical Journal*, **642**, 1256–1266.

Ferreira, S. E., Potgieter, M. S., Burger, R. A., Heber, B., and Fichtner, H. (2001a), The modulation of Jovian and galactic electrons in the heliosphere. II. Radial transport of a few MeV electrons, *Journal of Geophysical Research*, **106**, 29313–29322.

Ferreira, S. E. S., Potgieter, M. S., Burger, R. A., Heber, B., and Fichtner, H. (2001b), The modulation of Jovian and galactic electrons in the heliosphere. I. Latitudinal transport of a few-MeV electrons, *Journal of Geophysical Research*, **106**(A11), 24979–24988.

Ferreira, S. E. S., Potgieter, M. S., Heber, B., Fichtner, H., and Kissmann, R. (2003a), Transport of a few-MeV Jovian and galactic electrons at solar maximum, *Advances in Space Research*, **32**, 669–674.

Ferreira, S. E. S., Potgieter, M. S., Moeketsi, D. M., Heber, B., and Fichtner, H. (2003b), Solar Wind Effects on the Transport of 3–10 MeV Cosmic-Ray Electrons from Solar Minimum to Solar Maximum, *Astrophysical Journal*, **594**, 552–560.

Fichtner, H. (2001), Anomalous Cosmic Rays: Messengers from the Outer Heliosphere, *Space Science Reviews*, **95**, 639–754.

Fisk, L. A. (1976), Solar modulation of galactic cosmic rays. IV—Latitude-dependent modulation, *Journal of Geophysical Research*, **81**(10), 4646–4650.

Fisk, L. A. (1979), Mechanisms for energetic particle acceleration in the solar wind, in *AIP Conf. Proc. 56: Particle Acceleration Mechanisms in Astrophysics*, pp. 63–79.

Fisk, L. A. (1996), Motion of the footpoints of heliospheric magnetic field lines at the Sun: Implications for recurrent energetic particle events at high heliographic latitudes, *Journal of Geophysical Research*, **101**, 15547–15554.

Fisk, L. A., Gloeckler, G., and Zurbuchen, T. H. (2006), Acceleration of Low-Energy Ions at the Termination Shock of the Solar Wind, *Astrophysical Journal*, **644**, 631–637.

Fisk, L. A., and Jokipii, J. R. (1999), Mechanisms for Latitudinal Transport of Energetic Particles in the Heliosphere, *Space Science Reviews*, **89**, 115–124.

Fisk, L. A., Kozlovsky, B., and Ramaty, R. (1974), An Interpretation of the Observed Oxygen and Nitrogen Enhancements in Low-Energy Cosmic Rays, *Astrophysical Journal*, **190**, L35.

Florinski, V., and Zank, G. P. (2006), Particle acceleration at a dynamic termination shock, *Geophysical Research Letters*, **33**, L15110.

Forsyth, R. J., Balogh, A., and Smith, E. J. (2002), The underlying direction of the heliospheric magnetic field through the Ulysses first orbit, *Journal of Geophysical Research (Space Physics)*, **107**, pp. SSH19-1.

Fujii, Z., and McDonald, F. B. (2005), The spatial distribution of galactic and anomalous cosmic rays in the heliosphere at solar minimum, *Advances in Space Research*, **35**, 611–616.

Garcia-Munoz, M., Mason, G. M., and Simpson, J. A. (1973), A New Test for Solar Modulation Theory: The 1972 May–July Low-Energy Galactic Cosmic-Ray Proton and Helium Spectra, *Astrophysical Journal*, **182**, L81.

Giacalone, J., and Jokipii, J. R. (1999), The Transport of Cosmic Rays across a Turbulent Magnetic Field, *Astrophysical Journal*, **520**, 204–214.

Giacalone, J., Jokipii, J. R., and Matthaeus, W. H. (2006), Structure of the Turbulent Interplanetary Magnetic Field, *Astrophysical Journal*, **641**, L61–L64.

Gleeson, L. J., and Axford, W. I. (1967), Cosmic rays in the interplanetary medium, *Astrophysical Journal*, **149**, L115.

Gloeckler, G., Geiss, J., Balsiger, H., Fisk, L. A., Galvin, A. B., Ipavich, F. M., Ogilvie, K. W., von Steiger, R., and Wilken, B. (1993), Detection of interstellar pick-up hydrogen in the solar system, *Science*, **261**, 70–73.

Hattingh, M., and Burger, R. A. (1995), A new simulated wavy neutral sheet drift model, *Advances in Space Research*, **16**, 213.

Heber, B. (2001), Modulation of galactic and anomalous cosmic rays in the inner heliosphere, *Advances in Space Research*, **27**, 451–460.

Heber, B., and Cummings, A. (2001), Anomalous cosmic ray observations in the inner and outer heliosphere, in *The Outer Heliosphere: The Next Frontiers* (K. Scherer, H. Fichtner, H. J. Fahr, and E. Marsch, eds.), COSPAR Colloquia Series 11, Pergamon Press, Amsterdam, p.173.

Heber, B., and Marsden, R. G. (2001), Cosmic Ray Modulation over the Poles at Solar Maximum: Observations, *Space Science Reviews*, **97**, 309–319.

Heber, B., and Potgieter, M. S. (2000), Galactic cosmic ray observations at different heliospheric latitudes, *Advances in Space Research*, **26**(5), 839–852.

Heber, B., and Potgieter, M. S. (2006), Cosmic Rays at High Latitudes, *Space Science Reviews*, **127**, 117–194.

Heber, B., Raviart, A., Paizis, C., Bialk, M., Dröge, W., Ducros, R., Ferrando, P., Kunow, H., Müller-Mellin, R., Rastoin, C., Röhrs, K., and Wibberenz, G. (1993), Modulation of galactic cosmic ray particles observed on board the Ulysses spacecraft, in *Proc. 23rd Int. Cosmic Ray Conf., Calgary, Canada*, pp. 461–464, University of Calgary.

Heber, B., Dröge, W., Ferrando, P., Haasbroek, L., Kunow, H., Müller-Mellin, R., Paizis, C., Potgieter, M., Raviart, A., and Wibberenz, G. (1996a), Spatial variation of >40 MeV/n nuclei fluxes observed during Ulysses rapid latitude scan, *Astronomy & Astrophysics*, **316**, 538–546.

Heber, B., Dröge, W., Kunow, H., Müller-Mellin, R., Wibberenz, G., Ferrando, P., Raviart, A., and Paizis, C. (1996b), Spatial variation of >106 MeV proton fluxes observed during the Ulysses rapid latitude scan: Ulysses COSPIN/KET results, *Geophysical Research Letters*, **23**, 1513–1516.

Heber, B., Ferrando, P., Raviart, A., Wibberenz, G., Müller-Mellin, R., Kunow, H., Sierks, H., Bothmer, V., Posner, A., Paizis, C., and Potgieter, M. S. (1999), Differences in the temporal variation of galactic cosmic ray electrons and protons: Implications from Ulysses at solar minimum, *Geophysical Research Letters*, **26**(14), 2133–2136.

Heber, B., Ferrando, P., Raviart, A., Paizis, C., Posner, A., Wibberenz, G., , Müller-Mellin, R., Kunow, H., Potgieter, M. S., Ferreira, S. E. S., Burger, R. A., Fichtner, H., and Schlickeiser, R. (2002a), 3–20 MeV electrons in the inner three-dimensional heliosphere at solar maximum: Ulysses COSPIN/KET observations, *Astrophysical Journal*, **579**, 888–894.

Heber, B., Wibberenz, G., Potgieter, M. S., Burger, R. A., Ferreira, S. E. S., Müller-Mellin, R., Kunow, H., Ferrando, P., Raviart, A. C., Lopate, C., McDonald, F. B., and Cane, H. V. (2002b), Ulysses COSPIN/KET observations: Charge sign dependence and spatial gradients during the 1990–2000 A > 0 solar magnetic cycle, *Journal of Geophysical Research*, **107**, doi:10.10.1029/200.

Heber, B., Sarri, G., Wibberenz, G., Paizis, C., Ferrando, P., Raviart, A., Posner, A., Müller-Mellin, R., and Kunow, H. (2003a), The Ulysses fast latitude scans: COSPIN/KET results, *Annales Geophysicae*, **21**, 1275–1288.

Heber, B., Clem, J. M., Müller-Mellin, R., Kunow, H., Ferreira, S. E. S., and Potgieter, M. S. (2003b), Evolution of the galactic cosmic ray electron to proton ratio: Ulysses COSPIN/KET observations, *Geophysical Research Letters*, **30**, 61.

Henize, V. K., Ferreira, S. E. S., and Potgieter, M. S. (2003), Modeling a Few-MeV Jovian and Galactic Electron Spectra in the Inner Heliosphere, in *Int. Cosmic Ray Conf.*, p. 3819.

Henize, V. K., Potgieter, M. S., Ferreira, S. E. S., Moeketsi, D. M., and Heber, B. (2005), Causes of a few-MeV electron variations during solar maximum, in *35th COSPAR Scientific Assembly*, p. 904.

Hovestadt, D., Vollmer, O., Gloeckler, G., and Fan, C. Y. (1973), Measurement of Elemental Abundance of Very Low Energy Solar Cosmic Rays, in *Int. Cosmic Ray Conf.*, p. 1498.

Izmodenov, V., Malama, Y. G., Gloeckler, G., and Geiss, J. (2003), Effects of Interstellar and Solar Wind ionized Helium on the Interaction of the Solar Wind with the Local Interstellar Medium, *Astrophysical Journal*, **594**, L59–L62.

Jokipii, J. R. (1986), Particle acceleration at a termination shock. I—Application to the solar wind and the anomalous component, *Journal of Geophysical Research*, **91**(10), 2929–2932.

Jokipii, J. R. (1989), The physics of cosmic-ray modulation, *Advances in Space Research*, **9**, 105–119.

Jokipii, J. R. (1996), Theory of multiply charged anomalous cosmic rays, *Astrophysical Journal*, **466**, L47–L50.

Jokipii, J. R., and Kopriva, D. A. (1979), Effects of particle drift on the transport of cosmic rays. III—Numerical models of galactic cosmic-ray modulation, *Astrophysical Journal*, **234**, 384–392.

Jokipii, J., and Kóta, J. (1989), The polar heliospheric magnetic field, *Geophysical Research Letters*, **16**, 14.

Jokipii, J. R., and Kóta, J. (2000), Galactic and Anomalous Cosmic Rays in the Heliosphere, *Astronomy & Astrophysics Supplements*, **274**, 77–96.

Jokipii, J. R., and Wibberenz, G. (1998), Epilogue: Cosmic rays in the active heliosphere, *Space Science Reviews*, **83**, 365–368.

Jones, G. H., Balogh, A., and Smith, E. J. (2003), Solar magnetic field reversal as seen at Ulysses, *Geophysical Research Letters*, **30**, 2.

Kissmann, R., Fichtner, H., and Ferreira, S. E. S. (2004), The influence of CIRs on the energetic electron flux at 1 AU, *Astronomy & Astrophysics*, **419**, 357–363.

Klecker, B. (1999), Anomalous cosmic rays: Our present understanding and open questions, *Advances in Space Research*, **23**, 521–530.

Klecker, B., Oetliker, M., Blake, J., Hovestadt, D., Mason, G., Mazur, J., and McNab, M. (1997), *Proc. 25th Int. Cosmic Ray Conf.*, Vol. 7, pp. 333–336.

Klecker, B., Mewaldt, R. A., Bieber, J. W., Cummings, A. C., Drury, L., Giacalone, J., Jokipii, J. R., Jones, F. C., Krainev, M. B., Lee, M. A. *et al.* (1998), Anomalous cosmic rays, *Space Science Reviews*, **83**, 259–308.

Kóta, J., and Jokipii, J. R. (1991), The role of corotating interaction regions in cosmic-ray modulation, *Geophysical Research Letters*, **18**, 1797–1800.

Lange, D., Fichtner, H., and Kissmann, R. (2006), Time-dependent 3D modulation of Jovian electrons: Comparison with Ulysses/KET observations, *Astronomy & Astrophysics*, **449**, 401–410.

Langner, U. W., and Potgieter, M. S. (2004), Solar wind termination shock and heliosheath effects on the modulation of protons and antiprotons, *Journal of Geophysical Research (Space Physics)*, **109**(A18), 1103.

Langner, U. W., Potgieter, M. S., and Webber, W. R. (2003), Modulation of cosmic ray protons in the heliosheath, *Journal of Geophysical Research (Space Physics)*, (A10), 14.

Langner, U. W., Potgieter, M. S., Fichtner, H., and Borrmann, T. (2006), Modulation of anomalous protons: Effects of different solar wind speed profiles in the heliosheath, *Journal of Geophysical Research (Space Physics)*, **111**(A10), 1106.

le Roux, J. A. (2001), Anomalous cosmic rays: Current and future theoretical developments, in *The Outer Heliosphere: The Next Frontiers* (K. Scherer, H. Fichtner, H. J. Fahr, and E. Marsch, eds.), COSPAR Colloquia Series 11, Pergamon Press, Amsterdam, p. 163.

le Roux, J. A., Fichtner, H., and Zank, G. P. (2000), Self-consistent acceleration of multiply reflected pickup ions at a quasi-perpendicular solar wind termination shock: A fluid approach. *Journal of Geophysical Research*, **105**, 12557–12578.

le Roux, J. A., and Potgieter, M. S. (1995), The simulation of complete 11 and 12 year modulation cycles for cosmic rays in the heliosphere using a drift model with global merged interaction regions, *Astrophysical Journal*, **442**, 847–851.

le Roux, J. A., Potgieter, M. S., and Ptuskin, V. S. (1996), A transport model for the diffusive shock acceleration and modulation of anomalous cosmic rays in the heliosphere, *Journal of Geophysical Research*, **101**, 4791–4804.

le Roux, J. A., Zank, G. P., Li, G., and Webb, G. M. (2005), Nonlinear Energetic Charged Particle Transport and Energization in Enhanced Compressive Wave Turbulence near Shocks, *Astrophysical Journal*, **626**, 1116–1130.

Lee, M. A. (1983), The association of energetic particles and shocks in the heliosphere, *Rev. Geophys. Space Phys.*, **21**, 324.

Lee, M. A., and Fisk, L. A. (1982), Shock acceleration of energetic particles in the heliosphere. *Space Science Reviews*, **32**, 205–228.

Leipold, M., Fichtner, H., Heber, B., Groepper, P., Lascar, S., Burger, F., Eiden, M., Niederstadt, T., Sickinger, C., Herbeck, L., Dachwald, B., and Seboldt, W. (2003), Heliopause Explorer—A sailcraft mission to the outer boundaries of the solar system, in *Low-Cost Planetary Missions*, ESA SP-542, pp. 367–375, ESA, Noordwijk, The Netherlands.

Leske, R. A., Mewaldt, R. A., Christian, E., Cohen, C., Cummings, A. C., Slocum, P., Stone, E., von Rosenvinge, T., and Wiedenbeck, M. (2000), Observations of anomalous cosmic ray at 1 au, in *AIP Conf. Proc. 528: ACE 2000 Symposium*, Vol. 528, pp. 293–300.

L'Heureux, J., Fan, C., and Meyer, P. (1972), The quiet-time spectra of cosmic-ray electrons of energies between 10 and 200 MeV observed on OGO-5, *Astrophysical Journal*, **171**, 363–372.

Liewer, P. C., Mewaldt, R. A., Ayon, J. A., and Wallace, R. A. (2000), NASA's Interstellar Probe Mission, in *AIP Conf. Proc. 504: Space Technology and Applications International Forum*, pp. 911–916.

Lopate, C. (1991), Jobain and galactic electrons (2–30 MeV) in the heliosphere from 1 to 50 AU, in *Proc. 22nd ICRC*, p. 415.

MacLennan, C. G., and Lanzerotti, L. J. (1998), Low energy anomalous ions at northern heliolatitudes, *Geophysical Research Letters*, **25**, 3473–3476.

Malama, Y. G., Izmodenov, V. V., and Chalov, S. V. (2006), Modeling of the heliospheric interface: Multi-component nature of the heliospheric plasma, *Astronomy & Astrophysics*, **445**, 693–701.

Matthaeus, W. H., Qin, G., Bieber, J. W., and Zank, G. P. (2003), Nonlinear Collisionless Perpendicular Diffusion of Charged Particles, *Astrophysical Journal*, **590**, L53–L56.

McComas, D. J., Barraclough, B. L., Funsten, H. O., Gosling, J. T., Santiago-Muñoz, E., Skoug, R. M., Goldstein, B. E., Neugebauer, M., Riley, P., and Balogh, A. (2000), Solar wind observations over Ulysses' first full polar orbit, *Journal of Geophysical Research*, **105**(14), 10419–10434.

McComas, D. J., Elliott, H. A., Gosling, J. T., Reisenfeld, D. B., Skoug, R. M., Goldstein, B. E., Neugebauer, M., and Balogh, A. (2002). Ulysses' second fast-latitude scan: Complexity near solar maximum and the reformation of polar coronal holes, *Geophysical Research Letters*, **29**, 41.

McDonald, F. B., Cline, T. L., and Simnett, G. M. (1972), Multivarious temporal variations of low-energy relativistic cosmic-ray electrons, *Journal of Geophysical Research*, **77**, 2213–2231.

McDonald, F. B., Teegarden, B. J., Trainor, J. H., and Webber, W. R. (1974), The Anomalous Abundance of Cosmic-Ray Nitrogen and Oxygen Nuclei at Low Energies, *Astrophysical Journal*, **187**, L105–L108.

McKibben, R. (1998), *Three-Dimensional Solar Modulation of Cosmic Rays and Anomalous Components in the Inner Heliosphere*, p. 21, Kluwer.

McKibben, R. B. (2001), Cosmic rays at all latitudes in the inner heliosphere, in *The Heliosphere near Solar Minimum. The Ulysses Perspective* (A. Balogh, R. G. Marsden, and E. J. Smith, eds.), pp. 327–371, Springer-Praxis, Chichester, UK, ISBN 1-85233-204-2.

McKibben, R. B. (2005), Cosmic-ray diffusion in the inner heliosphere, *Advances in Space Research*, **35**, 518–531.

McKibben, R. B., O'Gallagher, J. J., Simpson, J. A., and Tuzzolino, A. J. (1973), Preliminary PIONEER-10 Intensity Gradients of Galactic Cosmic Rays, *Astrophysical Journal*, **181**, L9.

McKibben, R. B., Connell, J. J., Lopate, C., Simson, J. A., and Zhang, M. (1996), Observations of galactic cosmic rays and the anomalous helium during the Ulysses passage from the south pole to the north pole, *Astronomy & Astrophysics*, **316**(2), 547–554.

McKibben, R. B., Connell, J. J., Lopate, C., Anglin, J. D., Balogh, A., Dalla, S., Sanderson, T. R., Marsden, R. G., Hofer, M. Y., Kunow, H., Posner, A., and Heber, B. (2003), ULYSSES COSPIN observations of cosmic rays and solar energetic particles from the South Pole to the North Pole of the Sun during solar maximum, *Annales Geophysicae*, **21**, 1217–1228.

Mewaldt, R., Cummings, J., Leske, R., Selesnick, R., Stone, E., and von Rosenvinge, T. (1996), A study of the composition and energy spectra of anomalous cosmic rays using the geomagnetic field, *Geophysical Research Letters*, **23**, 617–620.

Möbius, E., Hovestadt, D., Klecker, B., Scholer, M., and Gloeckler, G. (1985), Direct observation of He(+) pick-up ions of interstellar origin in the solar wind, *Nature*, **318**, 426–429.

Moeketsi, D. M., Potgieter, M. S., Ferreira, S. E. S., Heber, B., Fichtner, H., and Henize, V. K. (2005), The heliospheric modulation of 3–10 MeV electrons: Modeling of changes in the solar wind speed in relation to perpendicular polar diffusion, *Advances in Space Research*, **35**, 597–604.

Moraal, H. (2001), The discovery and early development of the field of anomalous cosmic rays, in *The Outer Heliosphere: The Next Frontiers* (K. Scherer, H. Fichtner, H. J. Fahr, and E. Marsch, eds.), COSPAR Colloquia Series 11, Pergamon Press, Amsterdam, p. 147.

Moraal, H., Gleeson, L. J., and Webb, G. M. (1979), Effects of charged particle drifts on the modulation of the intensity of galactic cosmic rays, in *Proc. 16th ICRC*, Vol. 3, pp. 14.

Moskalenko, I. V., Strong, A. W., Ormes, J. F., and Potgieter, M. S. (2002), Secondary Antiprotons and Propagation of Cosmic Rays in the Galaxy and Heliosphere, *Astrophysical Journal*, **565**, 280–296.

Ndiitwani, D. C., Ferreira, S. E. S., Potgieter, M. S., and Heber, B. (2005), Modelling cosmic ray intensities along the Ulysses trajectory, *Annales Geophysicae*, **23**, 1061–1070.

Paizis, C., Heber, B., Raviart, A., Ducros, R., Ferrando, P., Rastoin, C., Kunow, H., Müller-Mellin, R., Sierks, H., and Wibberenz, G. (1995), Latitudinal effects of galactic cosmic rays onboard the ULYSSES spacecraft, in *Proc. 24th Int. Cosmic Ray Conf., Rome, Italy*, Vol. 4, p. 756.

Paizis, C., Heber, B., Raviart, A., Ferrando, M. P. P., and Müller-Mellin, R. (1997), Compton getting factor and latitude variation of cosmic rays, in *Proc. 25th Int. Cosmic Ray Conf., Durban*, p. 93.

Paizis, C., Heber, B., Ferrando, P., Raviart, A., Falconi, B., Marzolla, S., Potgieter, M. S., Bothmer, V., Kunow, H., Müller-Mellin, R., and Posner, A. (1999), Amplitude evolution and rigidity dependence of the 26-day recurrent cosmic ray decreases: COSPIN/KET results, *Journal of Geophysical Research*, **104**(A12), 28241.

Parhi, S., Burger, R. A., Bieber, J. W., and Matthaeus, W. H. (2001), Challenges for an ab initio theory of cosmic ray modulation, in *American Geophysical Union, Spring Meeting 2001*, Abstract #SH22F-06, p. 22.

Parker, E. N. (1958), Dynamics of the Interplanetary Gas and Magnetic Fields, *Astrophysical Journal*, **128**, 664.

Parker, E. N. (1961), The stellar-wind regions, *Astrophys. J.*, **134**, 20.

Parker, E. N. (1963), *Interplanetary Dynamical Processes*, Wiley-Interscience, New York.

Parker, E. N. (1965), The passage of energetic particles through interplanetary space, *Planet Space Science*, **13**, 949.

Perko, J. S., and Burlaga, F. L. (1990), Simulation of Voyager Cosmic-Ray Count Data Using Only Simultaneous Magnetic-Field Measurements, in *Int. Cosmic Ray Conf.*, p. 157.

Perko, J. S., and Fisk, L. A. (1983), Solar modulation of galactic cosmic rays. V—Time-dependent modulation, *Journal of Geophysical Research*, **88**(17), 9033–9036.

Pesses, M. E., Eichler, D., and Jokipii, J. R. (1981), Cosmic ray drift, shock wave acceleration, and the anomalous component of cosmic rays, *Astrophysical Journal*, **246**, L85–L88.

Pick, M., and Lario, D. (2007), Energetic particles in the inner solar system, in *Physics of the Heliosphere*, Springer-Verlag, Berlin.

Potgieter, M. S. (1989), Heliospheric terminal shock acceleration and modulation of the anomalous cosmic-ray component, *Advances in Space Research*, **9**, 21–24.

Potgieter, M. S. (1996), Heliospheric modulation of galactic electrons: Consequences of new calculations for the mean free path of electrons between 1 MeV and 10 GeV, *Journal of Geophysical Research*, **101**, 24411–24422.

Potgieter, M. S. (1997), The heliospheric modulation of galactic cosmic rays at solar minimum, *Advances in Space Research*, **19**, 883–892.

Potgieter, M. S. (1998), The modulation of galactic cosmic rays in the heliosphere: Theory and models, *Space Science Reviews*, **83**, 147–158.

Potgieter, M. S. (in press), Challenges from beyond the termination shock, *Space Science Reviews*.

Potgieter, M. S., and le Roux, J. A. (1989), A numerical model for a cosmic ray modulation barrier in the outer heliosphere, *Astronomy & Astrophysics*, **209**, 406–410.

Potgieter, M. S., and le Roux, J. A. (1992), The simulated features of heliospheric cosmic-ray modulation with time-dependent drift model I. General effects of the changing neutral sheet over the period 1985–1990, *Astrophysical Journal*, **386**, 336–346.

Potgieter, M. S., and le Roux, J. A. (1994), The Long-Term Heliospheric Modulation of Galactic Cosmic Rays according to a Time-dependent Drift Model with Merged Interaction Regions, *Astrophysical Journal*, **423**, 817.

Potgieter, M. S., and Moraal, H. (1985), A drift model for the modulation of galactic cosmic rays, *Astrophysical Journal*, **294**, 425–440.

Potgieter, M. S., and Moraal, H. (1988), Acceleration of cosmic rays in the solar wind termination shock. I—A steady state technique in a spherically symmetric model, *Astrophysical Journal*, **330**, 445–455.

Potgieter, M. S., le Roux, J. A., Burlaga, L. F., and McDonald, F. B. (1993), The role of merged interaction regions and drafts in the heliospheric modulation of cosmic rays beyond 20 AU—A computer simulation, *Astrophysical Journal*, **403**, 760–768.

Potgieter, M., Haasbroek, L., Ferrando, P., and Heber, B. (1997), The modeling of the latitude dependence of cosmic ray protons and electrons in the inner heliosphere, *Advances in Space Research*, **19**(6), 917–920.

Pyle, K. R., and Simpson, J. A. (1977), The Jovian relativistic electron distribution in interplanetary space from 1 to 11 AU—Evidence for a continuously emitting point source, *Astrophysical Journal*, **215**, L89–L93.

Rastoin, C. (1995), *Les électrons de Jupiter et de la Galaxie dans l'heliosphère d'après l'experience KET à bord de la sonde spatiale ULYSSE*, PhD thesis, Saclay.

Reinecke, J. P. L., and Potgieter, M. S. (1994), An explanation for the difference in cosmic ray modulation at low and neutron monitor energies during consecutive solar minimum periods, *Journal of Geophysical Research*, **99**, 14761–14767.

Richardson, I. G., Cane, H. V., and Wibberenz, G. (1999), A 22-year dependence in the size of near-ecliptic corotating cosmic ray depressions during five solar minima, *Journal of Geophysical Research*, **104**, 12549–12562.

Richardson, I. V. (2004), Corotating Interaction Regions, *Space Science Reviews*, **111**, 267–376.

Scherer, K and Fahr, H. J. (2003), Solar cycle induced variations of the outer heliospheric structures, *Geophysical Research Letters*, **30**, 17.

Scherer, K., and Ferreira, S. E. S. (2005), A heliospheric hybrid model: Hydrodynamic plasma flow and kinetic cosmic ray transport, *Astrophysics and Space Sciences Transactions*, **1**, 17–27.

Scherer, K., Fichtner, H., and Stawicki, O. (2002), Shielded by the wind: The influence of the interstellar medium on the environment of Earth, *Journal of Atmospheric and Terrestrial Physics*, **64**, 795–804.

Shalchi, A., and Schlickeiser, R. (2004), Quasilinear perpendicular diffusion of cosmic rays in weak dynamical turbulence, *Astronomy & Astrophysics*, **420**, 821–832.

Shalchi, A., Bieber, J. W., Matthaeus, W. H., and Schlickeiser, R. (2006), Parallel and Perpendicular Transport of Heliospheric Cosmic Rays in an Improved Dynamical Turbulence Model, *Astrophysical Journal*, **642**, 230–243.

Simpson, J. A. (2000), The Cosmic Ray Nucleonic Component: The Invention and Scientific Uses of the Neutron Monitor (Keynote Lecture), *Space Science Reviews*, **93**, 11–32.

Simpson, J. A. (2001), The cosmic radiation, in *The Century of Space Science* (J. A. M. Bleeker, J. Geiss, and M. C. E. Huber, eds.), Kluwer Academic, Dordrecht, The Netherlands, p. 117.

Simpson, J. A., Zhang, M., and Bame, S. (1996), A Solar Polar North-South Asymmetry for Cosmic-Ray Propagation in the Heliosphere: The ULYSSES Pole-to- Pole Rapid Transit, *Astrophysical Journal*, **465**, L69.

Simpson, J. A., Hamilton, D., Lentz, G., McKibben, R. B., Mogro-Campero, A., Perkins, M., Pyle, K. R., Tuzzolino, A. J., and O'Gallagher, J. J. (1974), Protons and Electrons in Jupiter's Magnetic Field: Results from the University of Chicago Experiment on Pioneer 10, *Science*, **183**, 306–309.

Smith, E. J. (2001), The heliospheric current sheet, *Journal of Geophysical Research*, **106**(15), 15819–15832.

Steenberg, C. D., and Moraal, H. (1996), An Acceleration/Modulation Model for Anomalous Cosmic-Ray Hydrogen in the Heliosphere, *Astrophysical Journal*, **463**, 776.

Teegarden, B. J., McDonald, F., Trainor, J. H., Webber, W. R., and Roelof, E. (1974), Interplanetary MeV-electrons of Jovian origin, *Journal of Geophysical Research*, **79**, 3615–3622.

Thomas, B. T., and Smith, E. J. (1981), The structure and dynamics of the heliospheric current sheet, *Journal of Geophysical Research*, **86**(15), 11105–11110.

Trattner, K. J., Marsden, R. G., Bothmer, V., Sanderson, T. R., Wenzel, K.-P., Klecker, B., and Hovestadt, D. (1996), ULYSSES COSPIN/LET: Latitudinal gradients of anomalous cosmic ray O, N and Ne, *Astronomy & Astrophysics*, **316**, 519–527.

Vasyliunas, V. M., and Siscoe, G. L. (1976), On the flux and the energy spectrum of interstellar ions in the solar system, *Journal of Geophysical Research*, **81**, 1247–1252.

Wibberenz, G., Richardson, I. G., and Cane, H. V. (2002), A simple concept for modeling cosmic ray modulation in the inner heliosphere during solar cycles 20–23, *Journal of Geophysical Research (Space Physics)*, **107**, 51.

Wiedenbeck, M. E. (2000), Cosmic-Ray Isotopic Composition Results from the ACE Mission, in *AIP Conf. Proc. 516: 26th Int. Cosmic Ray Conf., ICRC XXVI* (B. L. Dingus, D. B. Kieda, and M. H. Salamon, eds.), p. 301.

Witte, M., Rosenbauer, H., Banaszkiewicz, M., and Fahr, H. (1993), The Ulysses neutral gas experiment—Determination of the velocity and temperature of the interstellar neutral helium, *Advances in Space Research*, **13**, 121–130.

Zank, G. P. (1999), Interaction of the solar wind with the local interstellar medium: A theoretical perspective, *Space Science Reviews*, **89**, 413–688.

Zank, G. P., and Müller, H.-R. (2003), The dynamical heliosphere, *Journal of Geophysical Research (Space Physics)*, **108**, 71.

Zhang, M. (1997), A linear relationship between the latitude gradient and 26 day recurrent variation in the fluxes of galactic cosmic rays and anomalous nuclear components. I. Observations, *Astrophysical Journal*, **488**, 841–853.

Zurbuchen, T. H., Schwadron, N. A., and Fisk, L. A. (1997), Direct observational evidence for a heliospheric magnetic field with large excursions in latitude, *Journal of Geophysical Research*, **102**(11), 24175–24182.

7

Overview: The heliosphere then and now

Steven T. Suess

7.1 INTRODUCTION

Understanding of the integrated Sun–heliosphere system has been transformed by Ulysses, the only mission to explore the heliosphere in three dimensions and overcome the limitations of measurements restricted to the vicinity of the ecliptic plane. Ulysses' three orbits (O-I, O-II, O-III) have been very favorably aligned with respect to sunspot minimum and maximum conditions during solar cycle 23, giving rapid spatiotemporal cuts through the heliosphere at the extremes of sunspot activity during the fast-latitude scans, and more leisurely cuts through the heliosphere during the slow-latitude scans in the long rising and falling portions of the sunspot cycle (Figure 7.1). The first fast-latitude scan (FLS-I) took place in 1994–1995, at solar minimum and the start of cycle 23. The rising phase of cycle 23 took place during the second half of O-I and the first half of O-II. FLS-II occurred at the maximum of cycle 23. The falling phase of cycle 23 took place during the second half of O-II and the first half of O-III. By assembling the measurements through all the phases of cycle 23, it has been possible to characterize the "four-dimensional" heliosphere (space + time). A simple graphic representation of this characterization is a dial plot of solar wind speed such as those for O-I and O-II shown in Figure 7.2. The global viewpoint has knitted together the measurements of Ulysses with those of all the other missions that make up "The Great Observatory" (TGO) of heliospheric missions.

Stepping back to 1990, before the launch of Ulysses, SOHO, and ACE, and before Voyagers 1/2 neared the termination shock, global conditions in the heliosphere are now seen to have been poorly known. To be sure, it was known that the solar wind exhibits pronounced variations in heliographic latitude, but this result was based on remote observations—interplanetary radio scintillations (IPS), comet tails, etc.—and inferences based on in-ecliptic observations. *In situ* measurements had been limited to a narrow region near the ecliptic plane, with the exceptions of those by Voyagers 1 and 2 as they traveled out into the distant heliosphere at $\sim$34°N and

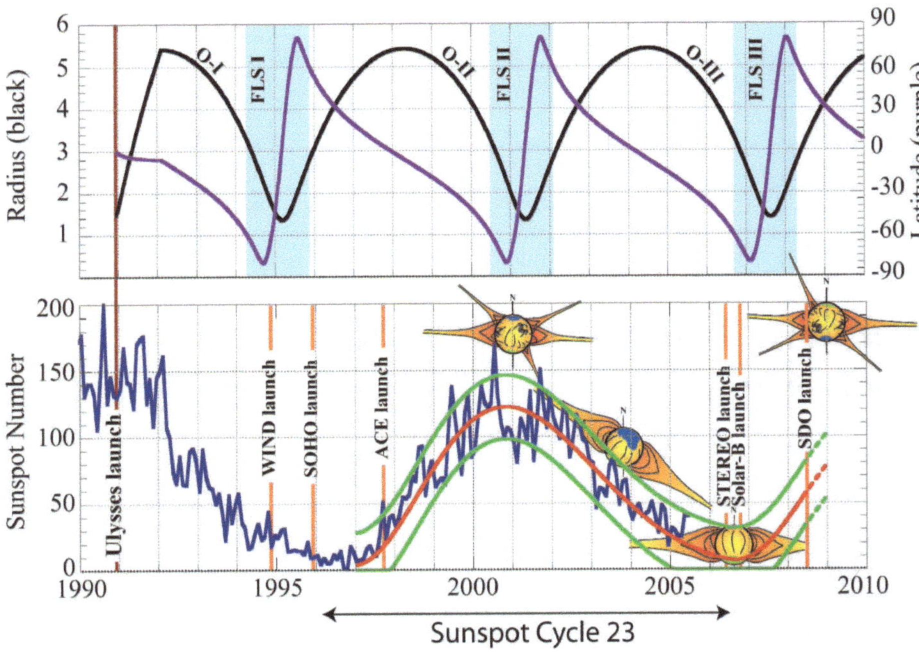

Figure 7.1. Sunspot cycle and Ulysses' radius and heliographic latitude through 2008. The orbits and fast-latitude scans are labelled O-I through O-III and FLS-I through FLS-III, respectively. Schematics of the Sun that are overlayed onto the observed and predicted sunspot cycle show the appearance of the corona through the cycle: spiky and disordered at maximum, tilted dipole with large coronal holes in the declining phase, and axial dipole with very large coronal holes at minimum.

~26°S, respectively. With Ulysses, our knowledge of the heliosphere was expected to expand and it has done so even more than expected. Original mission objectives were to investigate for the first time, as a function of latitude, the properties of the solar wind, the structure of the Sun–wind interface, the heliospheric magnetic field, solar radio bursts, and plasma waves, solar X-rays, solar and galactic cosmic rays, and both interstellar and interplanetary neutral gas and dust. All this has now been done and, along the way, there have been many unexpected discoveries. For derived science, this means that results and discoveries often involve connections: the solar–interplanetary magnetic field connection; the complex connections between interstellar neutral atoms, dust, pickup ions, and anomalous cosmic rays; the global solar wind–termination shock–heliosheath connection; the sub-Parker spiral–acceleration of particles at the termination shock connection; the magnetic field deviation–energetic particle drift and transport connection, and so on in a far from exhaustive list.

This chapter is an overview of how the heliosphere has come to be viewed as a consequence of observations carried out over the past 15 years. We begin with a quick look at the heliosphere as it was known *circa* 1992, when Ulysses passed Jupiter and

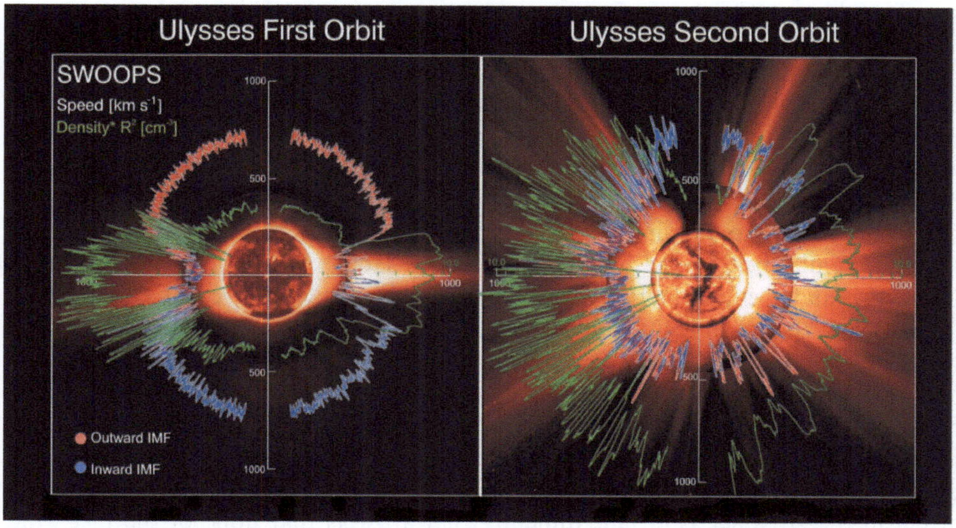

Figure 7.2. Dial plots of solar wind speed and density, with co-temporal coronal images, during O-I and O-II, with start dates shown in Figure 7.1. Time runs counter-clockwise from 9 o'clock, along with heliographic latitude. The gaps at the north and south poles reflect the maximum Ulysses latitude of 80.2°. The speed scale is [0, 1000] km/s and the density scale is [0, 10] cm^{-3}. The blue/red color of the solar wind speed values indicates the magnetic field polarity. Images of the Sun are SOHO/EIT on the disk, SOHO/LASCO above $2R_{\odot}$, and MSLO at 1–$2R_{\odot}$.

left the ecliptic plane. A description focusing only on the contributions of Ulysses through the end of O-I is contained in Balogh, Marsden, and Smith (2001).

7.2 THE KNOWN HELIOSPHERE IN 1992

In situ measurements of the fields and particles in the interplanetary medium began in the late 1950s (Gringauz *et al.*, 1960), shortly after Parker published his theory for the magnetized, steady solar wind (Parker, 1958). Observations immediately confirmed the existence of a supersonic wind and also the Archimedian spiral magnetic field predicted by Parker. Photospheric magnetic fields had been observed and measured since the early 20th century and the changes in coronal morphology over the solar cycle have been known far longer from eclipse observations (Billings, 1966). These two pieces of information were combined to deduce that the heliospheric magnetic field (HMF) should carry the imprint of the solar magnetic field and that the 3-D solar wind should reflect the changing corona over the solar cycle. Many suggestions were made for the nature of this imprint and one of the first discoveries from interplanetary missions was the sector structure of the HMF (Wilcox and Ness, 1965), which is the imprint on the HMF of the often dipolar appearance of the solar magnetic field and its solar cycle evolution. This was the beginning of the modern paradigm for the 4-D heliosphere.

7.2.1 The solar wind and the heliospheric magnetic field

Early measurements established that the solar wind can be broadly sorted into slow wind ($\lesssim$500 km/s), fast wind ($\gtrsim$650 km/s), and transients (Balogh, Marsden, and Smith, 2001; White, 1977). Skylab X-ray images of the corona made it clear that fast wind generally originates in coronal holes while slow wind comes from the streamer belt (Zirker, 1977). The coronal holes were associated with large magnetically unipolar regions on the Sun, making it possible to infer their locations using relatively simple potential field models of the corona (e.g., Hoeksema, Wilcox, and Scherrer, 1982) or, under more limited conditions, MHD models of coronal expansion (Steinolfson, Suess, and Wu, 1982).

A standard picture quickly emerged. Fast solar wind from a single coronal hole had a single predominant magnetic polarity, with high-speed streams being separated by a sector boundary across which the field changes direction. There are usually 2 to 4 sectors per 25.5 day (sidereal) solar rotation, depending on time during the solar cycle. In slow wind, the magnetic polarity is generally mixed, a consequence of its origin in the streamer belt. One difficulty that arose is that none of the fast wind seemed to be a thermal wind of the type modeled by Parker. Instead, additional momentum and/or heating is required above the base of the corona. The highly variable slow wind was found to be more nearly like a thermal wind.

The Sun and the solar cycle

These developments led to a standard description of how the Sun, the solar wind, and the HMF change together over the solar cycle. This was introduced in Chapter 1, the solar cycle was described in Chapter 2, and the description is summarized here. Near sunspot minimum, the white light corona is oblate, with a bulging streamer belt at the equator and large coronal holes over each pole. The streamer belt lay at the base of the heliospheric current sheet (HCS) dividing opposite magnetic polarities in the interplanetary medium (Figures 2.12 and 2.13). During the rapid rise to solar maximum, the coronal holes shrink and the shape of the streamer belt becomes more irregular. At sunspot maximum, the corona is highly structured, due to the presence of numerous active regions, and coronal holes are small and located haphazardly over the Sun, or absent altogether. As the cycle then continues, coronal holes again appear and grow, but are irregular in shape. Nevertheless, the Sun's magnetic field can still be approximated by a tilted dipole during this period. The resulting high-speed solar wind streams resulting from the wobbling dipole, as the Sun rotates, produce strong corotating interaction regions. The new polar coronal holes have the opposite magnetic polarity relative to those in the preceding sunspot cycle. This is now the second half of the 22-year Hale magnetic solar cycle. Finally, the dipole again becomes axially aligned as sunspot minimum is approached.

During most of the sunspot cycle, measurements of the solar magnetic field suggested the dominant component is a dipole (Hoeksema, Wilcox, and Scherrer, 1982). This is reflected in the sector structure, the polarity of the radial magnetic field in the interplanetary medium in the plane of the ecliptic, which generally contains two sectors as might be expected for a tilted dipole field. At sunspot maximum, the sector structure is often difficult to distinguish in HMF ecliptic measurements, which is a consequence of the large number of active regions on the Sun. On the other hand, at sunspot minimum there are often four sectors. This does not imply a large quadrupolar component at minimum. Instead, it is simply a consequence of small ripples in the heliospheric current sheet and the small angle between the ecliptic plane and the heliographic equator. This simple picture of heliosphere morphology is represented by the images of the corona at the different stages of the solar sunspot cycle that are shown in Figures 7.1, 2.11, 2.12, and 2.13.

The quasi-dipolar nature of the Sun's magnetic field over the majority of the solar cycle has the consequence for the 3-D heliosphere that the average solar wind speed increases with magnetic latitude. Shortly after solar maximum, when coronal holes have grown over the magnetic poles, an average speed increase with latitude was expected and interplanetary scintillation measurement showed this to be the case (Coles *et al.*, 1978).

The large morphological changes in the corona, the reversal of the dipole field, and the now-known relationship between coronal holes and high-speed wind implied major restructuring of the solar wind and heliosphere over the solar cycle. It also raised the questions of how mass flux, momentum flux, and net magnetic flux changes over the solar cycle. Models at that time (and, to a lesser degree, today) were unable to attack these questions and it was the goal of Ulysses to provide the answers.

Corotating interaction regions

In the declining phase of the solar cycle, polar coronal holes lie over magnetic poles that are tilted away from the rotation axis. The resulting solar wind is not at all axisymmetric. As the Sun rotates, the mid- and low-latitude heliosphere is exposed to alternating high- and low-speed solar wind, even extending into the opposite hemisphere, as sources of high-speed solar wind in coronal holes rotate beneath sources of slow wind in the streamer belt. This corresponds to the schematic of the corona in ~2003–2004 in Figure 7.1. Chapter 2 introduced this concept in terms of coronal holes tending to lie in large, unipolar magnetic field regions and the HCS providing a tracer for the organization of the heliosphere. There it was described how simple PFSS models of the coronal magnetic field could be used to estimate the location, size, and strength of coronal holes and the resulting corotating interaction regions (CIRs). The dynamic interaction that leads to these CIRs is conceptually straightforward. Once plasma leaves the Sun, high-speed wind overtakes slow wind, leading to a steepening velocity profile. The overall region of steepened front edge and long return to low speed is the corotating interaction region. The speed difference between fast and slow wind is roughly a factor of 2 so the interaction takes place predominantly between 1 AU and a few AU. A diagram of this interaction and the result from a simple 1-D model are shown in Figure 7.3. The in-ecliptic signature of CIRs was well-known from Skylab studies (Hundhausen, p. 225 in Zirker, 1977).

CIR evolution with distance was also well-studied by 1992. There were abundant data from Pioneers 10/11 and Voyagers 1/2 in the outer heliosphere, Helios 1/2 in the inner heliosphere, and many near-Earth spacecraft. The steepening, formation of

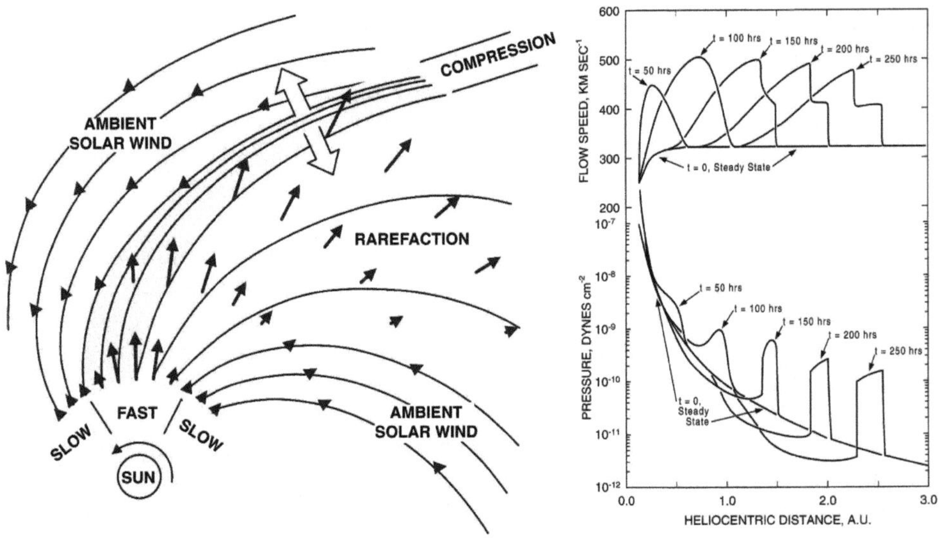

Figure 7.3. Left: Schematic of a corotating interaction region in the heliographic equatorial plane. Solid lines represent magnetic field lines while the length of the arrows is a measure of the flow speed (Pizzo, 1978). Right: Flow speed and pressure from a simple 1-D simulation showing the formation and evolution of an interaction region with distance from the Sun (Hundhausen, 1973).

shock ensembles, and some MHD effects were all understood. In particular, models, in combination with abundant in-ecliptic data, had produced a rudimentary understanding of the 3-D nature of CIRs. A good review and summary of what was known up to 1992 is given by Forsyth and Gosling (p. 107 in Balogh, Marsden, and Smith, 2001). But, there were several questions about the 3-D morphology of CIRs that Ulysses was expected to answer. These related to the properties of driven nonradial flows, influence of CIRs on the average properties of the solar wind, and the evolution of the shock ensembles produced in the CIRs. There were also questions about how far poleward the influence of CIRs extended into the volume of the heliosphere that mapped solely back into the polar coronal holes. A particular question was how effective CIRs are in accelerating particles.

Coronal mass ejections

Coronal mass ejections (CMEs) are relatively dense clouds of plasma ejected from the outer atmosphere of the Sun—a class of corona transients. CMEs and their interplanetary counterparts, ICMEs, first came under intense study during the time of the Skylab, SMM, and Solrad missions in the 1970s and 1980s. The origins of CMEs are still not well understood, even after more than 30 years of study. There has been steady progress, but many of the details are hidden in small-scale phenomena at the Sun that are only now becoming observable with current and upcoming missions such as Hinode and SDO. A discussion of their origins and interplanetary signatures is given by Forsyth and Gosling (2001). A global picture of CMEs in the inner heliosphere—their statistical properties and physical features—has been given by Gopalswamy (2004). The STEREO mission, launched in late 2006, is designed specifically to greatly expand knowledge of ICMEs in the inner heliosphere.

Remote observations of CMEs often start with their detection with a white light coronagraph. The ejected matter can be followed outward through the corona by observing the changes in corona brightness. Although coronagraphs have not been continuously operating, the gaps have been relatively small since ~1970 and now there is a large body of data indicating the statistical properties of CME occurrence over the solar cycle and how CMEs are related to various forms of solar activity (Webb and Howard, 1994). There was in 1992 an equally large body of information on the morphology of CMEs, specifically on the typical CME having a three-part structure consisting of a bright core, cavity, and bright outer shell (Hundhausen, 1997).

In situ observation of ICMEs has developed more slowly because the signatures of CMEs are not always unambigous. What is certain is that the ejecta sometimes has an identifiable signature in the bi-directional streaming of ~100 keV electrons along the magnetic field, plasma β, the existence of a "magnetic cloud", proton temperature, composition, ionization state, and several other parameters. But, it is possible that none of these signatures is always present. Much about ICMEs remained unknown or unclear in 1992. Ulysses would be able to add, literally, a new dimension to these studies. Due to the nature of the solar cycle and its synchronization with the orbital location of Ulysses, CIRs would become the main feature of interest in O-I while CMEs would take that place in O-II.

7.2.2 Solar wind composition and ionization state

The composition of solar wind plasma and the ionization state of the various species is a subject that was very poorly known before Ulysses, which carried a new generation instrument for the study of heavy ions that was able to unambiguously determine an ion's charge and mass. Research had been almost exclusively limited to data from spectrometers whose measurements determined only the mass/charge, with consequent overlap or confusion of different ion species. This was an important obstacle to understanding the source mechanism for the solar wind, in that the way in which solar matter is continuously fed into the corona was not known. There was no consensus about the mechanism of heating plasma in the corona to $>10^6$ K, and the acceleration and source of momentum driving the coronal material to supersonic velocity could not be uniquely determined.

It was already known by 1992 that elemental abundances in the solar wind are fractionated relative to the solar abundances, by atom–ion separation in the upper chromosphere and by ion–ion separation in the corona, where also the charge states of the ions become frozen-in (Geiss and Bochsler, p. 173 in Marsden, 1986). Because of this, it was anticipated that solar wind composition and charge states could be used to study conditions and processes in the corona. But, existing data did not give good coverage over the changing solar wind conditions. Nevertheless, some hints and suggestions had begun to emerge. It was evident that composition and charge states were significantly different in fast and slow wind and CMEs, as expected from the three different parts of the corona these types of solar wind originate. It had already been demonstrated that elements with low first-ionization potential (FIP) are enriched, relative to the solar surface, in the corona, solar wind, and solar flare particle populations. It had also long been known that helium and heavier ions travel faster than hydrogen (with velocity increments limited by the Alfvén speed) and that helium and heavier ions have freeze-in temperatures which are proportional to mass (*ibid.*). All of these results can now be seen to have been qualitative in comparison with modern measurements, therefore limiting quantitative analysis of coronal source mechanisms.

There is another completely different component of solar wind composition and ionization state that was virtually unexplored prior to Ulysses. This is the contribution of pickup ions—ions resulting from the ionization of neutrals coming from the interstellar medium, comets, dust, planets, and other sources. As a new ion is created through photoionization of collisional ionization, it is incorporated into the solar wind ("picked up") and carried along by the HMF. Although these ions were known to exist through, for example, measurements of solar radiation backscattered off interstellar neutral atoms that entered the heliosphere, they had not been measured directly. They have since been found to be a powerful tool for diagnosing processes regulating the global heliosphere.

7.2.3 Energetic particles and cosmic rays

Solar wind particles carry energies of a few tens of KeV per nucleon. Between this energy and a few tens to hundreds of MeV lie energetic particles of heliospheric origin. These particles come from the Sun and solar activity, shock waves in the corona and interplanetary medium, and several of the planets. There are also particles known as anomalous cosmic rays (ACRs) which originally enter the heliosphere as neutral particles and are later ionized and accelerated to energies of hundreds of KeV/nucleon up to several MeV/nucleon. Finally, cosmic rays entering the heliosphere from the local interstellar medium dominate the population above $\sim$100 MeV, depending on the time in the solar cycle. These populations have often been studied separately because only the higher energy particles can be detected at the surface of the Earth, because of the differences in their origins, and because the three populations can often be easily separated by energy. Cosmic rays have the longest observational history because they could be studied before the space age. Here, energetic particles refer to those of heliospheric origin while cosmic rays refer to particles originating outside the heliosphere.

Energetic particles

The various sources of the energetic particles that fill the heliosphere has been a major topic in solar system plasma physics. The sources are important in their own right and continue to present interesting problems. In addition, the particles can be used as probes and tracers of the features of the interplanetary medium, a use to which Ulysses is particularly suited (Lanzerotti and Sanderson, p. 259 in Balogh, Marsden, and Smith, 2001).

Aurora and geomagnetic activity long ago led to the conclusion that energetic particles streamed from the Sun to Earth. But, the advent of *in situ* measurements opened up a wide variety of associated investigations. The studies focused many different aspects of the particles. Some of these are composition and ionization state, dispersion in energy during transit from the Sun to Earth, timing relative to flare electromagnetic radiation, interplanetary propagation, and transport. A great body of theory has developed on the acceleration of particles in solar flares. Forman *et al.* (p. 249 in Sturrock *et al.*, 1986) give a comprehensive review of the state of knowledge of acceleration and propagation of solar flare energetic particles in the mid-1980s, with some accompanying discussion of shock acceleration. Acceleration in CIRs was also well-known (Lanzerotti and Sanderson, p. 259 in Balogh, Marsden, and Smith, 2001).

Many 3-D effects and phenomena would have to await Ulysses before they could be studied. But, radial gradients and propagation over the solar cycle, and the solar cycle dependence of CIR acceleration near the ecliptic had been characterized with the help of Voyagers 1/2, Helios 1/2, and several near-Earth spacecraft. It was, of course, known that Earth was a source. Less clear was the role of other planets as sources.

Anomalous cosmic rays

A source at or near the heliospheric termination shock was proposed long ago for ACRs (Fisk, Kozlovsky, and Ramaty, 1974; Pesses, Jokipii, and Eichler, 1981). The concept and theory for how these particles can then be detected in the inner heliosphere depends on the details of the processes of transport. ACRs are observed to have a solar and Hale (22-year solar magnetic cycle) cycle dependence in their spectra, with which the modulation theory must be consistent (Cummings and Stone, p. 51 in Fisk *et al.*, 1998). ACRs had been observed with AMPTE, which had tentatively confirmed that the particles were singly ionized, as required in the original concept. Ulysses and Voyagers 1/2 have both confirmed the external source of ACRs and raised the possibility that the acceleration mechanism is different than proposed in the original theory.

Cosmic rays

A vast body of data on cosmic rays has accumulated in the more than 100 years since their discovery. Higher energy cosmic rays can be monitored at ground level and a network of neutron monitors continues making synoptic measurements today. It was soon clear that they originate from outside the helioshere and that their modulation depends on the solar cycle. Parker (1963) derived the equation describing the convection, drifts, adiabatic cooling, and diffusion of cosmic rays in the heliosphere and models were developed based on various assumptions for the important processes (McKibben, 1986; Jokipii and Kóta, 1985).

The observed solar cycle modulation of cosmic rays led to the conclusion that diffusion and convection in the solar wind could be the dominant process. Cosmic rays interact with the solar wind and heliospheric magnetic field, being convected by the solar wind and diffusing as they are scattered by irregularities. This means that local measurements of the intensity at any given point depend on integrated effects over the complete trajectories of the particles in the heliosphere. Lacking global measurements prior to 1992, assumptions were made.

It was known before 1992 that adiabatic cooling and gradient drifts of cosmic rays in the HMF were important. The importance of drifts was deduced by the shape of the modulation depending on which half of the Hale magnetic cycle the observations were made. The gradient drift reverses with the polarity of the HMF, causing a change in the shape of the modulation over a solar activity cycle. This was observed in ground-level data (Chapter 1). The contribution of Ulysses to this study was expected to be a confirmation of the theory because the drifts depend on the 3-D structure of the field (Jokipii, 1986).

7.2.4 Interstellar and interplanetary neutral gas

Neutral atoms enter the heliosphere relatively unimpeded by the heliospheric interface. Their orbits are modified by the Sun's gravitational field and they are eventually photoionized or collisionally ionized. Neutral hydrogen is essentially fully ionized outside a few AU while neutral He can enter into the inner heliosphere, even inside 1 AU, at solar minimum. These atoms resonantly backscatter solar radiation which can, in turn, be detected to measure the properties of the incoming atoms. Starting in the late 1970s, several spacecraft have been equipped with instruments to measure the backscattered radiation (Lallement *et al.*, 1992; Fahr, 1986; Bertaux, Lallement, and Chassefière, 1986). By combining the observed backscattered radiation with ever-improving models, it was possible to derive a fairly good picture of the incoming neutral gas, including the density, flow speed, temperature, and flow direction. This information was vital for developing models of the heliospheric interface, for which the theory had already undergone significant development by 1992 (Zank, 1999).

7.2.5 Interstellar and interplanetary dust

There are many sources of dust in the heliosphere, very generally corresponding to sources for energetic particles in the sense that there are several internal sources and also there is dust entering from the interstellar medium. Internal sources include comets and planets and at least one unknown source. The existence of dust was known long ago from observations of the gegenschein and zodiacal light. These observations identified dust relatively close to the Sun relative to the size of the orbit of Ulysses (i.e., inside 1–2 AU).

Interplanetary measurements had been made on a few spacecraft prior to 1992. However, the data were sparse and very little was known about the global distribution. This was a rapidly developing field of study with the advent of Ulysses and Galileo, which carried almost identical detectors. The theory of dust dynamics in the heliosphere had also not received a lot of attention prior to Ulysses and Galileo. In particular, interstellar dust grains are generally charged and are affected by the Lorentz force as they move through the heliosphere (Landgraph *et al.*, 2003). The measurements made beyond 1 AU starting in ~1990 would effectively open up the new field of dust dynamics in the heliosphere.

7.3 THE KNOWN HELIOSPHERE AFTER A SOLAR ACTIVITY CYCLE WITH ULYSSES

Ulysses' contribution to a growing awareness that energetic particles are often far more mobile in latitude than expected serves to introduce the changes the mission has brought about. This observation is linked to similar results for cosmic rays, pickup ions, dust, and at least occasional direct magnetic field connections across latitude. Unexpected magnetic field topology has been a pervasive theme for Ulysses. For example, long-lasting deviations from the classical interplanetary spiral are due to movement of field line footpoints across the Sun and are one component of new theories for the origin of the solar wind that also involve the character of solar wind abundances and ionization state.

Results on the origin of the solar wind and properties of the local interstellar medium illustrate that Ulysses' measurements point both inward, towards the Sun, and outwards, towards the universe. Ulysses is analyzing sources of the solar wind and HMF while it is providing a description of the global solar wind at the time Voyager 1 and Voyager 2 are in the heliosheath.

This overview begins with the *global view* of the heliosphere after a solar cycle with Ulysses, including latitudinal gradients. But then it continues with sections on several individual topics that often represent the Sun and the heliosphere as an integrated system, including linking solar and heliospheric magnetic fields, the heliosphere and the interstellar medium, composition and ionization state, HMF deviations, locations of particle acceleration, dust dynamics, CMEs in 3-D, and the developing synergy with modeling and explaining the physics of the corona and links to the inner heliosphere.

There a dense forest of results from Ulysses since 1992, both from the mission alone and from it as a member of the Great Observatory. An overview cannot cover them all. Instead, a few interesting results are highlighted and reference is made to relevant chapters for more details. It is effectively a random walk through the dense forest of results.

7.3.1 The global view

The global structure of solar wind speed during solar minimum (O-I) and maximum (O-II) intervals of the solar cycle can be said to be illustrated by Figure 7.2, using Figure 7.1 to interpret time *versus* latitude. At sunspot minimum, the solar wind is divided into fast, smooth wind over the poles and slow, irregular wind above the streamer belt. Density is inversely correlated with speed. Perhaps the most striking feature found in O-I is the abrupt change in speed at the edge of the streamer belt. This has led to the solar wind in the few years before and at minimum being described as bimodal, in contrast to the smooth transition between fast and slow wind that had been anticipated. The thickness and character of the transition at the Sun before and during solar minimum has still not been well characterized. Dynamic interactions in CIRs smooth out the transition with increasing distance from the Sun (Figure 7.3). At maximum, there continues to be a characteristic slow wind much like that over the streamer belt at minimum. Superimposed on this is a large variance due to interplanetary CMEs (ICMEs), which are much more common in the few years before and after maximum (Gopalswamy, 2004).

There is a corresponding variation in solar wind ionization state and composition between fast and slow wind or, equivalently, in latitude that was first carefully documented during O-I. The ionization state of, for example, Fe or O, is found to be higher in slow wind than in fast wind while the abundance of low first-ionization potential (FIP) elements such as Fe or Si is found to be more nearly equal to photospheric abundances than in slow wind (low FIP elements are overabundant in slow wind). This is illustrated in Figure 7.4. The abundance difference has come to be known as the FIP effect. These two features have different origins. The ionization state is fixed in the corona as the solar wind flows outward through the transition height between collisional and collisionless regimes. Higher temperatures result in higher ionization states being "frozen" into the solar wind and the observation implies a result counterintuitive to some earlier expectations, that slow wind originates from hotter regions in the corona than fast wind. The FIP effect must occur lower in the solar atmosphere. Generally, it is thought to be a consequence of processes occurring near the large temperature gradient at the base of the transition region.

The FIP effect has become a component of a new hypothesis relating the origin of solar wind to the reconnection of emerging loops in the photosphere. Observations by SOHO/MDI and TRACE show that small and large magnetic loops (bipoles) are continuously emerging within supergranules. It was realized that flux emergence and reconnection with the existing HMF can be responsible for and constrain the heating of the solar corona and the acceleration of the solar wind (Fisk, Zurbuchen, and Schwadron, 1999; Schwadron and McComas, 2005). In this scenario, fast wind originates from small loops emerging within CHs, while slow wind originates from reconnection with larger loops in regions outside CHs. The time spent by plasma within a loop is relevant in enhancing the strength of the FIP bias. A consequence is the ionization state of the solar wind and speed can be used to estimate the temperature ("freeze-in temperature") and magnetic field strength at the origin. This was used on data from FLS-I to show there was a north–south asymmetry in coronal temperatures at the time, and a north–south field strength asymmetry consistent with *in situ* HMF results (Section 7.3.2).

Other solar wind parameters also exhibit characteristic latitudinal variations to a greater or lesser degree. Of particular interest for computing the shape of the helio-

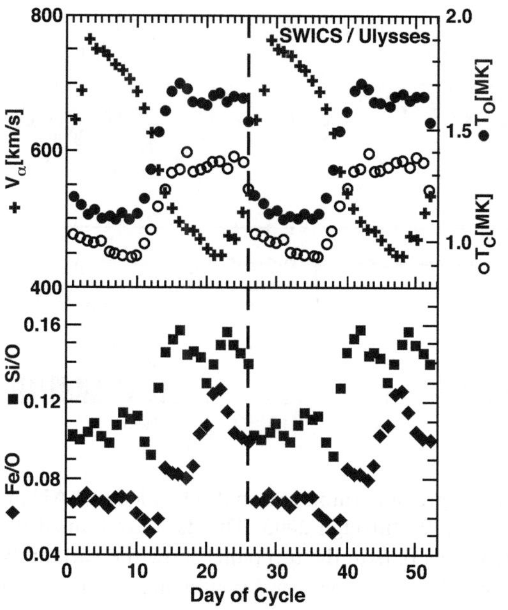

Figure 7.4. Top: Solar wind speed (+), oxygen (filled circles), and carbon (open circles) "freeze-in temperatures" in a CIR. Time is in days of the CIR, which had a duration of ~26 days. Data were superposed over several rotations. Bottom: Abundances of low FIP elements Fe and Si relative to O (Geiss *et al.*, 1995).

spheric termination shock and heliopause is the solar wind dynamic pressure (Izmodenov, 2004). Data show that the dynamic pressure of the solar wind was larger over the poles in O-I (Phillips *et al.*, 1995). Later, at sunspot maximum in O-II, the solar wind speed was, on average, lower and very spiky. A lot of the variance was, as above, due to CMEs.

The dynamic pressure is important for determining the overall size and shape of the heliosphere through the pressure balance that exists at the heliopause. The larger pressure over the poles in O-I implies that the termination shock was farther from the Sun at high latitudes. However, the more recent data in Figure 7.5 show the result to largely be a temporal effect (McComas *et al.*, 2003). These data place the result in broader context, with Figure 7.5 showing a rapid increase by a factor of ~2 during 1991, followed by an overall decrease from 1991 to 2001 that is larger than the latitudinal variation seen in 1995. A similar trend has been observed in in-ecliptic data in previous solar cycles. The implication is that the dynamic pressure of the solar wind undergoes a factor-of-2 change over a solar cycle, largely independent of latitude. Interestingly, the most recent Ulysses data indicate the pressure has again increased, but not nearly so much as in the earlier cycle. This could be a case of insufficient data to precisely identify the statistical character of the variation.

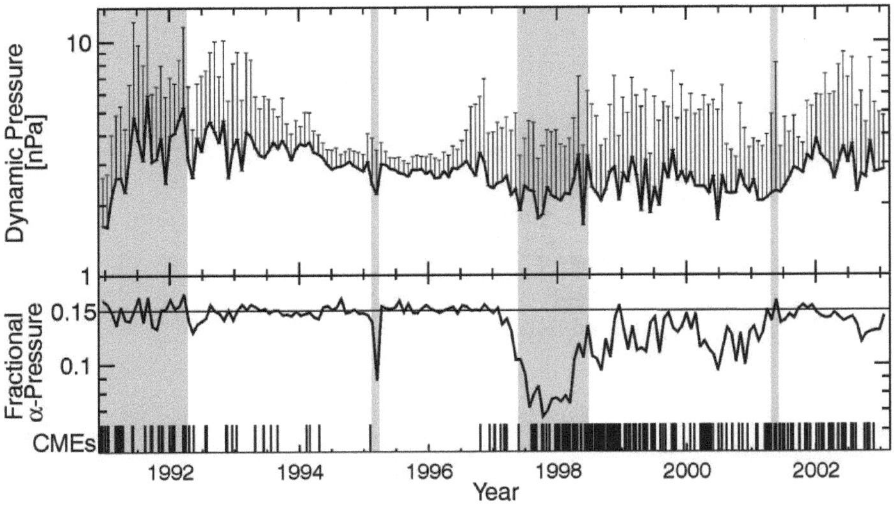

Figure 7.5. Total dynamic pressure (momentum flux) scaled to 1 AU and the fraction of alpha-to-proton pressures from 1990 through 2003. The data are binned by the solar rotation and mean values are plotted; $+1\sigma$ bars in the top panel show the variability. Gray bands highlight low-latitude (<10°) intervals, as can be seen by referring to Figure 7.1.

However, solar cycle 23 differs from 22 in having a far weaker polar field strength (Figure 2.6). There is still a lot to learn about cycle-to-cycle variations of the Sun.

A remarkable global result is that the strength of the radial component of the HMF (B_r), in one-rotation averages and mapped to a constant radius, is very nearly independent of latitude at both sunspot minimum and sunspot maximum (Smith *et al.*, 2003—see also Chapter 4). Although there was a prediction of this result for coronal hole flow (Suess *et al.*, 1977; Suess and Smith, 1996), its applicability to the global solar wind was not appreciated. The implications of the result are far-reaching. For example, it gives the opportunity to estimate the total open flux from the Sun using a local measurement. The result may offer possibilities for estimating pre–space age solar wind properties by using magnetospheric proxies for the HMF strength. The result for coronal physics is the implication that streamers are primarily confined by pressure from the expanding (diverging) magnetic field in adjacent coronal holes, as opposed to curvature forces inside the streamers. This phenomenon will receive continuing attention into solar cycle 23 because of the reduced polar field strengths so far exhibited in this cycle (Figure 2.6) and the question whether the HMF strength will be correspondingly smaller.

7.3.2 Coronal and heliospheric magnetic fields

The accessibility of global data has produced several discoveries on the relationship between the coronal magnetic field and the HMF. A few examples will be given here, with a more extensive discussion being given in Chapters 2 and 4.

An important Ulysses goal was to establish how the HMF responds to the changing solar magnetic field over the solar cycle. It has already been noted that there has been little change in the average strength of the HMF in cycle 22. But, the solar magnetic field reverses over each sunspot cycle and this process must somehow be reflected in the HMF. It had long been known that the HMF was that of an equatorial dipole magnetic field at the Sun at sunspot minimum and that of a tilted dipole during the declining phases of the solar cycle, as suggested by the images superimposed on Figure 7.1. What about the time around sunspot maximum, when the photospheric field seemed to be dominated by a relatively chaotic distribution of sunspot groups? When Ulysses magnetometer data was extrapolated back to the Sun, it was found to derive from a dipole field with its axis lying in the equator and corotating with the Sun (Jones, Balogh, and Smith, 2003). This result implies that the solar magnetic field reverses more by rotation in latitude rather than by the dipole strength going through zero. In fact, this can be seen in the Wilcox Solar Observatory data shown in Figure 2.15 (top) and had been suggested earlier from those data and could also be seen, in retrospect, in the source surface data shown in Figure 2.12. The significance is that the solar dipole undergoes reversal by a process different, or more complex, than that suggested in the traditional Babcock–Leighton picture of the 22-year Hale cycle. It is rewarding to find similar behaviors in the photospheric field and the HMF.

During FLS-I, an additional feature of the HMF was revealed that also modifies the traditional picture of the heliosphere. This is that there was a north–south asymmetry in the HMF, with the field being stronger in the south than in the north. Naturally, for flux conservation, the HCS was displaced southward—by about 10° (Chapter 4). A corresponding asymmetry was also found in cosmic rays (Chapter 6), ionization state, and many other solar wind parameters (Chapters 3 and 4). It has long been known that the Sun is not perfectly north–south symmetric, but it was unknown whether the asymmetry was reflected in properties of the interplanetary medium. The asymmetry during FLS-I can be understood by inspecting Figure 2.6 which shows the solar polar field strength measured at WSO. During FLS-I (1994–1995), the field strength at the Sun was larger in the south than in the north, just as seen at Ulysses (Gloeckler, Zurbuchen, and Geiss, 2003).

Observing a north–south asymmetry at sunspot minimum, the question naturally arises of what the asymmetry might have been at maximum. Again, the Sun is never perfectly symmetric at maximum. In particular, the polar coronal holes typically appear in one hemisphere before the other and this happened at the beginning of cycle 23. Figure 2.10 shows the early north CH development in terms of CH boundaries determined from He 10830 Å observations. There were no polar holes in late 2000. In late 2001, there was a prominent northern hole and still no southern hole. The image on the right shows the coronal (Fe XVII 195 Å) hole from SOHO/EIT in November 2001, with a prominent coronal hole in the north and nothing visible in the south. Looking back at the polar field strengths in Figure 2.6, it can be seen that the reversal in the north was more rapid and the field gained greater strength than in the south—apparently explaining the early formation of the northern CH. In the solar wind, the consequence of this situation is that Ulysses passed over the north pole just as the northern CH was forming, encountered typical fast wind from the CH, and was sporadically immersed in fast wind from this growing CH as it moved back towards the equator (Figure 7.1). In the right panel of Figure 7.2, the green line on the right side shows the solar wind speed being predominantly above 500 km/s during the entire passage of Ulysses from north back to the equator.

The "sub-Parker spiral"

One of the earliest predictions for the HMF was that it would be drawn into an Archimedian spiral (Parker, 1958). This was such an elegant concept that it is almost a test of the accuracy of the magnetometer on Ulysses that the spiral be confirmed, rather than an investigation of the spiral itself. Nevertheless, extended intervals were found in which the spiral angle was less than predicted. The explanation for this lay in the hypothesis that the HMF footpoints move around in the photosphere, moving from one flux element to another (Fisk, Zurbuchen, and Schwadron, 1999; Schwadron and McComas, 2005). While doing this, some footpoints will move across CH boundaries, causing them to contain both fast wind and slow wind. The hypothesis is compared very favorably with Ulysses CIR observastions during O-I. The phenomenon is diagrammed in Figure 7.6, which compares the cases without (top) and with (bottom) footpoint motion.

High-latitude Alfvénic fluctuations

As in the ecliptic, high-latitude fast solar wind was found to be permeated by Alfvénic fluctuations. This feature takes on importance because the presence of magnetic field fluctuations enhances cosmic ray diffusion. In general, fluctuations are similar to those in low-latitude fast wind, which is not surprising. But, noting this, Jokipii *et al.* (1995) argued that the fluctuations would be kinematically distorted with increasing distance from the Sun so that the effective power decrease with increasing distance would be $\sim r^{-2}$. The theoretical predictions were found to be well supported by the observed fluctuations. The effect is larger power levels far from the Sun than otherwise would be the case for waves. This result is fundamental to the explanation of cosmic ray gradients, which are even then not fully understood.

Interstellar dust

Interstellar dust entering the heliosphere is directly detected by Ulysses/DUST (Krüger *et al.*, 2006). Results imply the gas-to-dust mass ratio of the interstellar medium is a factor of 4–5 larger than derived from UV and optical extinction curves, which is still unexplained. Because the dust is generally negatively charged, it undergoes $\mathbf{v} \times \mathbf{B}$ drift as it moves through the heliosphere. The drift direction changes direction over the Hale cycle, with the reversal of the HMF, similar to cosmic rays. This leads to a Hale cycle variation that has been modeled to help understand the

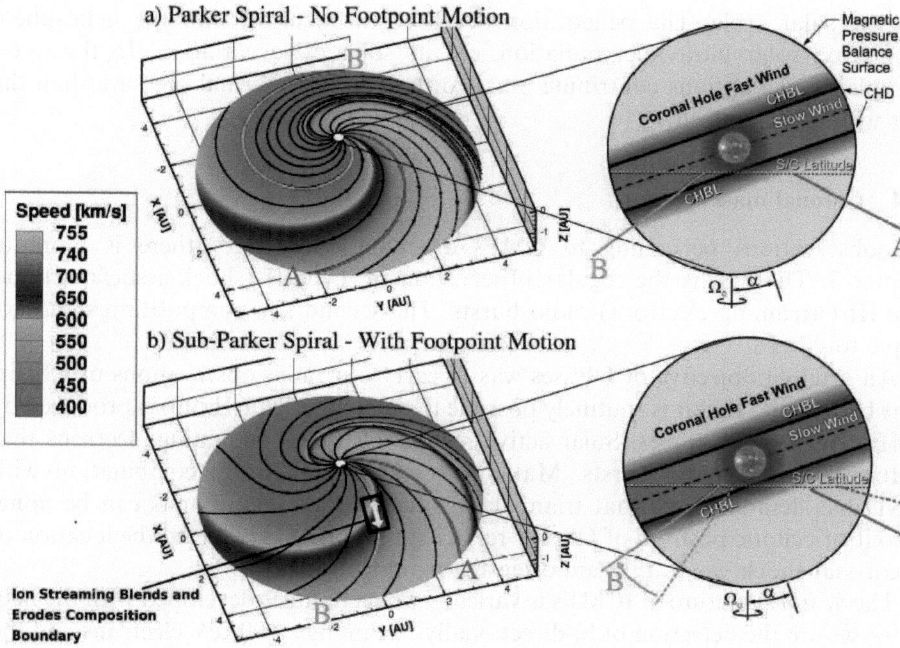

Figure 7.6. Sub-Parker spirals (Schwadron and McComas, 2005): magnetic fields in a CIR without (a) and with (b) footpoint motion at the Sun. In (b) differential rotation of magnetic field footpoints creates a magnetic connection across stream interfaces, causing magnetic field lines to cross stream interfaces. With footpoint motion, field lines are connected across the stream interface.

native dust population in the local interstellar medium (Landgraph *et al.*, 2003). The model helps to explain the large (3×) changes that have been observed at Ulysses.

7.3.3 Composition and ionization state

Ulysses carries a new class of instrument that produced the result shown in Figure 7.4. Identical or similar instruments are now carried on other missions, including ACE and STEREO A/B. They analyze the composition and ionization state of low-energy plasma in the energy-per-charge range from <1 to >50 keV/e. This covers the thermal plasma and more. What has been found is that this is a rich depository of information on coronal processes, particle acceleration in the solar wind, and various kinds of pickup ions. Besides the inference of a north–south temperature asymmetry mentioned above, a great deal has been gained by studying ions "picked up" by the solar wind when atoms of various sources are collisionally or photoionized. Sources include interstellar neutrals, inner heliosphere dust, and comets. These sources are not, themselves, dependent on the solar cycle. But, the detectability of the pickup ions does depend both on solar activity and on distance from the Sun. Interstellar pickup ions are one tool for analyzing the heliospheric interface, which moves in and out over the solar cycle. The penetration depth of the neutrals into the heliosphere depends on solar ultraviolet radiation and its solar cycle variation. In the outer heliosphere, pickup ions contribute more to the ambient thermal pressure than the solar wind plasma itself.

7.3.4 Coronal mass ejections

Two observations pertaining to CMEs are mentioned here—there is more in Chapter 3. The first is the regular observation of Type II (shock-associated) and Type III (streaming electrons) radio bursts. The second is a composition signature unique to CMEs.

An original objective of Ulysses was to carry out radio observations of CMEs. It has been found that it is routinely possible to detect the Type II bursts produced by ICME-driven shock waves. Solar active regions also emit streaming electrons that produce Type III radio bursts. Making these observations in combination with WIND has demonstrated that triangulation tracking of these bursts can be done. The out-of-ecliptic position of Ulysses relative to WIND has revealed the location of the coronal shock waves that are often the main source of SEPs.

The *in situ* signature of ICMEs is varied. Two techniques developed with the help of Ulysses are the detection of bi-directionally streaming ~100 keV electrons and the detection of enhanced high-ionization state Fe. The former is due to electrons of coronal temperatures streaming in both directions along the HMF, which can only happen on field lines which are connected back to the Sun at both ends or which are closed loops. The latter is the signature of hot material deposited in the ICME flux rope during the eruption at the Sun. With these two identifiers, it has become much more possible to distinguish ambient solar wind plasma that has been dynamically disturbed from ejecta coming from the solar eruption.

7.3.5 Energetic particles

In considering energetic particles, it is important to differentiate between transport in slow wind and in fast wind. The magnetic field in fast wind is more turbulent than in slow wind, so propagation in fast wind should be much more difficult. However, the magnetic field in the slow wind tends to be full of discontinuities, channels, and other features which, depending on the size and thickness of the discontinuities and the energy of the particles, affect propagation in various ways. In general, slow wind contains magnetic field structures such as CIRs, with their associated forward and reverse shocks, and CMEs, with their associated interplanetary shocks, all of which again affect the propagation and accelerate particles. The field in fast solar wind tends to be homogeneous at these scales and devoid of large-scale discontinuities.

Of equal importance for energetic particle propagation is the change in the character of the global heliosphere between solar minimum and maximum. The propagation depends to varying degrees, depending on energy, on conditions across distances up to the scale of the heliosphere itself, as will be seen when cosmic rays are discussed.

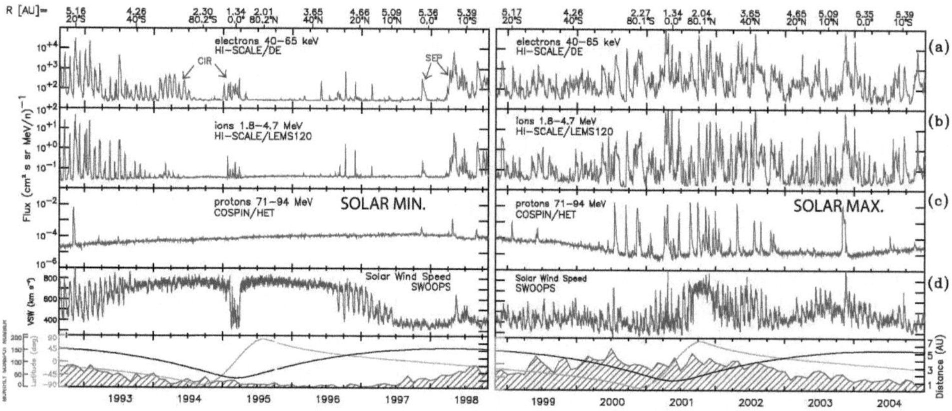

Figure 7.7. 12.5-year plot of selected energetic particle data, showing observations over O-I and O-II. Included are 40–65 keV electrons, 1.8–4.7 MeV ions, and 71–94 MeV protons. These are all of heliospheric origin (Lario and Pick, see Chapter 5).

Figure 7.7 very neatly presents an overview of energetic particle (*versus* cosmic ray) fluxes in O-I and O-II, in the 4-D context described in Section 7.2. It is used here as the basis for the discussion. A summary of Ulysses' energetic particle observations around solar maximum can be found in Sanderson (2004) and around solar minimum in Lanzerotti and Sanderson (2001), in addition to the overall review in Chapter 5. Starting from the bottom of Figure 7.7, first is the heliographic latitude and radius of the spacecraft, the solar wind speed, the 71–94 MeV proton particle intensity, the 1.8–4.7 MeV/n ion intensity, and the 40–65 keV electron intensity. Along the bottom is also plotted the sunspot number. The 1.8–4.7 MeV/n ion intensity is reasonably representative of SEPs accelerated close to the Sun, either by flares or by the CMEs when close to the Sun.

In O-I, at mid-latitudes, significant increases were observed due to CIRs. These are seen in 1993–1994 as periodic solar wind speed and corresponding particle (40–65 keV electron) flux enhancements. At the very highest latitudes few, if any, increases were observed due to either CIRs or CMEs. However, it was very surprising to find that after the CIR signature in solar wind speed disappeared with increasing southern latitude in 1994, there were still periodic electron increases. The observation of CIR-associated increases above the HCS and the CIRs themselves poses an interesting problem that is still being investigated. It is one example of the class of problems dealing with unexpected mobility of energetic particles and cosmic rays across latitude. Interpretation of these particle enhancements has been made in terms of the connection of the spacecraft to the CIR shocks, combined with the motions of the footpoints of the field lines that connect to Ulysses that were discussed earlier in connection with the sub-Parker spiral.

Particle increases due to CIRs were again observed during FLS-I in early 1995 as the spacecraft rapidly crossed through the ecliptic, but again, over the northern poles, no increases were observed. Particle increases associated with CIRs were again seen in

1996 as the spacecraft descended slowly down to low latitudes towards the end of O-I, though this time with not the same regularity as during the period in 1992 and 1993. Most of 1997 was dominated by CIR-associated particle increases, whilst most of 1998 was dominated by CME-associated particle increases as the level of activity of the Sun began to increase in solar cycle 23.

O-II began as the level of solar activity of cycle 23 was increasing. During most of the orbit, energetic particles in the heliosphere were dominated by the presence of CMEs.

In the early part of O-II, the high-energy particle flux (71–94 MeV protons) was dominated by the background cosmic ray flux, which slowly dropped during the course of the period 1998 to early 2000. From mid-2000 onwards, there were many intense increases in this proton particle intensity and, even more so, in 1.8–4.7 MeV ions, the result of an increase in flare and CME activity on the Sun as solar maximum was approached. This activity peaked during 2001 and 2002. Activity increased again for a short period towards the end of 2003, coincident with the October/November 2003 events on the Sun. Many more energetic particles, both high-energy SEP particles and locally accelerated low-energy particles, were observed over the poles than in O-I. For most of O-II, the solar wind speed was low, except for a short 3–4 month period at the end of 2001. Over the poles, the magnetic field was much more disturbed than during the first orbit.

Multi-spacecraft particle observations during solar maximum showed large SEP events are simultaneously observed by spacecraft widely separated in longitude and latitude. Particle anisotropies, lack of *in situ* evidence for field line excursions from low to high latitude, and solar observations show that global coronal activity may populate the entire inner heliosphere with energetic particles during major SEP events, explaining these observations. However, the actual process involved in the coronal activity is not well defined.

The reservoir effect

Reservoirs are observed during isolated major SEP events and during sequences of events . It is the phenomenon of similar decaying intensities after the passage of the associated interplanetary transient, observed at widely spaced spacecraft, in longitude and, especially, in latitude. It presents the appearance of a global reservoir of energetic particles filling the heliosphere after the passage of a transient. It has been interpreted as a consequence of the forming of a merged interaction region (MIR) from several ICMEs. The interpretation means that particle reservoirs form behind HMF structures that are able to reflect and confine energetic particles, and only observed when these structures are beyond the spacecraft. More-efficient-than-normal cross-field diffusion may also be involved.

Overexpanding CMEs

During O-II, Ulysses observed "over-expanding" ICMEs in the northern high-speed solar wind. Highest intensities of energetic particles (1.8–4.8 MeV ions) were observed in association with the passage of ICMEs and not with the shocks. This contrasts with typical in-ecliptic 1 AU observations, which is probably a consequence of the associated magnetic field topology. ICMEs propagating into high-speed solar wind streams are not able to drive strong shocks that efficiently accelerate energetic particles. Particle confinement within the ICMEs is responsible for the slower decay of the intra-ICME intensities with respect to those outside the ICMEs.

The interpretation of these phenomena depends on the ICME flux rope penetrating into the region of fast wind, rather than simply a pressure disturbance propagating into the fast wind. The magnetic field topology of ICMEs can sometimes be modeled with a flux rope—a magnetic cloud. However, the 3-D geometry of these flux ropes is still something of a mystery. It may be possible to use overexpanding ICMEs to help define the magnetic topology.

Energetic particle summary

During the highest northern-latitude parts of O-I, Ulysses was immersed in the fast solar wind. Fast solar wind tends mainly to be homogeneous, and devoid of large-scale discontinuities, but is much more turbulent than the slow solar wind, and so particles propagate to high latitudes in high-speed solar wind with some difficulty. Energetic particle events observed during this time had smooth time–intensity profiles, near-isotropic particle angular distributions at all energies at the onset, flow directions during the rising phase of the events along the field, and no evidence for any net flow across the field lines. These particles propagated to the highest heliographic latitudes traveling along magnetic field lines and not across them. Observations do not allow drawing conclusions about propagation closer to the Sun, but most likely, to reach the high latitudes, particles must either diffuse across field lines closer to the Sun, or else there is some large-scale distortion of the magnetic field lines.

At high southern latitudes during O-II, Ulysses was continually in the slow solar wind. Most of the particle events observed at this time occurred at the same time as some other pre-existing and unrelated structure, such as a CME from a previous solar flare, or a CIR, passed over the spacecraft. Particle propagation was dominated by the presence of these structures, the frequent occurrence of which meant that it was quite rare to find an event where the event was unaffected by one.

The forward and reverse shocks and the stream interfaces of the CIRs, and interplanetary shocks and magnetic clouds of the CMEs all affected the particle propagation, sometimes accelerating the lower energy particles locally. These structures tended to be full of discontinuities, which again affected the propagation, the effect depending on the size and thickness of the discontinuities and the energy of the particles.

Table 7.1 is a summary of energetic particle characteristics in high-latitude fast and slow wind. The slow solar wind was observed during the southern polar pass and the fast solar wind region observed during northern polar pass, both in O-II.

Table 7.1. Summary of energetic particle observations during FLS-II, contrasting high-latitude fast- and slow-wind results.

Characteristic	Southern polar pass	Northern polar pass
Solar wind	Slow	Fast
Time–intensity profiles	Irregular	Smooth
Structures at onset	Frequent	Rare
Event onset times	Rapid	Delayed
High-latitude propagation	Modified by draping	Direct alongfield
Anisotropy at onsets	Large	Nearly isotropic
Particles inside CME	Intensities depressed	Intensities elevated
Particle flow directions	Field-aligned	Field-aligned (non–field aligned near structures)

7.3.6 Cosmic rays

Solar minimum

Ulysses' observations during O-I and FLS-I, at solar minimum, were well suited to complement other investigations of spatial structure and the consequence of spatial structure for our understanding of particle propagation in the inner heliosphere. It was found that the proton spectrum was highly modulated so that Ulysses did not observe the local interstellar spectra at polar latitudes. The variation of the electron intensities was dominated by temporal changes, not an intensity change correlated with Ulysses latitude. The latitudinal gradient of electrons was consistent with it being zero. The latitudinal gradient of cosmic ray protons was small and showed a maximum at $\sim$2 GV. These observations were not expected and are still not very well explained in models. It makes the modeling of cosmic rays much more demanding but also more interesting.

It has long been known that recurrent variations in energetic particle intensities are associated with CIRs; normally, a CIR causes a depression in the cosmic ray flux while MeV/n nuclei and keV electron intensity increases, centered around the forward and reverse shock. From latitudes above 40°, Ulysses was embedded in fast solar wind originating from the southern polar coronal hole, and corotating forward and reverse shock waves disappeared. In contrast to what had been expected, recurrent particle increases and galactic cosmic ray decreases were observed up to polar latitudes. Even more surprising was the fact that the 40–65 keV electrons were delayed from the 0.5–1.0 MeV protons by up to 4 days. Two competing proposals were put forward, as described in Chapter 6. One explanation is that because of enhanced latitudinal diffusion the temporal variation at high heliolatitudes is determined by the interaction regions at low latitudes. The delay is then explained by the details of the diffusion process, especially perpendicular diffusion.The second explanation relies on the analysis of the Ulysses magnetic field data which showed that the polar magnetic field was dominated by strong variations. Systematic modifications of the standard Parker theory of the heliospheric magnetic field would then cause a meridional field component along which particles may move easily to polar latitudes. Since both proposals explain the observations equally well, only magnetic field measurements in the distant, high-latitude heliosphere, where the systematic effects will be larger than the statistical variation, may prove which one is correct.

A real surprise of the Ulysses mission was the observation that the galactic cosmic ray flux was not symmetric to the heliographic equator. This went along with corresponding observations of north–south asymmetry in the solar wind, inferred coronal electron temperature, and magnetic field.

Solar maximum

Ulysses' O-II in 1998–2003 was well suited to investigate the latitudinal variation at solar maximum, when heliospheric magnetic structure is characterized by a highly inclined current sheet that is distorted by coronal mass ejections at all latitudes. Gradient measurements at that time exhibited fluctuations of the order of 2 and larger so that no unambiguous evidence could be found for latitudinal gradients. In contrast to solar-minimum conditions, the cosmic ray distribution was almost spherically symmetric; this is consistent with model computations. From this the important conclusion was made that since the latitudinal gradients were positive at solar minimum, the total modulation is relatively higher at polar latitudes than in the ecliptic. It was concluded that less than a 10% contribution by $\mathbf{E} \times \mathbf{B}$ drift is required at extreme solar maximum to explain the observations while the magnetic field reversed. The proton-to-electron intensity ratios turned out to be an excellent indicator of when the magnetic field actually reversed; the reversal must have occurred between late 2000 and mid-2001.

The "reservoir effect" was an important result for cosmic rays, as well as for energetic particles. For 35–70 MeV and 70–95 MeV protons, essentially all large events at Earth produce comparable intensity increases at Ulysses during FLS-II. Nearly equal particle intensities at Ulysses and close to Earth occurred after 3–4 days. A similar reservoir effect was observed in the 1970s, but without FLS-II it would have been very difficult to distinguish between the high-latitude and low-latitude measurements. It was concluded that either an acceleration front for energetic particles in large events extends over a broad range in latitude and longitude, or that mechanisms exist to transport particles efficiently across the mean magnetic field close to the Sun.

Insights on particle propagation in a turbulent astrophysical plasma

The Ulysses mission to high heliolatitudes has led to several insights concerning propagation and modulation theory, in particular on the relative importance of the various diffusion coefficients. It was concluded that to obtain agreement between current modulation models and Ulysses observations, enhancing $\kappa_{\perp\theta}$ latitudinally and changing the rigidity dependence of $\kappa_{\perp\theta}$ differently from $\kappa_{\|}$ was essential. The latitudinal enhancement is related to the different solar wind regimes observed around solar minimum.

The observed electron-to-proton ratios (implicitly also containing the radial and latitudinal gradients) indicated that large particle drifts were occurring during solar minimum but diminished significantly toward solar maximum when less than 10% drifts were required in models to explain the observed values.

These combined observational and modeling studies initiated new projects concerning diffusion and turbulence theories which have become known as the *ab initio* approach to modulation theory and modeling, with exceptional progress being made in the past few years.

Cosmic rays in O-III

In O-III, with a second solar minimum, but now with A<0, drifts will be reversed. Of interest will be corotating particle variations at high latitude. In particular, the behavior of cosmic ray electrons will be studied. As noted above, this has so far defied satisfactory explanation. With the launch of STEREO in 2006, a network of spacecraft will be in place for spatial and temporal characterization of SEPs, CIR and CME-accelerated particles, and cosmic ray modulation in the inner heliosphere. The objective for O-III is to measure latitude gradients of GCRs and ACRs when A<0. Open questions are: (i) Will the decrease continue to high latitude or only to middle latitudes? (ii) Will the magnitude of the gradients be comparable with those in 1994–1995? (iii) How does the latitude structure of CIRs compare with that of GCRs?

7.3.7 The heliosphere–interstellar medium interface

The global solar wind during the epoch of Voyagers 1/2 in the heliosheath

Voyager 1 passed the heliospheric termination shock late in 2004 and Voyager 2 is expected to do the same by late 2008. Predictions that this would happen were made using IMP-8 data in the ecliptic and models of the interface between the LISM and the solar wind (Izmodenov, Gloeckler, and Malama, 2003). The distance to the termination shock (TS) is 90–110 AU in this prediction and as observed (Figure 7.8). It takes solar wind on the order of 1 year to reach the termination shock. Therefore, solar wind passing Ulysses on 2007 will determine the distance to the TS in 2008.

The Voyagers are exploring the heliosheath in 2005–2010, corresponding to Ulysses observations in 2004–2009. Figure 7.9 shows how the time of FLS-III overlays the encounters of Voyagers 1/2 with the TS. The IBEX mission is expected to be launched in summer 2008 and will remotely analyze the heliospheric interface during late 2008 and 2009. Ulysses is well-positioned to sample the global solar wind during the epoch of Voyagers 1/2 in the heliosheath and IBEX observations. This is happening at the same time as the end of cycle 23 and beginning of cycle 24. Figure 7.8 is a prediction for how the distance to the TS varied over cycle 23. Figure 7.5 shows how the dynamic pressure varied, in latitude and time, at Ulysses. What this means is that the distance and probably also the shape of the termination shock will change over the solar cycle. The dependence of dynamic pressure on latitude is not yet known from the results in Figure 7.5 because, without a model of the source of the solar wind, it is impossible to separate latitudinal and temporal changes. There was

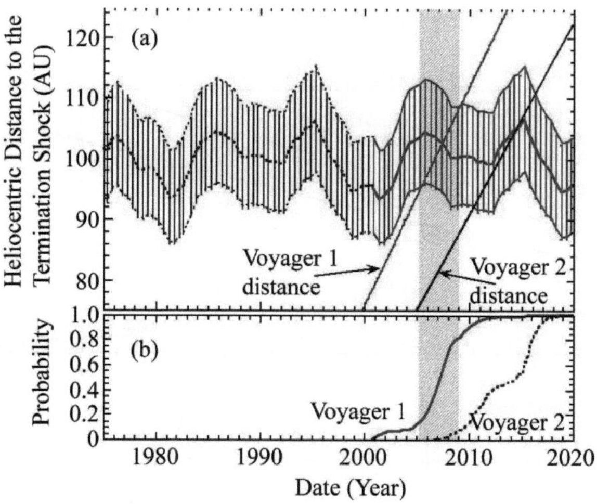

Figure 7.8. Location of the termination shock and probability of shock encounter by Voyagers 1/2 as a function of time. Shown in the left in (a) is the calculated mean and $\pm 1 - \sigma$ limits of the termination shock distance found using IMP-8 solar wind speed and density. Shown in the right is the calculated mean and $\pm 1 - \sigma$ limits of the termination shock distance using a repeat of the previous 20-year sequence (Izmodenov, Gloeckler, and Malama, 2003).

apparently little latitudinal variation in cycle 23, but recently there has been a temporal variation that exceeds the latitudinal variation and it is not obvious that the temporal variation is uniform over latitude. This behavior became more clear with new data during FLS-III.

Interstellar neutral gas

The Ulysses GAS instrument detects interstellar neutral helium entering the heliosphere. This had never been directly measured. The GAS instrument determines the direction from which helium atoms enter the heliosphere, their density, and the angular dispersion—the last parameter being used to determine the temperature. Estimates of all these parameters were already available from, for example, solar backscattered EUV measurements (Fahr, p. 421 in Marsden, 1986; Bertaux *et al.*, p. 435 in Marsden, 1986). But, they would be far more precisely known after the direct measurements. No solar cycle variation of the neutral atoms was expected, but their properties are a major factor in modeling the shape and dynamics of the interface between the heliosphere and the interstellar medium. This is because some interstellar neutral hydrogen undergoes charge exchange while interstellar neutral helium is essentially unaffected.

7.3.8 Summary

1 The heliosphere, from the photosphere to the heliopause, and outward to beyond the heliospheric bow shock, is a coupled system that responds to a variety of

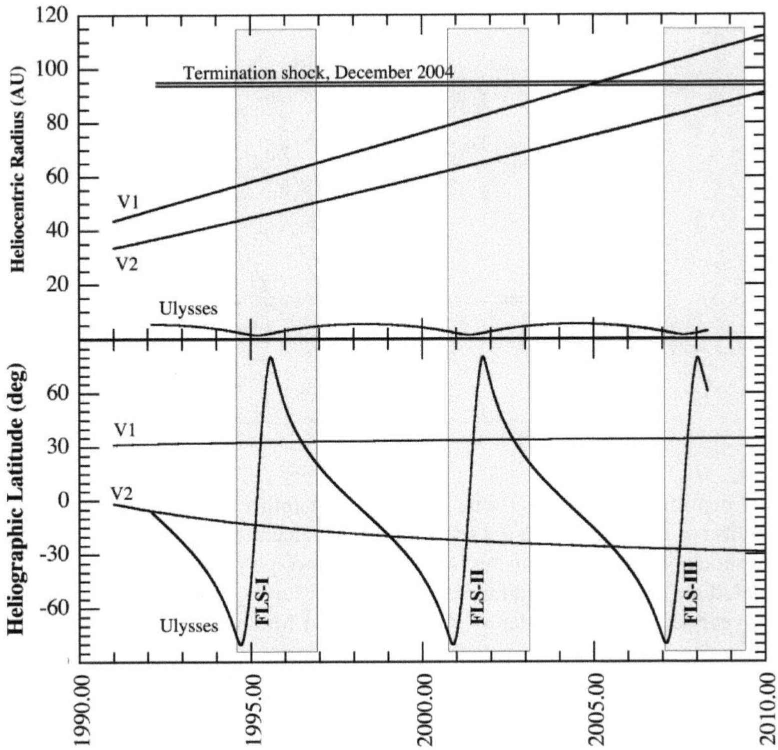

Figure 7.9. Orbits of Ulysses, Voyager 1, and Voyager 2 from 1991 through 2010 (Ulysses begins at passage of Jupiter and ends at the beginning of April 2008). Voyager = solid lines, Voyager 2 = dashed lines, Ulysses = thin lines.

variations in the solar activity cycle and Hale magnetic cycle. The Ulysses mission has provided important information on this response at high heliographic latitudes. As its final contribution, the mission, which is coming to an end, points towards what measurements are needed in the future. On the one hand, it is shown that high-latitude measurements in the inner heliosphere will continue to return new information on how the Sun and the solar dynamo work, how the heliospheric magnetic field responds to the dynamo, and on the effect of solar activity on the inner heliosphere. On the other hand, specific requirements have been raised for measurements of the latitudinal variation of cosmic rays and energetic particles in the outer heliosphere. The purpose of such measurements would be to determine how cosmic rays enter the heliosphere. This, in turn, leads to insight on how the heliosphere has interacted with the local interstellar medium over its 4.5×10^9 year lifetime.

2 High-latitude measurements in the heliosphere have been extraordinarily and unexpectedly productive in understanding the physics of the integrated Sun and heliosphere system.

7.4 ACKNOWLEDGMENTS

The preparation of this chapter and this book has been supported by the Ulysses project and the Ulysses/SWOOPS instrument team.

7.5 REFERENCES

Balogh, A., R. G. Marsden, and E. J. Smith (eds.) (2001), *The Heliosphere near Solar Minimum: The Ulysses Perspective*, Springer/Praxis, Chichester, UK.

Balogh, A., T. J. Beek, R. J. Forsyth, P. C. Hedgecock, R. J. Marquedant, E. J. Smith, D. J. Southwood, and B. T. Tsurutani (1992), The magnetic field investigations on the Ulysses mission: Instrumentation and preliminary scientific results, *Astron. and Astrophys.*, **92**, 221.

Bame, S. J., D. J. McComas, B. L. Barraclough, J. L. Phillips, K. J. Sofaly, J. C. Chavez, B. E. Goldstein, and R. K. Sakurai (1992), The Ulysses solar wind plasma experiment, *Astron. and Astrophys.*, **92**, 237.

Bame, S. J., B. E. Goldstein, J. T. Gosling, J. W. Harvey, D. J. McComas, M. Neugebauer, and J. L. Phillips (1993), Ulysses observations of a recurrent high speed solar wind stream and the heliomagnetic streamer belt, *Geophys. Res. Letters*, **20**, 2323.

Bertaux, J. L., R. Lallement, and E. Chassefière (1986), Interstellar gas parameters and solar wind anisotropies deduced from H and He observations in the solar system, in *The Sun and the Heliosphere in Three Dimensions* (R. G. Marsden, ed.), p. 435, Astrophysics and Space Science Library Vol. 123, D. Reidel, Dordrecht, The Netherlands.

Billings, D. E. (1966), *A Guide to the Solar Corona*, Academic Press, New York.

Coles, W. A., B. J. Rickett, V. H. Rumsery, J. J. Kaufman, D. G. Turley, S. Ananthakrishnan, J. W. Armstrong, J. K. Harmons, S. L. Scott, and D. G. Sime (1978), Solar cycle changes in the polar solar wind, *Nature*, **286**, 239.

Fahr, H. (1986), Neutral interstellar gases in the heliosphere: New aspects of the problem, in *The Sun and the Heliosphere in Three Dimensions* (R. G. Marsden, ed.), p. 421, Astrophysics and Space Science Library Vol. 123, D. Reidel, Dordrecht, The Netherlands.

Fisk, L. F., B. Kozlovsky, and R. Ramaty (1974), An interpretation of the observed oxygen and nitrogen enhancements in low energy cosmic rays, *Astrophys. J. Lett.*, **190**, L35.

Fisk, L. F., T. H. Zurbuchen, & N. A. Schwadron (1999), On the coronal magnetic field: Consequences of large-scale motions, *Astrophys. J.*, **521**, 868.

Fisk, L. F. J. R. Jokipii, G. M. Simnett, R. von Steiger, & K.-P. Wenzel (eds.) (1998), *Cosmic Rays in the Heliosphere*, Kluwer Academic, Dordrecht, The Netherlands.

Forsyth, R. J., and Gosling, J. T. (2001), Corotating and transient structures in the heliosphere, in *The Heliosphere near Solar Minimum: The Ulysses Perspective* (A. Balogh, R. G. Marsden, and E. J. Smith, eds.), pp. 107–166, Springer-Praxis, Chichester, UK.

Geiss, J., and P. Bochsler (1986), Solar wind composition and what we expect to learn from out-of-ecliptic measurements, in *The Sun and the Heliosphere in Three Dimensions* (R. G. Marsden, ed.), p. 173, Astrophysics and Space Science Library Vol. 123, D. Reidel, Dordrecht, The Netherlands.

Geiss, J., G. Gloeckler, R. von Steiger, H. Balsiger, L. A. Fisk, A. B. Galvin, F. Ipavich, J. F. McKenzie, K. W. Ogilvie, and B. Wilken (1995), The southern high-speed stream—results from the SWICS instrument on ULYSSES, *Science*, **268**(5213), 1033.

Gloeckler, G., T. H. Zurbuchen, and J. Geiss (2003), Implications of the observed anticorrelation between solar wind speed and coronal temperature, *J. Geophys. Res.*, **108**(A4), 1158, doi:1029/2002JA009286.

Gopalswamy, N. (2004), A global picture of CMEs in the inner heliosphere, in *The Sun and the Heliosphere as an Integrated System* (G. Poletto and S. T. Suess, eds.), p. 201, Kluwer Academic, Dordrecht, The Netherlands.

Gosling, J. T., D. J. McComas, J. L. Phillips, L. A. Weiss, V. J. Pizzo, B. E. Goldstein, and R. J. Forsyth (1994), A new class of forward–reverse shock pair in the solar wind, *Geophys. Res. Lett.*, **21**, 2271.

Gringauz, K. I., Bezrukikh, V.. V., Ozerov, V. D., and Rybchinskiy, R. E. (1960), Study of the interplanetary ionized gas, high energy electrons, and solar corpuscular radiation by means of three electrode traps for charged particles on the second Soviet cosmic rocket, *Doklady* (Engl. trans.), **5**, 361–364.

Hoeksema, J. T., J. M. Wilcox, and P. H. Scherrer (1982), Structure of the heliospheric current sheet in the early portion of sunspot cycle 21, J. Geophys. Res., 87, 10331, 1982.

Hundhausen, A. J. (1973), Nonlinear model of high-speed solar wind streams, *J. Geophys. Res.*, **78**, 1528.

Hundhausen, A. J. (1977), An interplanetary view of coronal holes, in *Coronal Holes and High Speed Streams* (J. B. Zirker, ed.), p. 225, Colorado Associated Universities Press.

Hundhausen, A. J. (1997), Coronal mass ejections, in *Cosmic Winds and the Heliosphere* (J. R. Jokipii, C. P. Sonett, and M. S. Giampapa, eds.), p. 259, University of Tucson Press, Tucson, AZ.

Izmodenov, V. V. (2004), The heliospheric interface: Models and observations, in *The Sun and the Heliosphere as an Integrated System* (G. Poletto and S. T. Suess, eds.), p. 23, Kluwer Academic, Dordrecht, The Netherlands.

Izmodenov, V., G. Gloeckler, and Y. Malama (2003), When will Voyager 1 and 2 cross the temination shock?, *Geophys. Res. Lett.*, **30**(7), 1351, doi: 1029/2002GL016127.

Jokipii, J. R. (1986), Effects of three-dimensional heliospheric structures on cosmic-ray modulation, in *The Sun and the Heliosphere in Three Dimensions* (R. G. Marsden, ed.), p. 375, Astrophysics and Space Science Library Vol. 123, D. Reidel, Dordrecht, The Netherlands.

Jokipii, J. R., and J. Kóta (1985), Cosmic rays near the heliospheric current sheet. II—An ensemble approach to comparing theory and observation, *J. Geophys. Res.*, **91**, 2885.

Jokipii, J. R., J. Kóta, J. Giacalone, T. S. Horbury, and E. J. Smith, Interpretation and consequences of large-scale magnetic variances observed at high heliographic latitude, *Geophys. Res. Lett.*, **22**(23), 3385.

Jones, G. H., A. Balogh, and E. J. Smith (2003), Solar magnetic field reversal as seen at Ulysses, *Geophys. Res. Lett.*, **30**, Cite ID 8028.

Krüger, H., Altobelli, N., Anweiler, B., Dermott, S. F., Dikarev, V., Graps, A. L., Grün, E., Gustafson, B. A., Hamilton, D. P., Hanner, M. S. *et al.* (2006), Five years of Ulysses dust data: 2000–2004, *Planet. & Space Sci.*, **54**, 932.

Lallement, R., J. L. Bertaux, B. R. Sander, and E. Chassefière (1992), The interplanetary H glow as seen by Voyager during the cruise, in *Solar Wind Seven* (E. Marsch and R. Schwenn, eds.), p. 209, COSPAR Colloquia Series Vol. 3, Pergamon Press.

Landgraph, M., H. Krüger, N. Altobelli, and E. Grün (2003), Penetration of the heliosphere by the interstellar dust stream during solar maximum, *J. Geophys. Res.*, **108**(A10), 8030.

Lanzerotti, L. J., and T. R. Sanderson (2001), Energetic particles in the heliosphere, in *The Heliosphere near Solar Minimum* (A. Balogh, R. G. Marsden, and E. J. Smith, eds.), p. 259, Springer-Praxis, Chichester, UK.

Marsden, R. G. (ed.) (1986), *The Sun and the Heliosphere in Three Dimensions*, Astrophysics and Space Science Library Vol. 125, D. Reidel, Dordrecht, The Netherlands.

McComas, D. J., H. A. Elliott, N. A. Schwadron, J. T. Gosling, R. M. Skoug, and B. E. Goldstein (2003), The three-dimensional solar wind around solar maximum, *Geophys. Res. Lett.*, **30**(10), 1517.

McKibben, R. B. (1986), Modulation of galactic cosmic rays in the heliosphere, in *The Sun and the Heliosphere in Three Dimensions* (R. G. Marsden, ed.), p. 361, Astrophysics and Space Science Library Vol. 123, D. Reidel, Dordrecht, The Netherlands.

Parker, E. N. (1958), Dynamics of the interplanetary gas and magnetic fields, *Astrophys. J.*, **128**, 664.

Parker, E. N. (1963), *Interplanetary Dynamical Processes*, Interscience Publishers.

Pesses, M. E., J. R. Jokipii, and D. Eichler (1981), Cosmic ray drift, shock wave acceleration, and the anomalous component of cosmic rays, *Astrophys. J. Lett.*, **246**, L85.

Phillips, J. L., S. J. Bame, A. Barnes, B. L. Barraclough, W. C. Feldman, B. E. Goldstein, J. T. Gosling, G. W. Hoogeveen, D. J. McComas, M. Neugebauer, and S. T. Suess (1995), Ulysses solar wind plasma observations from pole to pole, *Geophys. Res. Lett.*, **22**(23), 3301.

Pizzo, V. J. (1978), A three-dimensional model of corotating streams in the solar wind. I—Theoretical foundations, *J. Geophys. Res.*, **83**, 5563.

Reisenfeld, D. B., J. T. Gosling, R. J. Forsyth, P. Riley, and O. C. St. Cyr (2003), Properties of high-latitude CME-driven disturbances during Ulysses second northern polar passage, *Geophys. Res. Lett.*, **30**(19), 8031.

Sanderson, T. R. (2004), Propagation of energetic particles to high latitudes, in *The Sun and the Heliosphere as an Integrated System* (G. Poletto and S. T. Suess, eds.), p. 259, Kluwer Academic, Dordrecht, The Netherlands.

Schwadron, N. A., and D. J. McComas (2005), The sub-Parker spiral structure of the heliospheric magnetic field, *Geophys. Res. Lett.*, **32**, L03112, doi:10.1029/2004GL021579.

Smith, E. J., Marsden, R. G., Balogh, A., Gloeckler, G., Geiss, J., McComas, D. J., McKibben, R. B., MacDowall, R. J., Lanzerotti, L. J., Krupp, N. *et al.* (2003), The Sun and Heliosphere at Solar Maximum, *Science*, **302**(5648), 1165.

Steinolfson, R. S., S. T. Suess, and S. T. Wu (1982), The steady global corona, *Astrophys. J.*, **255**, 730.

Sturrock, P. A., T. E. Holzer, D. M. Mihalas, and R. K. Ulrich (eds.) (1986), *Physics of the Sun, Volume II: The Solar Atmosphere*, D. Reidel, Dordrecht, The Netherlands.

Suess, S. T., and E. J. Smith (1996), Latitudinal dependence of the radial IMF component: Coronal imprint, *Geophys. Res. Lett.*, **23**(22), 3267.

Suess, S. T., A. K. Richter, C. R. Winge, and S. F. Nerney (1977), Solar polar coronal hole—A mathematical simulation, *Astrophys. J.*, **217**, 296.

Webb, D. F., and R. A. Howard (1994), The solar cycle variation of coronal mass ejections and the solar wind mass flux, *J. Geophys. Res.*, **99**, 4201.

White, O. R. (ed.) (1977), *The Solar Output and Its Variation*, Colorado Associated University Press, Boulder, CO.

Wilcox, J. M., and N. F. Ness (1965), Quasi-stationary Corotating Structures in the Interplanetary Medium, *J. Geophys. Res.*, **70**, 5793.

Zank, G. P., Interaction of the solar wind with the local interstellar medium: A theoretical perspective, Space Science Rev., 89(3/4), 413, 1999.

Zirker, J. B. (ed.) (1977), *Coronal Holes and High Speed Streams*, Colorado Associated University Press, Boulder, CO.

Index

MIX
Papier aus verantwortungsvollen Quellen
Paper from responsible sources
FSC® C105338

If you have any concerns about our products,
you can contact us on
ProductSafety@springernature.com

In case Publisher is established outside the EU,
the EU authorized representative is:
Springer Nature Customer Service Center GmbH
Europaplatz 3, 69115 Heidelberg, Germany

Printed by Libri Plureos GmbH
in Hamburg, Germany